S0-BLI-292

Gravitational Solitons

This book gives a self-contained exposition of the theory of gravitational solitons and provides a comprehensive review of exact soliton solutions to Einstein's equations.

The text begins with a detailed discussion of the extension of the inverse scattering method to the theory of gravitation, starting with pure gravity and then extending it to the coupling of gravity with the electromagnetic field. There follows a systematic review of the gravitational soliton solutions based on their symmetries. These solutions include some of the most interesting in gravitational physics, such as those describing inhomogeneous cosmological models, cylindrical waves, the collision of exact gravity waves, and the Schwarzschild and Kerr black holes.

This work will equip the reader with the basic elements of the theory of gravitational solitons as well as with a systematic collection of nontrivial applications in different contexts of gravitational physics. It provides a valuable reference for researchers and graduate students in the fields of general relativity, string theory and cosmology, but will also be of interest to mathematical physicists in general.

VLADIMIR A. BELINSKI studied at the Landau Institute for Theoretical Physics, where he completed his doctorate and worked until 1990. Currently he is Research Supervisor by special appointment at the National Institute for Nuclear Physics, Rome, specializing in general relativity, cosmology and nonlinear physics. He is best known for two scientific results: firstly the proof that there is an infinite curvature singularity in the general solution of Einstein equations, and the discovery of the chaotic oscillatory structure of this singularity, known as the BKL singularity (1968–75 with I.M. Khalatnikov and E.M. Lifshitz), and secondly the formulation of the inverse scattering method in general relativity and the discovery of gravitational solitons (1977–82, with V.E. Zakharov).

ENRIC VERDAGUER received his PhD in physics from the Autonomous University of Barcelona in 1977, and has held a professorship at the University of Barcelona since 1993. He specializes in general relativity and quantum field theory in curved spacetimes, and pioneered the use of the Belinski–Zakharov inverse scattering method in different gravitational contexts, particularly in cosmology, discovering new physical properties in gravitational solitons. Since 1991 his main research interest has been the interaction of quantum fields with gravity. He has studied the consequences of this interaction in the collision of exact gravity waves, in the evolution of cosmic strings and in cosmology. More recently he has worked in the formulation and physical consequences of stochastic semi-classical gravity.

CAMBRIDGE MONOGRAPHS ON
MATHEMATICAL PHYSICS

General editors: P. V. Landshoff, D. R. Nelson, D. W. Sciama, S. Weinberg

J. Ambjørn, B. Durhuus and T. Jonsson   *Quantum Geometry: A Statistical Field Theory Approach*
A. M. Anile   *Relativistic Fluids and Magneto-Fluids*
J. A. de Azcárraga and J. M. Izquierdo   *Lie Groups, Lie Algebras, Cohomology and Some Applications in Physics*†
V. Belinski and E. Verdaguer   *Gravitational Solitons*
J. Bernstein   *Kinetic Theory in the Early Universe*
G. F. Bertsch and R. A. Broglia   *Oscillations in Finite Quantum Systems*
N. D. Birrell and P. C. W. Davies   *Quantum Fields in Curved Space*†
S. Carlip   *Quantum Gravity in 2 + 1 Dimensions*
J. C. Collins   *Renormalization*†
M. Creutz   *Quarks, Gluons and Lattices*†
P. D. D'Eath   *Supersymmetric Quantum Cosmology*
F. de Felice and C. J. S. Clarke   *Relativity on Curved Manifolds*†
P. G. O. Freund   *Introduction to Supersymmetry*†
J. Fuchs   *Affine Lie Algebras and Quantum Groups*†
J. Fuchs and C. Schweigert   *Symmetries, Lie Algebras and Representations: A Graduate Course for Physicists*
A. S. Galperin, E. A. Ivanov, V. I. Ogievetsky and E. S. Sokatchev   *Harmonic Superspace*
R. Gambini and J. Pullin   *Loops, Knots, Gauge Theories and Quantum Gravity*†
M. Göckeler and T. Schücker   *Differential Geometry, Gauge Theories and Gravity*†
C. Gómez, M. Ruiz Altaba and G. Sierra   *Quantum Groups in Two-dimensional Physics*
M. B. Green, J. H. Schwarz and E. Witten   *Superstring Theory, volume 1: Introduction*†
M. B. Green, J. H. Schwarz and E. Witten   *Superstring Theory, volume 2: Loop Amplitudes, Anomalies and Phenomenology*†
S. W. Hawking and G. F. R. Ellis   *The Large-Scale Structure of Space-Time*†
F. Iachello and A. Aruna   *The Interacting Boson Model*
F. Iachello and P. van Isacker   *The Interacting Boson–Fermion Model*
C. Itzykson and J.-M. Drouffe   *Statistical Field Theory, volume 1: From Brownian Motion to Renormalization and Lattice Gauge Theory*†
C. Itzykson and J.-M. Drouffe   *Statistical Field Theory, volume 2: Strong Coupling, Monte Carlo Methods, Conformal Field Theory, and Random Systems*†
J. I. Kapusta   *Finite-Temperature Field Theory*†
V. E. Korepin, A. G. Izergin and N. M. Boguliubov   *The Quantum Inverse Scattering Method and Correlation Functions*†
M. Le Bellac   *Thermal Field Theory*†
N. H. March   *Liquid Metals: Concepts and Theory*
I. M. Montvay and G. Münster   *Quantum Fields on a Lattice*†
A. Ozorio de Almeida   *Hamiltonian Systems: Chaos and Quantization*†
R. Penrose and W. Rindler   *Spinors and Space-time, volume 1: Two-Spinor Calculus and Relativistic Fields*†
R. Penrose and W. Rindler   *Spinors and Space-time, volume 2: Spinor and Twistor Methods in Space-Time Geometry*†
S. Pokorski   *Gauge Field Theories*, 2nd edition
J. Polchinski   *String Theory, volume 1: An Introduction to the Bosonic String*
J. Polchinski   *String Theory, volume 2: Superstring Theory and Beyond*
V. N. Popov   *Functional Integrals and Collective Excitations*†
R. G. Roberts   *The Structure of the Proton*†
J. M. Stewart   *Advanced General Relativity*†
A. Vilenkin and E. P. S. Shellard   *Cosmic Strings and Other Topological Defects*†
R. S. Ward and R. O. Wells Jr   *Twistor Geometry and Field Theories*†

† Issued as a paperback

Gravitational Solitons

V. Belinski

National Institute for Nuclear Physics (INFN), Rome

E. Verdaguer

University of Barcelona

CAMBRIDGE
UNIVERSITY PRESS

PUBLISHED BY THE PRESS SYNDICATE OF THE UNIVERSITY OF CAMBRIDGE
The Pitt Building, Trumpington Street, Cambridge, United Kingdom

CAMBRIDGE UNIVERSITY PRESS
The Edinburgh Building, Cambridge CB2 2RU, UK
40 West 20th Street, New York, NY 10011-4211, USA
10 Stamford Road, Oakleigh, VIC 3166, Australia
Ruiz de Alarcón 13, 28014, Madrid, Spain
Dock House, The Waterfront, Cape Town 8001, South Africa

http://www.cambridge.org

© V. Belinski and E. Verdaguer 2001

This book is in copyright. Subject to statutory exception
and to the provisions of relevant collective licensing agreements,
no reproduction of any part may take place without
the written permission of Cambridge University Press.

First published 2001

Printed in the United Kingdom at the University Press, Cambridge

Typeface Times 11/13pt.   *System* LaTeX 2_ε [DBD]

A catalogue record of this book is available from the British Library

Library of Congress Cataloguing in Publication data

Belinski, V. (Vladimir), 1941–
Gravitational solitons / V. Belinski, E. Verdaguer.
p.   cm.
Includes bibliographic references and index.
ISBN 0 521 80586 4
1. Solitons. 2. Gravitational waves. I. Verdaguer, E. (Enric) II. Title.
QC174.26.W28 B39 2001
530.14–dc21   2001018488

ISBN 0 521 80586 4   hardback

Contents

Q C
174
.26
W28
B39
2001
PHYS

Preface

Solitons are some remarkable solutions of certain nonlinear wave equations which behave in several ways like extended particles: they have a finite and localized energy, a characteristic velocity of propagation and a structural persistence which is maintained even when two solitons collide. Soliton waves propagating in a dispersive medium are the result of a balance between nonlinear effects and wave dispersion and therefore are only found in a very special class of nonlinear equations. Soliton waves were first found in some two-dimensional nonlinear differential equations in fluid dynamics such as the Korteweg–de Vries equation for shallow water waves. In the 1960s a method, known as the Inverse Scattering Method (ISM) was developed [111] to solve this equation in a systematic way and it was soon extended to other nonlinear equations such as the sine-Gordon or the nonlinear Schrödinger equations.

In the late 1970s the ISM was extended to general relativity to solve the Einstein equations in vacuum for spacetimes with metrics depending on two coordinates only or, more precisely, for spacetimes that admit an orthogonally transitive two-parameter group of isometries [23, 24, 206]. These metrics include quite different physical situations such as some cosmological, cylindrically symmetric, colliding plane waves, and stationary axisymmetric solutions. The ISM was also soon extended to solve the Einstein–Maxwell equations [4]. The ISM for the gravitational field is a solution-generating technique which allows us to generate new solutions given a background or seed solution. It turns out that the ISM in the gravitational context is closely related to other solution-generating techniques such as different Bäcklund transformations which were being developed at about the same time [135, 224]. However, one of the interesting features of the ISM is that it provides a practical and useful algorithm for direct and explicit computations of new solutions from old ones. These solutions are generally known as soliton solutions of the gravitational field or gravitational solitons for short, even though they share only some, or none, of the properties that solitons have in other nonlinear contexts.

Among the soliton solutions generated by the ISM are some of the most relevant in gravitational physics. Thus in the stationary axisymmetric case the Kerr and Schwarzschild black hole solutions and their generalizations are soliton solutions. In the 1980s there was some active work on exact cosmological models, in part as an attempt to find solutions that could represent a universe which evolved from a quite inhomogeneous stage to an isotropic and homogeneous universe with a background of gravitational radiation. In this period there was also renewed activity in the head-on collision of exact plane waves, since the resulting spacetimes had interesting physical and geometrical properties in connection with the formation of singularities or regular caustics by the nonlinear mutual focusing of the incident plane waves. Some of these solutions may also be of interest in the early universe and the ISM was of use in the generation of new colliding wave solutions. In the cylindrically symmetric context the ISM also produced some solutions representing pulse waves impinging on a solid cylinder and returning to infinity, which could be of interest to represent gravitational radiation around a straight cosmic string. Also some soliton solutions were found illustrating the gravitational analogue of the electromagnetic Faraday rotation, which is a typical nonlinear effect of gravity. Some of this work was reviewed in ref. [288].

In this book we give a comprehensive review of the ISM in gravitation and of the gravitational soliton solutions which have been generated in the different physical contexts. For the solutions we give their properties and possible physical significance, but concentrate mainly on those with possible physical interest, although we try to classify all of them. The ISM provides a natural starting point for their classification and allows us to connect in remarkable ways some well known solutions.

The book is divided into eight chapters. In chapter 1 we start with an overview of the ISM in nonlinear physics and discuss in particular the sine-Gordon equation, which will be of use later. We then go on to generalize and adapt the ISM in the gravitational context to solve the Einstein equations in vacuum when the spacetimes admit an orthogonally transitive two-parameter group of isometries. We describe in detail the procedure for obtaining gravitational soliton solutions. The ISM is generalized to solve vacuum Einstein equations in an arbitrary number of dimensions and the possibility of generating nonvacuum soliton solutions in four dimensions using the Kaluza–Klein ansatz is considered. In chapter 2 we study some general properties of the gravitational soliton solutions. The case of background solutions with a diagonal metric is discussed in detail. A section is devoted to the topological properties of gravitational solitons and we discuss how some features of the sine-Gordon solitons can be translated under some restrictions to the gravitational solitons. Some remarkable solutions such as the gravitational analogue of the sine-Gordon breather are studied.

Chapter 3 is devoted to the ISM for the Einstein–Maxwell equations under the same symmetry restrictions for the spacetime. The generalization of the

ISM in this context was accomplished by Alekseev. This extension is not a straightforward generalization of the previous vacuum technique; to some extent it requires a new approach to the problem. Here we follow Alekseev's approach but we adapt and translate it into the language of chapter 1. To illustrate the procedure the Einstein–Maxwell analogue of the gravitational breather is deduced and briefly described.

In chapters 4 and 5 we deal with gravitational soliton solutions in the cosmological context. This context has been largely explored by the ISM and a number of solutions, some new and some already known, are derived to generalize isotropic and homogeneous cosmologies. Most of the cosmological solutions have been generated from the spatially homogeneous but anisotropic Bianchi I background metrics. Soliton solutions which have a diagonal form can be generalized leading to new solutions and connecting others. Here we find pulse waves, cosolitons, composite universes, and in particular the collision of solitons on a cosmological background. The last of these is described and studied in some detail, and compared with the soliton waves of nonlinear physics. In chapter 5 soliton metrics that are not diagonal or in backgrounds different from Bianchi I are considered. Nondiagonal metrics are more difficult to characterize and study but they present the most clear nonlinear features of soliton physics such as the time delay when solitons interact. Solutions representing finite perturbations of isotropic cosmologies are also derived and studied.

In chapter 6 we describe gravitational solitons with cylindrical symmetry. Mathematically most of the gravitational solutions in this context are easily derived from the cosmological solution of the two previous chapters but, of course, they describe different physics. In chapter 7 we describe the connection of gravitational solitons with exact gravitational plane waves and the head-on collision of plane waves. We illustrate the physically more interesting properties of the spacetimes describing plane waves and the head-on collision of plane waves with some simple examples. The interaction region of the head-on collision of two exact plane waves has the symmetries which allow the application of the ISM. We show how most of the well known solutions representing colliding plane waves may be derived as gravitational solitons.

Chapter 8 is devoted to the stationary axisymmetric gravitational soliton solutions. Now the relevant metric field equations are elliptic rather than hyperbolic, but the ISM of chapter 1 is easily translated to this case. We describe in detail how the Schwarzschild and Kerr metrics, and their Kerr–NUT generalizations are simply obtained as gravitational solitons from a Minkowski background. The generalized soliton solutions of the Weyl class, which are related to diagonal metrics in the cosmological and cylindrical contexts, are obtained and their connection with some well known solutions is discussed. Finally the Tomimatsu–Sato solution is derived as a gravitational soliton solution obtained by a limiting procedure from the general soliton solution.

In our view only some of the earlier expectations of the application of the ISM in the gravitational context have been partially fulfilled. This technique has allowed the generation of some new and potentially relevant solutions and has provided us with a unified picture of many solutions as well as given us some new relations among them. The ISM has, however, been less successful in the characterization of the gravitational solitons as the soliton waves of nongravitational physics. It is true that in some restricted cases soliton solutions can be topologically characterized in a mathematical sense, but this characterization is then blurred in the physics of the gravitational spacetime the solutions describe. Things like the velocity of propagation, energy of the solitons, shape persistence and time shift after collision have been only partially characterized, and this has represented a clear obstruction in any attempt to the quantization of gravitational solitons. We feel that more work along these lines should lead to a better understanding of gravitational physics at the classical and, even possibly, the quantum levels.

As regards to the level of presentation of this book we believe that its contents should be accessible to any reader with a first introductory course in general relativity. Little beyond the formulation of Einstein equations and some elementary notions on differential geometry and on partial differential equations is required. The rudiments of the ISM are explained with a practical view towards its generalization to the gravitational field.

We would like to express our gratitude to the collaborators and colleagues who over the past years have contributed to this field and from whom we have greatly benefited. Among our collaborators we are specially grateful to G.A. Alekseev, B.J. Carr, J. Céspedes, A. Curir, M. Dorca, M. Francaviglia, X. Fustero, J. Garriga, J. Ibáñez, P.S. Letelier, R. Ruffini, and V.E. Zakharov. We are also very grateful to W.B. Bonnor, J. Centrella, S. Chandrasekhar, A. Feinstein, V. Ferrari, R.J. Gleiser, D. Kitchingham, M.A.H. MacCallum, J.A. Pullin, H. Sato, A. Shabat and G. Neugebauer for stimulating discussions or suggestions.

Rome Vladimir A. Belinski
Barcelona Enric Verdaguer
September 2000

1

Inverse scattering technique in gravity

The purpose of this chapter is to describe the Inverse Scattering Method (ISM) for the gravitational field. We begin in section 1.1 with a brief overview of the ISM in nonlinear physics. In a nutshell the procedure involves two main steps. The first step consists of finding for a given nonlinear equation a set of linear differential equations (spectral equations) whose integrability conditions are just the nonlinear equation to be solved. The second step consists of finding the class of solutions known as soliton solutions. It turns out that given a particular solution of the nonlinear equation new *soliton* solutions can be generated by purely algebraic operations, after an integration of the linear differential equations for the particular solution. We consider in particular some of the best known equations that admit the ISM such as the Korteweg–de Vries and the sine-Gordon equations. In section 1.2 we write Einstein equations in vacuum for spacetimes that admit an orthogonally transitive two-parameter group of isometries in a convenient way. In section 1.3 we introduce a linear system of equations for which the Einstein equations are the integrability conditions and formulate the ISM in this case. In section 1.4 we explicitly construct the so-called n-soliton solution from a certain background or seed solution by a procedure which involves one integration and a purely algebraic algorithm which involves the so-called pole trajectories. In the last section we discuss the use of the ISM for solving Einstein equations in vacuum with an arbitrary number of dimensions, and the use of the Kaluza–Klein ansatz to find some nonvacuum soliton solutions in four dimensions.

1.1 Outline of the ISM

The ISM is an important tool of mathematical physics by means of which it is possible to solve a certain type of nonlinear partial differential equations using the techniques of linear physics. This book is not about the ISM, its main concern are the so-called soliton solutions, and these only in the context of general relativity. But since such solutions can be obtained by the ISM, it is of course of interest to have some familiarity with the method. However, mastering the ISM is by no means essential for reading this book because, first to find soliton solutions one does not require the full machinery of the ISM, and second the peculiarities of the gravitational case require specific techniques that will be explained in detail in the following sections.

In subsection 1.1.1 we give a brief summary of the ISM including relevant references to the literature. Terms such as Schrödinger equation, scattering data, and transmission and reflection coefficients are borrowed from quantum mechanics, thus readers familiar with that subject may gain some insight from this subsection. Some readers may prefer to have only a quick glance at subsection 1.1.1 and to look in more detail at subsection 1.1.2 where some familiar examples of fluid dynamics and of relativistic physics are discussed. Of particular interest for the purposes of this book is the last example discussed and the method of how to construct solitonic solutions by purely algebraic operations from a given particular solution.

In any case, the key points that should be retained from subsection 1.1.1 are the following. A nonlinear partial differential equation such as (1.1) for the function $u(z, t)$ is integrable by the ISM when the following occur. First, one must be able to associate to the nonlinear equation a linear eigenvalue problem such as (1.2), where the unknown function $u(z, t)$ plays the role of a 'potential' in the linear operator. Given an initial value $u(z, 0)$, (1.2) defines scattering data: this is the well known problem in quantum mechanics of scattering of a particle in a potential $u(z, 0)$ and includes the transmission and reflection coefficients and the energy eigenvalues. Second, it must be possible to provide an equation such as (1.3) for the time evolution of these data, such that the integrability conditions of the two equations (1.2) and (1.3) implies (1.1). In this case the nonlinear equation is integrable by the ISM and the solution $u(z, t)$ is found by computing the potential corresponding to the time-evolved scattering data. This last step is the inverse scattering problem and requires the solution of a usually nontrivial linear integral equation. Although the whole procedure is generally complicated there is a special class of solutions called *soliton solutions* for which the inverse scattering problem can be solved exactly in analytic form.

1.1.1 The method

Let us consider the nonlinear two-dimensional partial differential equation for the function $u(z, t)$

$$u_{,t} = F(u, u_{,z}, u_{,zz}, \ldots), \tag{1.1}$$

where t is the time variable, z is a space coordinate, and F is a nonlinear function. To integrate this equation, which is first order with respect to time, by the ISM one considers the scattering problem for the following stationary one-dimensional Schrödinger equation,

$$L\psi = \lambda\psi, \quad L = -\frac{d^2}{dz^2} + u(z, t),\qquad (1.2)$$

where the unknown function $u(z, t)$ plays the role of the potential. Here the time t in u is an external parameter that should not be confused with the conventional time in quantum mechanics, which appears in the time-dependent Schrödinger equation associated to (1.2). We assume also that $u(t, z)$ vanishes at $z \to \pm\infty$ fast enough (like $z^2u \to 0$ or faster).

Let $u(z, 0)$ be the Cauchy data at time $t = 0$ and consider the so-called direct scattering problem, which consists of finding the full set of scattering data $S(\lambda, 0)$ produced by the potential $u(z, 0)$. The scattering data $S(\lambda, 0)$ are the set of quantities that allow us to find the asymptotic values of the eigenfunctions $\psi(\lambda, z, 0)$ at $z \to -\infty$ through the given asymptotic values of $\psi(\lambda, z, 0)$ at $z \to +\infty$ for each value of the spectral parameter λ. This parameter is the energy of the scattered particle and positive values are the continuous spectrum for the problem (1.2). Moreover, a discrete set of negative eigenvalues of λ can also enter into the problem corresponding to the bound states of the particle in the potential u. Thus, the set $S(\lambda, 0)$ should contain the forward and backward scattering amplitudes for the continuous spectrum (in the one-dimensional problem these are the transmission and reflection coefficients, $T(\lambda)$ and $R(\lambda)$, respectively), and the negative eigenvalues λ_n of the discrete spectrum together with some coefficients, C_n, which link the asymptotic values of the eigenfunctions for the bound states $\psi_n(\lambda_n, z, 0)$ at $z \to \pm\infty$.

We can also consider the inverse of the problem just described. The task in this case is to reconstruct the potential $u(z)$ through a given set of scattering data $S(\lambda)$. This is the inverse scattering problem. It has been investigated in detail in the last forty years and the main steps of its solution are now well known. In principle, for any appropriate set of scattering data $S(\lambda)$ it is possible to reconstruct the corresponding potential $u(z)$. It is easy to see that one could solve the Cauchy problem for $u(z, t)$ using this technique. In fact, let us imagine that after constructing the scattering data $S(\lambda, 0)$ corresponding to the potential $u(z, 0)$ at $t = 0$ we could know the time evolution of S and are able to get from the initial values $S(\lambda, 0)$ the scattering data $S(\lambda, t)$ at any arbitrary time t. Then we can apply the inverse scattering technique to $S(\lambda, t)$ and reconstruct the potential $u(z, t)$ at any time. This would give the desired solution to the Cauchy problem.

This programme, however, is only attractive if such a 'miracle' can happen which means, for practical purposes, that we need some evolution equations for the scattering data $S(\lambda, t)$ that can be integrated in a simple way. It turns out that for a number of special classes of differential equations of nonlinear physics this is the case. This discovery was made by Gardner, Greene, Kruskal and Miura in 1967 in a famous paper [111] dedicated to the method of solving the Cauchy data problem for the Korteweg–de Vries equation. This was the beginning of a rapid development of the ISM and now we have a vast literature on the subject. One of the more recent books is ref. [231], and readers can also find textbook expositions, including historical reviews, in refs [84, 302]. The review article [259] and the book of collected papers book [247], which includes a good introductory guide through the literature, are also very useful.

Now let us look closer at the remarkable possibility of finding the exact time evolution for the scattering data. The fact is that for integrable cases (in the sense of the ISM) the eigenvalues of the associated spectral problem (1.2) are independent of time t and the eigenfunctions $\psi(\lambda, z, t)$ obey, besides (1.2), another partial differential equation which is of first order in time. This is the key point, since this additional evolution equation for the eigenfunctions allows us to find the exact time dependence of the scattering data. This equation can be written as

$$\dot{\psi} = A\psi, \qquad (1.3)$$

where the differential operator A depends on $u(z, t)$ and contains only derivatives with respect to the space coordinate z. This remarkable set of equations, namely, (1.2) and (1.3), is often called a *Lax pair*, or *Lax representation* of the integrable system, or *L–A pair* [186]. The existence of two equations for the eigenfunction ψ means that a selfconsistency condition must be satisfied. In each case it is easy to show that this condition coincides exactly with the original equation of interest, (1.1). Consequently, the problem can now be put into a slightly different form: all integrable nonlinear two-dimensional equations are the selfconsistency conditions for the existence of a joint spectrum and a joint set of eigenfunctions for two differential operators whose coefficients (which play the role of potentials) depend on $u(z, t)$ and, in general, on its derivatives. This was the basic point for a further generalization of the ISM to multicomponent fields $u(z, t)$ and to several families of differential operators. This work was largely due to Zakharov and Shabat (see ref. [231], chapter 3, and ref. [84], chapter 6, and references therein). Of course, only very special classes of nonlinear differential equations admit L–A pairs and still today there is no general approach on how to find these classes. Despite the existence of a number of powerful techniques each differential system needs individual and, often, sophisticated consideration.

Let us return to our problem (1.1). From what we have just said we know that this equation is integrable by the ISM if the time evolution of the scattering data can be found. However, it is important to understand the restricted sense of this integrability. In order to perform an actual integration we need to be able to solve the inverse scattering problem for the data $S(\lambda, t)$. In general this cannot be done in analytic form, because the inverse problem $S(\lambda, t) \rightarrow u(z, t)$ is based on complicated integral equations of the Gelfand, Levitan and Marchenko [231]. Also there is no possibility, in general, for analytic solutions of the direct scattering problem $u(z, 0) \rightarrow S(\lambda, 0)$. What can really be done in general is to find the explicit expression for the asymptotic values of the field $u(z, t)$ at $t \rightarrow +\infty$ directly through the initial Cauchy data. Of course, the possibility of even this restricted use of the ISM is very valuable because in many physical problems all we need to know is the late time asymptotic values of the field.

Soliton solutions. Another great advantage of the ISM is really remarkable: for each integrable equation (1.1) (or system of equations) there are special classes of solutions $u(z, t)$ for which the direct and inverse scattering problems can be solved exactly in analytic form! These are the so-called *soliton solutions*. We mentioned before that for the continuous spectrum of positive λs the scattering data consist of the backward and forward scattering amplitudes or the reflection and the transmission coefficients, $R(\lambda)$ and $T(\lambda)$ respectively. The reflection coefficient is identically zero for solitons, and this property is independent of time. It can be shown that if for some initial potential $u(z, 0)$ all the coefficients $R(\lambda, 0)$ vanish, then they will vanish at any time t due to the evolution equations of the scattering data. The solutions $u(z, t)$ of that kind are often called 'reflectionless potentials'. In such cases the values λ_n of the discrete spectrum and the coefficients $C_n(\lambda_n, t)$, the time evolution of which can be also easily found, determine all the structure of the ISM. It is well known that the values λ_n coincide with the simple poles of the transmission amplitude $T(\lambda)$, and the positions of these poles completely determine the analytical structure of the scattering data and the eigenfunctions of the spectral problem (1.2) in the complex λ-plane. The transmission amplitude and the behaviour of the eigenfunctions of (1.2) and (1.3) as functions of the spectral parameter λ in the complex λ-plane are completely determined by this simple pole structure. In this case even a first look at the equation of the ISM suffices to see that the main steps of the ISM for the solitonic case are purely algebraic. This is integrability in its simplest direct sense.

1.1.2 Generalization and examples

Although we have discussed the idea of the ISM with the example of the first-order differential equation with respect to time for a single function $u(z, t)$, the qualitative character of our previous statements also remains valid in any extended integrable case. The generalization to second order equations and to multicomponent fields $u(z, t)$ is straightforward. In these cases instead of (1.2) and (1.3) we have two systems of equations and the multicomponent analogue of the spectral problem (1.2) presents no difficulties [231]. For such extended versions of the ISM we need only a change in the terminology. The generalized version of (1.2) is no longer a Schrödinger equation, but some Schrödinger-type system, and the same for the inverse scattering transformation of Gelfand,

Levitan and Marchenko. In addition the parameter λ can no longer be the energy but is instead some spectral parameter, etc.

Further development of the ISM [312] showed that most of the known two-dimensional equations and their possible integrable generalizations can be represented as selfconsistency conditions for two matrix equations,

$$\psi_{,z} = U^{(1)}\psi, \quad \psi_{,t} = V^{(1)}\psi, \tag{1.4}$$

where the matrices $U^{(1)}$ and $V^{(1)}$ depend rationally on the complex spectral parameter λ and on two real spacetime coordinates z and t. The column matrix ψ is a function of these three independent variables also. Differentiating the first of these two equations with respect to t and the second one with respect to z we obtain, after equating the results, the consistency condition for system (1.4):

$$U^{(1)}_{,t} - V^{(1)}_{,z} + U^{(1)}V^{(1)} - V^{(1)}U^{(1)} = 0. \tag{1.5}$$

This condition should be satisfied for all values of λ and this requirement coincides explicitly with the integrable differential equation (or system) of interest. Let us see a few examples [231], which will be of special interest.

Korteweg–de Vries equation. If we choose

$$U^{(1)} = i\lambda \begin{pmatrix} 1 & 0 \\ 0 & -1 \end{pmatrix} + \begin{pmatrix} 0 & 1 \\ u & 0 \end{pmatrix}, \tag{1.6}$$

$$V^{(1)} = 4i\lambda^3 \begin{pmatrix} 1 & 0 \\ 0 & -1 \end{pmatrix} + 4\lambda^2 \begin{pmatrix} 0 & 1 \\ u & 0 \end{pmatrix}$$

$$+ 2i\lambda \begin{pmatrix} u & 0 \\ u_{,z} & -u \end{pmatrix} + \begin{pmatrix} -u_{,z} & 2u \\ 2u^2 - u_{,zz} & u_{,z} \end{pmatrix}, \tag{1.7}$$

then the left hand side of (1.5) becomes a fourth order polynomial in λ. All the coefficients of this polynomial, except one, vanish identically and we get from (1.5):

$$\begin{pmatrix} 0 & 0 \\ u_{,t} - 6uu_{,z} - u_{,zzz} & 0 \end{pmatrix} = 0, \tag{1.8}$$

which is the Korteweg–de Vries equation:

$$u_{,t} - 6uu_{,z} - u_{,zzz} = 0, \tag{1.9}$$

an equation of the form of (1.1). The function ψ in this case is the column

$$\psi = \begin{pmatrix} \psi_1 \\ \psi_2 \end{pmatrix}, \tag{1.10}$$

and from the first equation of (1.4) we have the following spectral problem:

$$\psi_{1,z} = i\lambda\psi_1 + \psi_2, \tag{1.11}$$

$$\psi_{2,z} = -i\lambda\psi_2 + u\psi_1, \tag{1.12}$$

which is equivalent to the Schrödinger equation (1.2). In fact, from the first equation (1.11) we can express ψ_2 in terms of ψ_1, and then substituting into the second, we get

$$-\psi_{1,zz} + u\psi_1 = \lambda^2\psi_1, \tag{1.13}$$

which coincides with (1.2) after a redefinition of the spectral parameter ($\lambda^2 \to \lambda$).

A second example appears when one is dealing with relativistic invariant second order field equations. From the mathematical point of view the physical nature of the variables z and t in (1.4) is irrelevant and we can interpret them as null (light-like) coordinates. But in order to avoid notational confusion, here and in the following, the variables t and z are always, respectively, time-like and space-like coordinates, and we introduce a pair of null coordinates ζ and η as

$$\zeta = \frac{1}{2}(z + t), \quad \eta = \frac{1}{2}(z - t). \tag{1.14}$$

Now, instead of (1.4) and (1.5) we have, in these new coordinates,

$$\psi_{,\zeta} = U^{(2)}\psi, \quad \psi_{,\eta} = V^{(2)}\psi, \tag{1.15}$$

$$U^{(2)}_{,\eta} - V^{(2)}_{,\zeta} + U^{(2)}V^{(2)} - V^{(2)}U^{(2)} = 0, \tag{1.16}$$

where $U^{(2)} = U^{(1)} + V^{(1)}$ and $V^{(2)} = U^{(1)} - V^{(1)}$.

Sine-Gordon equation. If we choose

$$\left.\begin{array}{l} U^{(2)} = i\lambda \begin{pmatrix} 1 & 0 \\ 0 & -1 \end{pmatrix} + \begin{pmatrix} 0 & u_{,\zeta} \\ u_{,\zeta} & 0 \end{pmatrix}, \\[16pt] V^{(2)} = \frac{1}{4i\lambda} \begin{pmatrix} \cos u & -i\sin u \\ i\sin u & -\cos u \end{pmatrix}, \end{array}\right\} \tag{1.17}$$

we get, from (1.16),

$$\begin{pmatrix} 0 & u_{,\zeta\eta} - \sin u \\ u_{,\zeta\eta} - \sin u & 0 \end{pmatrix} = 0, \tag{1.18}$$

which is the sine-Gordon equation:

$$u_{,\zeta\eta} = \sin u. \tag{1.19}$$

The function ψ is still the column (1.10) and the spectral problem that follows from the first of equations (1.15) gives

$$\psi_{1,\zeta} = i\lambda\psi_1 + \frac{i}{2}u_{,\zeta}\psi_2, \tag{1.20}$$

$$\psi_{2,\zeta} = -i\lambda\psi_2 + \frac{i}{2}u_{,\zeta}\psi_1. \tag{1.21}$$

After solving the direct scattering problem for this 'stationary' system it is easy to find the evolution of scattering data in the 'time' η. The inverse scattering transform then gives the solution for $u(\zeta, \eta)$ (see the details in ref. [231]).

In general the matrices U and V can have an arbitrary size $N \times N$ (the same follows for the column matrix ψ) as well as a more complicated dependence on the parameter λ. Each choice will give some complicated (in general) integrable system of differential equations. Most of them do not yet have a physical interpretation but a number of interesting possibilities arise.

Principal chiral field equation. Let us consider, first of all, the case when U and V are regular at infinity in the λ-plane and have simple poles only at finite values of the spectral parameter (we should not confuse these poles with the poles of the scattering data in the same plane). As was shown in ref. [312] in this case we can construct matrices U and V which vanish at $|\lambda| \to \infty$, due to the gauge freedom in the system (1.15)–(1.16). We shall restrict ourselves to the simplest case in which U and V have only one pole each. Without loss of generality we can choose the positions of these poles to be at $\lambda = \lambda_0$ and $\lambda = -\lambda_0$, where λ_0 is an arbitrary constant. Now for $U^{(2)}$ and $V^{(2)}$ we have

$$U^{(2)} = \frac{K}{\lambda - \lambda_0}, \quad V^{(2)} = \frac{L}{\lambda + \lambda_0}, \tag{1.22}$$

where the matrices K and L are independent of λ. Substitution of (1.22) into (1.16) shows that the left hand side of (1.16) vanishes if and only if the following relations hold:

$$K_{,\eta} - L_{,\zeta} = 0, \tag{1.23}$$

$$K_{,\eta} + L_{,\zeta} + \frac{1}{\lambda_0}(KL - LK) = 0. \tag{1.24}$$

Equation (1.24) suggests that we can represent K and L in terms of 'logarithmic derivatives' of some matrix g as

$$K = -\lambda_0 g_{,\zeta} g^{-1}, \quad L = \lambda_0 g_{,\eta} g^{-1}. \tag{1.25}$$

Then, (1.24) is simply the integrability condition of (1.25) for the matrix g, and (1.23) is the field equation for some integrable relativistic invariant model:

$$\left(g_{,\zeta} g^{-1}\right)_{,\eta} + \left(g_{,\eta} g^{-1}\right)_{,\zeta} = 0. \tag{1.26}$$

This matrix equation is associated with the model of the so-called principal chiral field and received much attention in the 1980s and 1990s. The first description of the integrability of this model in the language of the commutative representation (1.16) was given in ref. [312], but a more detailed description can be found in ref. [311] or in ref. [231]. The exact solution of the corresponding quantum chiral field model was investigated in refs [244] and [95].

From any solution $\psi(\zeta, \eta, \lambda)$ of the 'L–A pair' (1.15) one immediately gets a solution of the field equation (1.26) for g. In fact, from (1.15), (1.22) and (1.25) it follows that

$$\psi_{,\zeta}\psi^{-1} = \frac{K}{\lambda - \lambda_0} = \frac{-\lambda_0 g_{,\zeta} g^{-1}}{\lambda - \lambda_0} \longrightarrow g_{,\zeta} g^{-1}, \tag{1.27}$$

$$\psi_{,\eta}\psi^{-1} = \frac{L}{\lambda + \lambda_0} = \frac{\lambda_0 g_{,\eta} g^{-1}}{\lambda + \lambda_0} \longrightarrow g_{,\eta} g^{-1}, \tag{1.28}$$

when $\lambda \to 0$, which means that the matrix of interest equals the matrix eigenfunction $\psi(\zeta, \eta, \lambda)$ at the point $\lambda = 0$,

$$g(\zeta, \eta) = \psi(\zeta, \eta, 0). \tag{1.29}$$

The solution of the general Cauchy problem for (1.26) can be obtained in the framework of the classical ISM in the form we have explained. We can also use a more elegant and modern method, based on the Riemann problem in the theory of functions of complex variables, which was proposed by Zakharov and Shabat [231, 312]. Of course any method will lead us to integral equations of the Gelfand, Levitan and Marchenko type and the Zakharov and Shabat method is no exception. But what is important for us here is that the previous approach is the best suited for practical calculations in the solitonic case. In this book we will deal only with solitons and we will follow the commutative representation (1.15) and (1.16) of the ISM.

If we are interested only in the solitonic solutions of (1.26) we do not need to study the Riemann problem, the spectrum and the direct and inverse scattering transforms. All we need to know is one particular exact solution (g_0, ψ_0) of (1.26) and (1.15), which we will call the background solution or the seed solution, together with the number of solitons we wish to introduce on this background. We know already that in the solitonic case the poles of the transmission amplitude completely determine the problem. Since the transmission amplitude is just a part of the eigenfunction $\psi(\zeta, \eta, \lambda)$, such a function exhibits the same simple pole structure in some arbitrarily large, but finite, part of the λ-plane. Simple inspection shows that in this case $\psi(\zeta, \eta, \lambda)$ can be represented in the form

$$\psi = \chi \psi_0, \tag{1.30}$$

where $\psi_0(\zeta, \eta, \lambda)$ is the particular solution mentioned before and χ is a new matrix, called the *dressing matrix*, which can be normalized in such a way that it tends to the unit matrix, I, when $|\lambda| \to \infty$. Then the λ dependence of the χ matrix for the solitonic case is very simple:

$$\chi = I + \sum_{n=1}^{N} \frac{\chi_n}{\lambda - \lambda_n}, \tag{1.31}$$

where λ_n are arbitrary constants and the χ_n matrices are independent of λ. The number of poles in (1.31) corresponds to the number of solitons which we have added to the background (g_0, ψ_0). Of course the set of λ_n constitutes the discrete spectrum of the spectral problem (1.15), but this need not concern us here. After choosing any set of parameters λ_n and a background solution (g_0, ψ_0), we should substitute (1.30) and (1.31) into (1.15), and the matrices χ_n will be obtained by purely algebraic operations. After that, from (1.31), (1.30) and (1.29) we obtain the solution for $g(\zeta, \eta)$ in terms of the background solution g_0:

$$g = \chi(\zeta, \eta, 0)g_0 = g_0 - \left(\sum_{n=1}^{N} \lambda_n^{-1} \chi_n \right) g_0. \tag{1.32}$$

This is an example of the so-called *dressing technique* developed by Zakharov and Shabat. For the pure solitonic case it is straightforward to compute the new solutions from a given background solution.

1.2 The integrable ansatz in general relativity

If we wish to apply the two-dimensional ISM to the Einstein equations in vacuum

$$R_{\mu\nu} = 0, \tag{1.33}$$

where $R_{\mu\nu}$ is the Ricci tensor, we need to examine the particular case in which the metric tensor $g_{\mu\nu}$ depends on two variables only, which correspond to spacetimes that admit two commuting Killing vector fields, i.e. an Abelian two-parameter group of isometries. In this chapter we take these variables to be the time-like and the space-like coordinates $x^0 = t$ and $x^3 = z$ respectively. This corresponds to nonstationary gravitational fields, i.e. to wave-like and cosmological solutions of Einstein equations (1.33), and the two Killing vectors are both space-like. In any spacetime using the coordinate transformation freedom, $x^\mu = x^\mu(x'^\nu)$, we can fix the following constraints on the metric tensor

$$g_{00} = -g_{33}, \quad g_{03} = 0, \quad g_{0a} = 0. \tag{1.34}$$

Here, and in the following the Latin indices $a, b, c, \ldots$ take the values $1, 2$. In these coordinates the spacetime interval becomes

$$ds^2 = f(dz^2 - dt^2) + g_{ab}dx^a dx^b + 2g_{a3}dx^a dz, \tag{1.35}$$

where $f = -g_{00} = g_{33}$. If we now restrict ourselves to the case in which all metric components in (1.35) depend on t and z only, the Einstein equations for such a metric are still too complicated for the ISM or, more precisely, it is unknown at present whether the ISM can be applied in this case. The situation is different in the particular case in which $g_{a3} = 0$. Since it is not possible to eliminate the metric coefficients g_{a3} by any further coordinate transformation such a simplification should be considered as a real physical constraint. This corresponds to assuming the existence of 2-surfaces orthogonal to the group orbits, i.e. to assuming an orthogonally transitive group of isometries, which is a restriction on the two commuting Killing vectors. We should note that in the stationary axisymmetric case the two commuting Killing vectors already guarantee the existence of orthogonal 2-surfaces, provided some conditions on the nonsingular symmetry axis are satisfied [236, 179]. Therefore, from now on, we shall deal with the simplified block diagonal form of the metric (1.35):

$$ds^2 = f(t, z)(dz^2 - dt^2) + g_{ab}(t, z)dx^a dx^b. \qquad (1.36)$$

The stationary axisymmetric gravitational fields correspond to the analogue of this metric when the independent variables are both space-like. From the mathematical point of view the ISM for the stationary case presents no essential differences with respect to the present case and the solutions in such a case can be extracted from that case after appropriate complex transformations. However, due to the essential difference in the boundary conditions problem in the two cases it is better to consider the stationary metrics separately; we shall deal with this case in chapter 8.

The metric (1.36) was first considered in 1937 by Einstein and Rosen [90] for a diagonal matrix g_{ab}, when the Einstein equations (1.33) actually reduce to one linear equation in cylindrical coordinates. The inclusion of the off-diagonal component g_{12} changes the situation drastically, and converts the Einstein equations into an essentially nonlinear problem. In the language of the weak gravitational waves this corresponds to the appearance of a second independent polarization state of the wave. For the stationary analogue of the metric (1.36) such a generalization means (under reasonable boundary conditions) that rotation has been included. Equations for the metric (1.36) were first considered by Kompaneets [176], who noted some of their general properties. In the past, several authors using different simplifying assumptions have obtained a number of exact nontrivial solutions for a metric of the type (1.36) or its stationary analogue (most of these solutions are listed in ref. [179]), but a regular integration procedure was only found in 1978 [23].

From the physical point of view the metric (1.36) and its stationary analogue have many applications in gravitational theory. It suffices to say that to such a class belong the classical solutions of the Robinson–Bondi plane waves, the Einstein–Rosen cylindrical wave solutions and their two-polarization generalizations, the homogeneous cosmological models of Bianchi types I–VII including the Friedmann–Lemaître–Robertson–Walker models, the Schwarzschild and Kerr solutions, Weyl's axisymmetric solutions, etc. For many more contemporary results the reader can refer to refs [179, 180]. All this shows that in spite of its relative simplicity a metric of the type (1.36) encompasses a wide variety of physically relevant cases, and that a method for integrating the corresponding Einstein equations could significantly move forward some of our understanding of gravitational theory.

It turns out that this case can be successfully treated by means of some generalization of the Zakharov–Shabat form of the ISM. The Einstein equations for the metric (1.36) are most conveniently investigated in the null coordinates (ζ, η) introduced in (1.14). In what follows we shall always denote by g the two-dimensional real and symmetric matrix with elements g_{ab}, i.e. the two-dimensional block of the metric tensor (1.36):

$$g = \begin{pmatrix} g_{11} & g_{12} \\ g_{21} & g_{22} \end{pmatrix}. \tag{1.37}$$

For the determinant of this matrix it is convenient to introduce the notation

$$\det g = \alpha^2, \tag{1.38}$$

and we shall always consider that α is nonnegative: $\alpha \geq 0$. This is in agreement with the fact that the points $\alpha = 0$ usually (but not always) correspond to physical singularities and in such cases continuation of the solutions through these points is meaningless. It turns out that the R_{0a} and R_{3a} components of the Ricci tensor for the metric (1.36) are identically zero. The remaining system of the vacuum Einstein equations (1.33) for this metric decomposes into two sets. The first one follows from equations $R_{ab} = 0$ and with the use of the null coordinates (1.14) can be written in the form of a single matrix equation:

$$\left(\alpha g_{,\zeta} g^{-1}\right)_{,\eta} + \left(\alpha g_{,\eta} g^{-1}\right)_{,\zeta} = 0. \tag{1.39}$$

The second set follows from the equations $R_{00} + R_{33} = 0$ and $R_{03} = 0$, and gives the metric coefficient $f(t, z)$ in terms of the matrix g, solution of (1.39), via the relations:

$$(\ln f)_{,\zeta} (\ln \alpha)_{,\zeta} = (\ln \alpha)_{,\zeta\zeta} + \frac{1}{4\alpha^2} \operatorname{Tr} A^2, \tag{1.40}$$

$$(\ln f)_{,\eta} (\ln \alpha)_{,\eta} = (\ln \alpha)_{,\eta\eta} + \frac{1}{4\alpha^2} \operatorname{Tr} B^2, \tag{1.41}$$

where the matrices A and B are defined by

$$A = -\alpha g_{,\zeta} g^{-1}, \quad B = \alpha g_{,\eta} g^{-1}. \tag{1.42}$$

It is easy to see that the integrability condition for (1.40) and (1.41) with respect to f is automatically satisfied if g satisfies (1.39). The equation $R_{00} - R_{33} = 0$ can be written in the form

$$(\ln f)_{,\zeta\eta} = \frac{1}{4\alpha^2} \operatorname{Tr} AB - (\ln \alpha)_{,\zeta\eta}, \tag{1.43}$$

but this is not a new equation, it is just a consequence of the system (1.38)–(1.42) when α is not a constant. The special case in which α is constant does not deserve special treatment because it corresponds to flat Minkowski spacetime. In fact, it follows from (1.40) and (1.41) that in this case, $\operatorname{Tr} A^2 = \operatorname{Tr} B^2 = 0$, and it is easy to see that this can happen only for $g = \text{constant}$. In this specific case (1.40) and (1.41) do not determine the coefficient f and one needs (1.43) which, since $A = B = 0$, has the solution $f = \exp[f_1(\zeta) + f_2(\eta)]$, with arbitrary functions f_1 and f_2. But now a simple coordinate transformation $\zeta = \zeta(\zeta')$ and $\eta = \eta(\eta')$ reduces the new coefficient f to a constant.

It is remarkable that the basic set of Einstein equations for the metric (1.36), i.e. (1.39), is very similar to (1.26) for the principal chiral field. The difference is that in (1.39) we have the additional factor $\alpha = (\det g)^{1/2}$ instead of a constant. If one were to forget (1.40) and (1.41), then (1.39) would formally have nontrivial solutions even when α is constant and these would correspond to a subclass of solutions of chiral field theory. However, as we have just seen, such a special class of solutions has no relevance for the gravitational field.

Therefore, the technique described in the previous section requires some generalization in order to be applied to the gravitational field. As will be seen shortly, the general idea of the method remains the same: it is based on the study of the analytic structure of the eigenfunctions of the two operators (as functions of a complex parameter λ), which can be associated by a definite law to the system (1.38)–(1.39). In particular, for soliton solutions of (1.38)–(1.39), the structure of the poles of the corresponding functions in the λ-plane plays a fundamental role. For α not constant, (1.38)–(1.39) require the introduction of generalized differential operators entering into the 'L–A pair', which depend on the function $\alpha(\zeta, \eta)$, and which also contain derivatives with respect to the spectral parameter. For soliton solutions this leads to 'floating' poles of the eigenfunctions, and instead of stationary poles $\lambda_n = \text{constant}$ as in chiral field theory we now have pole trajectories $\lambda_n(\zeta, \eta)$.

The complete solution of the problem, i.e. the construction of the 'L–A pair' for (1.38)–(1.39) together with the general ISM for its integration and the procedure for computing the solitonic solution was presented by Belinski and Zakharov in ref. [23], where the first solitonic solution for the gravitational field was exhibited. In the next section we will follow the main lines of this paper together with paper [24] where the technique for the stationary analogue of metric (1.36) was developed. In connection with this we have to mention two important independent results which appeared at about the same time. The first is due to Maison [206], who constructed the linear eigenvalue problem in the spirit of Lax for the stationary analogue of metric (1.36). Due to the above discussed peculiarity of the gravitational equations his result was, of course, more sophisticated than the standard form of the Lax equations, but he posed the correct conjecture that the existence of the 'L–A pair' that he found entails the complete integrability of the system. The second result was

due to Harrison [135, 136] and Neugebauer [224], who derived the analogue
of the Bäcklund transformation for the stationary case of metric (1.36). By
means of the Bäcklund transformation it is possible to get from a given solution
a new solution, which usually can be seen as one soliton added to the given
background solution. The existence of the Bäcklund transformation implies
the complete integrability of the system. In his approach Harrison used the
'prolongation scheme' devised by Wahlquist and Estabrook (see ref. [247]).
Technically the constructions of Maison, Harrison and Neugebauer differ from
the ISM developed in refs [23] and [24], but practice has proved that this last
approach is more suitable for direct and explicit calculations. The equivalence of
the ISM, which in this context is also sometimes called *soliton transformation*,
with Harrison's Bäcklund transformations or Neugebauer's Bäcklund transformations was proved by Cosgrove [64, 65, 66].

1.3 The integration scheme

We now turn to a systematic investigation of (1.38)–(1.39). The trace of (1.39),
taking into account the condition (1.38), yields

$$\alpha_{,\zeta\eta} = 0. \tag{1.44}$$

Thus, the square root of the determinant of the matrix g satisfies a wave equation
(this result was already known to Einstein and Rosen [90]) with solutions

$$\alpha = a(\zeta) + b(\eta), \tag{1.45}$$

where $a(\zeta)$ and $b(\eta)$ are arbitrary functions. We shall later need a second
independent solution of (1.44) which we denote by $\beta(\zeta, \eta)$ and we choose it
in the form

$$\beta = a(\zeta) - b(\eta). \tag{1.46}$$

It should be understood that metric (1.36) admits, in addition, arbitrary
coordinate transformations $z' = f_1(z+t) + f_2(z-t), t' = f_1(z+t) - f_2(z-t)$,
which do not change the conformally flat form of the $f(dz^2 - dt^2)$ part. By
an appropriate choice of the functions f_1 and f_2 one can bring the functions
$a(\zeta)$ and $b(\eta)$ in (1.45) into a prescribed form. When this freedom is used
to write (α, β) as spacetime coordinates we say that the metric (1.36) has the
canonical form and (α, β) are called *canonical coordinates*. For instance, if the
variable $\alpha(\zeta, \eta)$ is time-like (corresponding to solutions of cosmological type)
the coordinates can be chosen in such a way that $\alpha = t$ and $\beta = z$; in this
case α and β are canonical coordinates. It is, however, more convenient to carry
through the analysis in a general form, without specifying the functions $a(\zeta)$
and $b(\eta)$ in advance, and turning to special cases when necessary.

It is easy to see that (1.39) is equivalent to a system consisting of (1.42) and
two first order matrix equations for the matrices A and B. From (1.39) and

(1.42) the first obvious equation for A and B is

$$A_{,\eta} - B_{,\zeta} = 0. \tag{1.47}$$

The second equation is easily derived as an integrability condition of (1.42) with respect to g. We obtain in this manner

$$A_{,\eta} + B_{,\zeta} + \alpha^{-1}[A, B] - \alpha_{,\eta}\alpha^{-1}A - \alpha_{,\zeta}\alpha^{-1}B = 0, \tag{1.48}$$

where the square brackets denote the commutator.

In close analogy with the ideas described in section 1.1 the main step now consists in representing (1.47) and (1.48) in the form of compatibility conditions of a more general overdetermined system of matrix equations related to an eigenvalue–eigenfunction problem for some linear differential operators. Such a system will depend on a complex spectral parameter λ, and the solutions of the original equations for the matrices g, A and B will be determined by the possible types of analytic structure of the eigenfunctions in the λ-plane. Although as we have already mentioned at present there is no general algorithm for the determination of such systems, this can be done [23] for the particular case of (1.38)–(1.39). To do so we introduce the following differential operators

$$D_1 = \partial_\zeta - \frac{2\alpha_{,\zeta}\lambda}{\lambda - \alpha}\partial_\lambda, \quad D_2 = \partial_\eta + \frac{2\alpha_{,\eta}\lambda}{\lambda + \alpha}\partial_\lambda, \tag{1.49}$$

where the symbol ∂ with a subscript denotes partial differentiation with respect to the corresponding variable and λ is a complex parameter independent of the coordinates ζ and η. It is easy to verify that the commutator of the operators D_1 and D_2 vanishes identically when α satisfies the wave equation. Thus taking (1.44) into account we have

$$[D_1, D_2] = 0. \tag{1.50}$$

We now introduce, as in section 1.1, a complex matrix function $\psi(\lambda, \zeta, \eta)$, which in this context is usually called the *generating matrix*, and consider the system of equations

$$D_1\psi = \frac{A}{\lambda - \alpha}\psi, \quad D_2\psi = \frac{B}{\lambda + \alpha}\psi, \tag{1.51}$$

where the matrices A and B are real and do not depend on the parameter λ (the same requirements are satisfied, of course, by the real function α). Then it turns out that the compatibility conditions for (1.51) coincide exactly with (1.47)–(1.48). In order to see this it is necessary to operate with D_2 on the first of equations (1.51) and with D_1 on the second one, and subtract the results. On account of the commutativity of D_1 and D_2 we get zero on the left hand side, while on the right hand side we get a rational function of λ which vanishes if, and only if, the conditions (1.47)–(1.48) are satisfied. It is easy to see that a solution of the system (1.51) guarantees not only that the equations satisfied by the matrices A and B are true, but also yields a solution of (1.42), i.e. the sought matrix $g(\zeta, \eta)$ which satisfies the original equations (1.38)–(1.39). The matrix $g(\zeta, \eta)$ is simply the value of the generating matrix $\psi(\lambda, \zeta, \eta)$ at $\lambda = 0$:

$$g(\zeta, \eta) = \psi(0, \zeta, \eta). \tag{1.52}$$

Indeed, in this case (1.51) for $\lambda = 0$ (for solutions which are regular in the neighbourhood of $\lambda = 0$) exactly duplicate (1.42). The matrix $g(\zeta, \eta)$ must, of course, be real and symmetric; later we shall formulate additional restrictions to the solutions of (1.51) which guarantee these requirements.

The procedure of integration assumes knowledge of at least one particular solution. Let $g_0(\zeta, \eta)$ be a particular solution of (1.38)–(1.39), then by means of (1.42) one can determine the matrices $A_0(\zeta, \eta)$ and $B_0(\zeta, \eta)$, and integrating (1.51) one can obtain the corresponding generating matrix $\psi_0(\lambda, \zeta, \eta)$. We now make the substitution

$$\psi = \chi \psi_0 \tag{1.53}$$

in (1.51), and obtain the following equations for the dressing matrix $\chi(\lambda, \zeta, \eta)$:

$$D_1 \chi = \frac{1}{\lambda - \alpha}(A\chi - \chi A_0), \quad D_2 \chi = \frac{1}{\lambda + \alpha}(B\chi - \chi B_0). \tag{1.54}$$

There are additional conditions to be imposed on the dressing matrix χ in order to ensure the reality and symmetry of the matrix g. The first consists of requiring the reality of χ on the real axis of the complex λ-plane (the matrix ψ must also satisfy this condition). This implies

$$\overline{\chi}(\overline{\lambda}) = \chi(\lambda), \quad \overline{\psi}(\overline{\lambda}) = \psi(\lambda), \tag{1.55}$$

where a bar denotes complex conjugation. Note that often for the sake of brevity we do not indicate the arguments ζ and η of some functions. The second condition is less trivial and is related to the following invariance property of the solutions of the system (1.54). Let us assume that the matrix $\chi(\lambda)$ satisfies (1.54). Replacing the argument λ in this matrix by α^2/λ, we obtain the new matrix $\chi'(\lambda)$:

$$\chi'(\lambda) = g\widetilde{\chi}^{-1}(\alpha^2/\lambda)g_0^{-1}, \tag{1.56}$$

where the tilde denotes transposition of the matrix. Direct verification suffices to convince oneself that the new matrix $\chi'(\lambda)$ also satisfies (1.54) if the matrix g is symmetric. We shall assume that $\chi'(\lambda) = \chi(\lambda)$ to guarantee the symmetry of the matrix g. Thus this condition takes the form

$$g = \chi(\lambda)g_0\widetilde{\chi}(\alpha^2/\lambda). \tag{1.57}$$

Moreover, it is necessary to require that when $\lambda \to \infty$ the dressing matrix $\chi(\lambda, \zeta, \eta)$ tends to the unit matrix,

$$\chi(\infty) = I. \tag{1.58}$$

Then these relations imply

$$g = \chi(0)g_0, \tag{1.59}$$

a result which also follows from conditions (1.52)–(1.53).

Thus, the problem now consists of solving (1.54) and determining the dressing matrix χ that satisfies the supplementary conditions (1.55) and (1.58). The following important point should be emphasized. The solution $g(\zeta, \eta)$ must also satisfy the requirement that $\det g = \alpha^2$. The function $\alpha(\zeta, \eta)$ is the same for the background solution g_0 and for the generalized g (recall that α is a given solution of the wave equation (1.44)) and, by definition, the background solution also satisfies the requirement $\det g_0 = \alpha^2$. Therefore, as follows from (1.59), one must impose on the matrix χ another restriction: $\det \chi(0) = 1$. However, it is more convenient not to worry about this condition during computations and to use a simple renormalization in the final results in order to obtain the correct functions. These (correct) functions will be called physical functions. It is easy to establish the legitimacy of this procedure from (1.39). In fact, if we obtain a solution of that equation with $\det g \neq \alpha^2$, the trace of (1.39) implies that $\det g$ satisfies the equation

$$\left(\alpha(\ln \det g)_{,\zeta}\right)_{,\eta} + \left(\alpha(\ln \det g)_{,\eta}\right)_{,\zeta} = 0. \tag{1.60}$$

We can then form the physical matrix $g^{(ph)}$ by

$$g^{(ph)} = \alpha(\det g)^{-1/2}g, \tag{1.61}$$

and it is easy to see that $g^{(ph)}$ satisfies (1.39) and also the condition $\det g^{(ph)} = \alpha^2$. The matrices A and B are also subject to appropriate transformations, namely,

$$A^{(ph)} = A - \alpha \left\{\ln[\alpha(\det g)^{-1/2}]\right\}_{,\zeta} I, \tag{1.62}$$

$$B^{(ph)} = B + \alpha \left\{\ln[\alpha(\det g)^{-1/2}]\right\}_{,\eta} I, \tag{1.63}$$

where A and B are defined in terms of g according to (1.42) and $A^{(ph)}$ and $B^{(ph)}$ are defined by the same formulas but in terms of the matrix $g^{(ph)}$.

1.4 Construction of the *n*-soliton solution

In the general case, in direct analogy with the theory of principal chiral fields, the determination of the matrix $\chi(\zeta, \eta, \lambda)$ amounts to solving the Riemann problem of analytic function theory which, in turn, is reduced to solving a singular matrix integral equation [23]. The general solution for χ represents the sum of the solitonic and the nonsolitonic parts. In this book we consider the purely solitonic solutions, i.e. when the nonsolitonic part is absent. This problem does not require the use of the Riemann problem and can be solved explicitly.

The existence of solutions of the soliton type is due to the presence in the λ-plane of points at which the determinant of χ has simple poles. Thus the purely solitonic solutions correspond to the case in which χ is representable as a rational function of the parameter λ with a finite number of simple poles, and is such that it tends to the unit matrix when $\lambda \rightarrow \infty$ as required by (1.58). From the reality condition for g, i.e. (1.55), it follows that these poles are either on the real axis of the complex λ-plane or come in pairs, i.e. for each complex pole $\lambda = \mu$ there is a corresponding complex conjugate pole $\lambda = \bar{\mu}$. From the symmetry condition (1.57) it follows that for each pole $\lambda = \mu$ there is a point $\lambda = \alpha^2/\mu$ of degeneracy of χ where the determinant of χ vanishes. The inverse matrix χ^{-1} has the same properties, as can easily be seen from (1.55) and (1.57). It thus follows that the dressing matrix χ has the form

$$\chi = I + \sum_{k=1}^{n} \frac{R_k}{\lambda - \mu_k}, \tag{1.64}$$

where the matrices R_k as well as the numerical functions μ_k no longer depend on λ. The reality condition discussed above implies that in the sum of (1.64) to each real term μ_k there should correspond a real matrix R_k, and that to each complex function μ_k there should correspond another function $\mu_{k+1} = \bar{\mu}_k$, and that $R_{k+1} = \bar{R}_k$. It can be seen from (1.64) and (1.59) that the solution of (1.39) for the matrix $g(\zeta, \eta)$ is

$$g(\zeta, \eta) = \left(I - \sum_{k=1}^{n} \mu_k^{-1} R_k \right) g_0. \tag{1.65}$$

Let us now determine the matrix R_k explicitly. For this it is necessary to substitute (1.64) into (1.54) and impose that these equations be satisfied at the poles $\lambda = \mu_k(\zeta, \eta)$. First, it can be seen that these equations explicitly determine the dependence of the positions of the poles on the coordinates ζ and η, i.e. the functions $\mu_k(\zeta, \eta)$. In fact, the right hand sides of (1.54) at the points $\lambda = \mu_k$ have only first order poles, whereas the left hand sides, $D_1\chi$ and $D_2\chi$, have second order poles. The requirement that the coefficients of the powers $(\lambda - \mu_k)^{-2}$ vanish on the left hand sides yields the following equations for the pole trajectories $\mu_k(\zeta, \eta)$:

$$\mu_{k,\zeta} = \frac{2\alpha_{,\zeta} \mu_k}{\alpha - \mu_k}, \quad \mu_{k,\eta} = \frac{2\alpha_{,\eta} \mu_k}{\alpha + \mu_k}. \tag{1.66}$$

These equations have the following invariance: if μ_k is a solution of (1.66), then α^2/μ_k is also a solution. The solutions of (1.66) are the roots of the quadratic equation

$$\mu_k^2 + 2(\beta - w_k)\mu_k + \alpha^2 = 0, \tag{1.67}$$

where w_k are arbitrary complex constants. For each given w_k, (1.67) yields two roots, μ_k and α^2/μ_k. If the modulus of the first root, $|\mu_k|$, is in the interval $[0, \alpha]$ the modulus of the second root, $|\alpha^2/\mu_k|$, is outside this interval. This enables us to introduce the notion of μ_k^{in} and μ_k^{out} corresponding to these two choices. All the poles μ_k^{in} of the χ matrix (1.64) are inside the circle $|\lambda| = \alpha$ in the complex λ-plane, and all the poles μ_k^{out} are outside this circle. These solutions for μ_k can be written in the form

$$\mu_k^{in} = (w_k - \beta)\left\{1 - [1 - \alpha^2(\beta - w_k)^{-2}]^{1/2}\right\}, \tag{1.68}$$

$$\mu_k^{out} = (w_k - \beta)\left\{1 + [1 - \alpha^2(\beta - w_k)^{-2}]^{1/2}\right\}, \tag{1.69}$$

and the branches of the square roots in these formulas should conform to the adopted definition of the solutions, i.e. $|\mu_k^{in}| < \alpha$ and $|\mu_k^{out}| > \alpha$. For example, in the case of real poles (real constants w_k) in the region where $1 - \alpha^2(\beta - w_k)^{-2} > 0$, the square roots take only positive values. The behaviour of μ_k^{in} and μ_k^{out} as functions of α and β is shown for this case in fig. 1.1 and fig. 1.2.

Let us rewrite (1.54) in the form

$$\frac{A}{\lambda - \alpha} = (D_1\chi)\chi^{-1} + \chi\frac{A_0}{\lambda - \alpha}\chi^{-1}, \tag{1.70}$$

$$\frac{B}{\lambda + \alpha} = (D_2\chi)\chi^{-1} + \chi\frac{B_0}{\lambda + \alpha}\chi^{-1}. \tag{1.71}$$

Since the left hand sides of these equations are regular at the poles $\lambda = \mu_k$, it is necessary that the residues of these poles on the right hand sides vanish at $\lambda = \mu_k$. This requirement leads to the following equations for the matrices R_k:

$$R_{k,\zeta}\chi^{-1}(\mu_k) + R_k\frac{A_0}{\mu_k - \alpha}\chi^{-1}(\mu_k) = 0, \tag{1.72}$$

$$R_{k,\eta}\chi^{-1}(\mu_k) + R_k\frac{B_0}{\mu_k + \alpha}\chi^{-1}(\mu_k) = 0, \tag{1.73}$$

where use has been made of the relation

$$R_k\chi^{-1}(\mu_k) = 0, \tag{1.74}$$

which follows from the identity $\chi\chi^{-1} = I$, at the poles $\lambda = \mu_k$. It can be seen from (1.74) that R_k and $\chi^{-1}(\mu_k)$ are degenerate matrices and their matrix elements can be written in the form

$$(R_k)_{ab} = n_a^{(k)}m_b^{(k)}, \quad [\chi^{-1}(\mu_k)]_{ab} = q_a^{(k)}p_b^{(k)}, \tag{1.75}$$

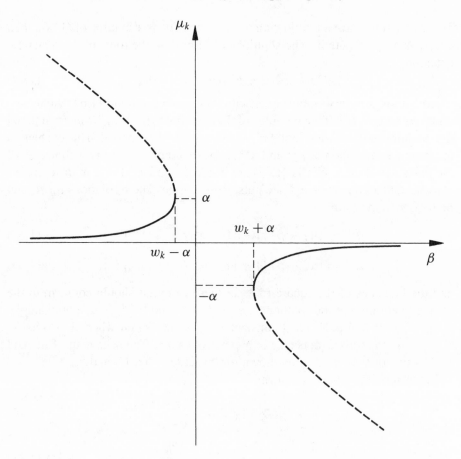

Fig. 1.1.   The behaviour of μ_k as a function of β for some fixed value of α for a real pole (real w_k). For definiteness we choose the value of the arbitrary constant w_k to be in the range $-\alpha < w_k < \alpha$. The smooth lines show μ_k for the case μ_k^{in} and the broken lines correspond to the case μ_k^{out}.

thus (1.74) implies that

$$m_a^{(k)} q_a^{(k)} = 0. \tag{1.76}$$

Here and in the following, summation will be understood to be over repeated vector and tensor indices a, b, c, d (recall that these take the values 1 and 2 only).

Substituting (1.75) into (1.72) and (1.73) we obtain the equations which determine the evolution of the vectors $m_a^{(k)}$:

$$\left[m_{a,\zeta}^{(k)} + m_b^{(k)} \frac{(A_0)_{ba}}{\mu_k - \alpha} \right] q_a^{(k)} = 0, \quad \left[m_{a,\eta}^{(k)} + m_b^{(k)} \frac{(B_0)_{ba}}{\mu_k + \alpha} \right] q_a^{(k)} = 0. \tag{1.77}$$

A solution of these equations is easily expressed in terms of a given particular

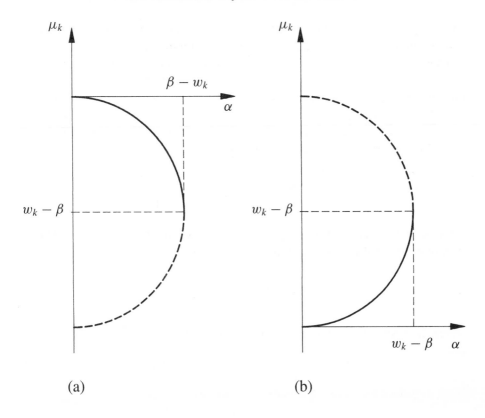

Fig. 1.2.   The behaviour, in the real-pole case, of μ_k as a function of α for some fixed value of β: (a) $\beta > w_k$, and (b) $\beta < w_k$. As in the previous figure the smooth line corresponds to μ_k^{in} and the broken line to μ_k^{out}.

solution ψ_0 of (1.51). Introducing the matrices

$$M_k = (\psi_0^{-1})_{\lambda=\mu_k} = \psi_0^{-1}(\mu_k, \zeta, \eta), \tag{1.78}$$

it is not difficult to see that these satisfy the equations

$$M_{k,\zeta} + M_k \frac{A_0}{\mu_k - \alpha} = 0, \quad M_{k,\eta} + M_k \frac{B_0}{\mu_k + \alpha} = 0. \tag{1.79}$$

Thus, a solution of (1.77) for the vectors $m_a^{(k)}$ will be

$$m_a^{(k)} = m_{0b}^{(k)}(M_k)_{ba} = m_{0b}^{(k)}[\psi_0^{-1}(\mu_k, \zeta, \eta)]_{ba}, \tag{1.80}$$

where the $m_{0b}^{(k)}$ are arbitrary complex constant vectors. In the solution (1.80) for the vectors $m_a^{(k)}$, there may also be arbitrary complex factors depending on the index k and the coordinates ζ and η. However, such factors reduce to an inessential renormalization of the vectors $m_a^{(k)}$ and disappear from the final expression for the matrices R_k; we therefore set them equal to 1.

There remains the task of determining the vectors $n_a^{(k)}$ and thus the matrices R_k. This can be done by means of the supplementary condition (1.57) that must be satisfied by the dressing matrix χ. Substituting (1.64) into (1.57) and considering the relation obtained in this way at the poles of the matrix $\chi(\alpha^2/\lambda)$, i.e. at the points $\lambda = \alpha^2/\mu_k$, we conclude that the matrices R_k satisfy the following system of n algebraic matrix equations:

$$R_k g_0 \left[I + \sum_{l=1}^{n} (\alpha^2 - \mu_k \mu_l)^{-1} \mu_k \tilde{R}_l \right] = 0, \tag{1.81}$$

where $k = 1, \ldots, n$. Substituting in this (1.75) for the matrices R_k we obtain a system of linear algebraic equations for the vectors $n_a^{(k)}$:

$$\sum_{l=1}^{n} \Gamma_{kl} n_a^{(l)} = \mu_k^{-1} m_c^{(k)} (g_0)_{ca}, \tag{1.82}$$

where the matrix Γ_{kl} is symmetric and its elements are

$$\Gamma_{kl} = -m_c^{(k)} m_b^{(l)} (g_0)_{cb} (\alpha^2 - \mu_k \mu_l)^{-1}. \tag{1.83}$$

If we introduce the symmetric matrix Π_{kl} inverse to Γ_{kl},

$$\sum_{m=1}^{n} \Pi_{km} \Gamma_{ml} = \delta_{kl}, \tag{1.84}$$

where δ_{kl} is the Kroneker symbol, we get from (1.82) for the vectors $n_a^{(k)}$,

$$n_a^{(k)} = \sum_{l=1}^{n} \mu_l^{-1} \Pi_{kl} L_a^{(l)}, \tag{1.85}$$

where

$$L_a^{(l)} = m_c^{(l)} (g_0)_{ca}. \tag{1.86}$$

Now, using (1.75), (1.85) and (1.86) we get, from (1.65), the metric components g_{ab}:

$$g_{ab} = (g_0)_{ab} - \sum_{k,l=1}^{n} \mu_k^{-1} \mu_l^{-1} \Pi_{kl} L_a^{(k)} L_b^{(l)}. \tag{1.87}$$

With this expression the matrix g is obviously symmetric. Let us now consider the reality condition. If all the functions $\mu_k(\zeta, \eta)$ are real the components g_{ab} are automatically real when we take all the arbitrary constants appearing in the solution to be real. In fact, the particular solution $\psi_0(\lambda, \zeta, \eta)$ is always assumed to satisfy the second of the conditions (1.55) and consequently $\psi_0(\lambda)$ is real on the real axis of the λ-plane, i.e. at the points $\lambda = \mu_k$. It can now be seen from (1.80) that the arbitrary constants $m_{0b}^{(k)}$ that occur in the vectors $m_a^{(k)}$ must be real, and then the vectors $m_a^{(k)}$ will also be real. It then follows that all the quantities from which the matrix g is made are real. If we now assume that there are also complex values among the functions $\mu_1, \mu_2, \ldots, \mu_n$, the conditions (1.55) then require that all the complex poles appear only as conjugate pairs; for each complex pole $\lambda = \mu$ its complex conjugate $\lambda = \bar{\mu}$ must also appear. Let us assume that there is such a pair of poles $\lambda = \mu_k$ and $\lambda = \mu_{k+1}$ with $\mu_{k+1} = \bar{\mu}_k$. To these poles there correspond vectors $m_a^{(k)}$ and $m_a^{(k+1)}$, which according to (1.80) are given by

$$m_a^{(k)} = m_{0b}^{(k)}[\psi_0^{-1}(\mu_k, \zeta, \eta)]_{ba}, \quad m_a^{(k+1)} = m_{0b}^{(k+1)}[\psi_0^{-1}(\mu_{k+1}, \zeta, \eta)]_{ba}. \quad (1.88)$$

A simple analysis shows that the matrix g will be real if for each such pair of complex conjugate poles the arbitrary constants $m_{0b}^{(k)}$ and $m_{0b}^{(k+1)}$ are taken to be conjugate to each other: $m_{0b}^{(k+1)} = \bar{m}_{0b}^{(k)}$. But since the function $\psi_0(\lambda, \zeta, \eta)$ satisfies the condition $\psi_0(\bar{\lambda}) = \bar{\psi}_0(\lambda)$, this means that the vectors $m_a^{(k)}$ and $m_a^{(k+1)}$ corresponding to each pair of conjugate poles are also conjugate to each other, i.e. $m_a^{(k+1)} = \bar{m}_a^{(k)}$. Accordingly, we can formulate the following rule to determine the choice of the arbitrary constants $m_{0b}^{(k)}$ in (1.80). To ensure that the matrix g is real, it is necessary to choose the arbitrary constants $m_{0b}^{(k)}$ in (1.80) so that the vectors $m_a^{(k)}$ corresponding to real poles $\lambda = \mu_k$ are real, and the vectors $m_a^{(k)}$ and $m_a^{(k+1)}$ corresponding to the pair of complex conjugate poles $\lambda = \mu_k$ and $\lambda = \mu_{k+1} = \bar{\mu}_k$ are complex conjugate to each other.

1.4.1 The physical metric components g_{ab}

Satisfying the requirements that g be real and symmetric is still not enough to have a physical solution. It must not be forgotten that g must also satisfy the supplementary condition (1.38); therefore we now calculate the determinant of the matrix g. The form (1.87) is not convenient for this calculation, and we use a different representation of our solution. We note that the process of perturbing the background solution g_0 and obtaining from it the n-soliton solution g, as described above, is formally equivalent to the introduction of the n solitons one at a time, in succession. The first step is to go from the background metric g_0 to the metric g_1 containing one soliton, corresponding to the presence in the matrix χ (which we call at this stage χ_1) of one pole only $\lambda = \mu_1$.

This one-soliton solution is easily obtained from the results given above. Now we have only one pole trajectory $\mu_1(\zeta, \eta)$ which is one of the functions (1.68) or (1.69) containing one arbitrary constant w_1. The vector $m_a^{(1)}$ follows from (1.80) for $k = 1$. From (1.85), (1.84) and (1.83) it is easy to get the vector $n_a^{(1)}$ and, after that, the matrix $(R_1)_{ab} = n_a^{(1)} m_b^{(1)}$. Substituting this matrix into (1.64) we can write the dressing matrix χ_1 in the form:

$$\chi_1 = I + \mu_1^{-1}(\lambda - \mu_1)^{-1}(\mu_1^2 - \alpha^2)P_1, \qquad (1.89)$$

where

$$(P_1)_{ab} = \frac{m_c^{(1)}(g_0)_{ca}m_b^{(1)}}{m_d^{(1)}(g_0)_{df}m_f^{(1)}}. \qquad (1.90)$$

We now get for the one soliton solution g_1,

$$g_1 = \chi_1(0)g_0 = [I - \mu_1^{-2}(\mu_1^2 - \alpha^2)P_1]g_0. \qquad (1.91)$$

It is not difficult to compute the determinant of g_1. First, we note the following remarkable properties of the matrix P_1, which follow easily from (1.90):

$$P_1^2 = P_1, \quad \text{Tr } P_1 = 1, \quad \det P_1 = 0. \qquad (1.92)$$

Using these properties and the general relation,

$$\det(I + F) = 1 + \text{Tr } F + \det F, \qquad (1.93)$$

which holds for any arbitrary 2×2 matrix F, we get from (1.91) that

$$\det g_1 = \mu_1^{-2}\alpha^2 \det g_0. \qquad (1.94)$$

We can now take the solution g_1 as a new background solution and repeat the operation of adding another soliton to it, corresponding to the pole $\lambda = \mu_2$. To do this we compute the matrix χ_1^{-1}, the inverse of the matrix χ_1 given in (1.89). Using the property $P_1^2 = P_1$ it is easy to check that

$$\chi_1^{-1} = I + \frac{\mu_1^2 - \alpha^2}{\alpha^2 - \lambda\mu_1} P_1. \qquad (1.95)$$

Now we form the new background generating matrix $\psi_1 = \chi_1\psi_0$, take its inverse ψ_1^{-1}, with the help of (1.95), calculate it at the point $\lambda = \mu_2$, and then find the corresponding vector $m'^{(2)}_a$,

$$m'^{(2)}_a = m'^{(2)}_{0b}[\psi_1^{-1}(\mu_2, \zeta, \eta)]_{ba}.$$

After that, we construct the matrix P'_2, in analogy with (1.90):

$$(P'_2)_{ab} = \frac{m'^{(2)}_c(g_1)_{ca}m'^{(2)}_b}{m'^{(2)}_d(g_1)_{df}m'^{(2)}_f},$$

which satisfies the same properties (1.92) as P_1. Then we construct the dressing matrix $\chi_2(\lambda)$ as

$$\chi_2 = I + \mu_2^{-1}(\lambda - \mu_2)^{-1}(\mu_2^2 - \alpha^2)P'_2,$$

and we get the two-soliton solution g_2:

$$g_2 = \chi_2(0)g_1 = [I - \mu_2^{-2}(\mu_2^2 - \alpha^2)P'_2][I - \mu_1^{-2}(\mu_1^2 - \alpha^2)P_1]g_0. \quad (1.96)$$

Repeating this process we get the n-soliton solution (1.87) in the form

$$g = \left(\prod_{k=1}^{n} [I - \mu_k^{-2}(\mu_k^2 - \alpha^2)P'_k] \right) g_0, \quad (1.97)$$

where $P'_1 \equiv P_1$ and all the matrices P'_k satisfy the same properties as the matrix P_1, i.e.

$$P'^2_k = P'_k, \quad \mathrm{Tr}\, P'_k = 1, \quad \det P'_k = 0. \quad (1.98)$$

Naturally the explicit form of the matrices P'_k quickly becomes cumbersome as k increases, and therefore this way of calculating the solutions is less convenient than the method described previously. But the representation of the solution in the form (1.97) is useful for the study of some particular questions and specially for calculating the determinant of the matrix g. The key point for this calculation is that since the matrices P'_k satisfy the properties (1.98), the contribution from each factor in (1.97) to the determinant of g can be calculated trivially, and the result is

$$\det g = \alpha^{2n} \left(\prod_{k=1}^{n} \mu_k^{-2} \right) \det g_0 = \alpha^{2n+2} \prod_{k=1}^{n} \mu_k^{-2}. \quad (1.99)$$

We still have to construct an n-soliton solution $g^{(ph)}$ which satisfies not only (1.39) but also the supplementary condition (1.38). For this 'physical' solution we have already derived the formula (1.61), consequently the final result for the n-soliton solution is

$$g^{(ph)} = \alpha \left(\alpha^{2n+2} \prod_{k=1}^{n} \mu_k^{-2} \right)^{-1/2} g = \alpha^{-n} \left(\prod_{k=1}^{n} \mu_k \right) g. \quad (1.100)$$

We should keep in mind that both signs are allowed in front of the matrix $g^{(ph)}$ due to the invariance of Einstein equations (1.38)–(1.42) with respect to the reflection $g^{(ph)} \rightarrow -g^{(ph)}$. This sign should be chosen separately for each case in order to ensure the correct signature of the metric. This completes the determination of the metric components g_{ab} for the n-soliton solution. We also note from (1.70)–(1.71) that one can obtain explicit expressions for the matrices A and B by equating the residues on the left and the right hand sides of these equations at the poles $\lambda = \alpha$ and $\lambda = -\alpha$; the result is:

$$A = 2\alpha\alpha_\zeta \left[\sum_{k=1}^{n} (\alpha - \mu_k)^{-2} R_k \right] \chi^{-1}(\alpha) + \chi(\alpha)A_0\chi^{-1}(\alpha), \quad (1.101)$$

$$B = 2\alpha\alpha_\eta \left[\sum_{k=1}^{n} (\alpha + \mu_k)^{-2} R_k \right] \chi^{-1}(-\alpha) + \chi(-\alpha)B_0\chi^{-1}(-\alpha). \quad (1.102)$$

The matrices R_k come from (1.75) and from (1.80) and (1.85). We then need to use expression (1.99) for the determinant of g in order to calculate the physical values of the matrices A and B, i.e. $A^{(ph)}$ and $B^{(ph)}$, in agreement with the prescriptions (1.62)–(1.63).

1.4.2 The physical metric component f

To complete the construction of the n-soliton solutions for the metric (1.36) we also need to calculate the metric coefficient f from (1.40)–(1.41) using the matrix g already found. Surprisingly enough the coefficient f in the general n-soliton case can also be calculated explicitly by algebraic operations only, like the metric components g_{ab}. Here it is convenient to perform the calculation of this coefficient in two stages. First we calculate the value of f which follows from (1.40)–(1.41) when we substitute in them the nonphysical solution g given by (1.87) which does not satisfy the condition $\det g = \alpha^2$. Then we use a simple procedure to find the physical value of the coefficient, $f^{(ph)}$, which is also obtained from the same equations (1.40)–(1.41) when $g^{(ph)}$ is substituted in them instead of g. Calculating the traces $\mathrm{Tr}\, A^2$ and $\mathrm{Tr}\, B^2$, using the expressions (1.101)–(1.102) for the matrices A and B, and substituting into (1.40)–(1.41), we find f by direct integration. It is a remarkable fact that this integration can actually be carried out. The key point in calculating the coefficient f corresponding to an n-soliton solution is to calculate it, first, for the one-soliton solution (the coefficient will be denoted in this case by f_1). From (1.89)–(1.94) after the necessary computations following the scheme indicated above, we get for the one-soliton solution,

$$f_1 = C_1 f_0 \left(\mu_1^2 - \alpha^2 \right)^{-1} \alpha \mu_1^2 \Gamma_{11}, \qquad (1.103)$$

where C_1 is an arbitrary constant, f_0 is the particular background solution for f, which corresponds to the solution g_0, and Γ_{11} is the single component of the matrix (1.83), which is a 1×1 matrix in this case ($k = 1, l = 1$):

$$\Gamma_{11} = \left(\mu_1^2 - \alpha^2 \right)^{-1} m_a^{(1)} m_b^{(1)} (g_0)_{ab}, \qquad (1.104)$$

where the vector $m_a^{(1)}$ follows from (1.80) for $k = 1$.

The next step in the calculation is to take the solution (g_1, f_1) as a new particular solution and repeat the operation, as we explained before in connection with the evaluation of g_2. Thus we get the coefficient f_2 which corresponds to the two-soliton solution with the poles $\lambda = \mu_1$ and $\lambda = \mu_2$. At this second step we already have to deal only with algebraic computations since the need for integration appears in the whole procedure only once, in the transition from the background solution (g_0, f_0) to the solution with one soliton (g_1, f_1). Omitting the details of these calculations, the result is

$$f_2 = C_2 f_0 \left(\mu_1^2 - \alpha^2 \right)^{-1} \left(\mu_2^2 - \alpha^2 \right)^{-1} \alpha^2 \mu_1^2 \mu_2^2 \left(\Gamma_{11} \Gamma_{22} - \Gamma_{12}^2 \right). \qquad (1.105)$$

Here C_2 is an arbitrary constant, f_0 is the background solution as in (1.103) and Γ_{11}, Γ_{22} and Γ_{12} are the components of the matrix (1.83). We now have three independent components of Γ_{kl}, since the indices k and l can take two values, 1 and 2.

Equations (1.103) and (1.105) suggest that in the general n-soliton case the coefficient f is given by the expression

$$f = C_n f_0 \alpha^n \left(\prod_{k=1}^{n} \mu_k^2 \right) \left[\prod_{k=1}^{n} (\mu_k^2 - \alpha^2) \right]^{-1} \det \Gamma_{kl}, \qquad (1.106)$$

where C_n is an arbitrary constant and $k, l = 1, 2, \ldots, n$. We see from (1.103) and (1.105) that this formula indeed holds for $n = 1$ and $n = 2$, and we can prove by induction that it holds for arbitrary n. The proof for the stationary analogue of metric (1.36) is given in the appendix of ref. [24]. For the nonstationary case, (1.36), the proof is almost identical and we do not give it here; it shows that (1.106) is indeed correct in general.

Now we must determine the physical value $f^{(ph)}$ of the coefficient f, i.e. the value that would be obtained from (1.40)–(1.41) if we had substituted in them the physical matrix $g^{(ph)}$ of (1.100) instead of g. For the matrices $A^{(ph)}$ and $B^{(ph)}$ we have the formulas (1.62) and (1.63). When we substitute the matrices $A^{(ph)}$ and $B^{(ph)}$ in (1.40)–(1.41) instead of A and B, we find that the physical coefficient $f^{(ph)}$ is given by the formula

$$f^{(ph)} = f \alpha^{1/2} F, \qquad (1.107)$$

where f is the value of the coefficient given by (1.106) and the function F is defined by the equations:

$$(\ln F)_{,\zeta} = -\frac{\alpha}{8\alpha_{,\zeta}} \left[(\ln \det g)_{,\zeta} \right]^2, \quad (\ln F)_{,\eta} = -\frac{\alpha}{8\alpha_{,\eta}} \left[(\ln \det g)_{,\eta} \right]^2.$$
$$(1.108)$$

Then substituting the expression (1.99) for $\det g$ we find that these equations are easily integrated (the essential ingredient for such integration comes from relations (1.66)), and we get

$$F = C_F \alpha^{-(n^2+2n+1)/2} \left(\prod_{k=1}^{n} \mu_k \right)^{n-1} \left[\prod_{k=1}^{n} (\mu_k^2 - \alpha^2) \right] \left[\prod_{k>l=1}^{n} (\mu_k - \mu_l)^{-2} \right],$$
$$(1.109)$$

where C_F is an arbitrary constant. From this and (1.106)–(1.107) we get the final expression for the physical value of the coefficient f:

$$f^{(ph)} = C_f f_0 \alpha^{-n^2/2} \left(\prod_{k=1}^{n} \mu_k \right)^{n+1} \left[\prod_{k>l=1}^{n} (\mu_k - \mu_l)^{-2} \right] \det \Gamma_{kl}, \qquad (1.110)$$

where C_f is an arbitrary constant which should be taken with the appropriate sign in order to ensure the correct sign for $f^{(ph)}$. For clarity we point out that the product

$$\prod_{k>l=1}^{n} (\mu_k - \mu_l)^{-2}$$

is equal to 1 for $n = 1$, to $(\mu_2 - \mu_1)^{-2}$ for $n = 2$, to $(\mu_3 - \mu_2)^{-2}(\mu_3 - \mu_1)^{-2}(\mu_2 - \mu_1)^{-2}$ for $n = 3$, and so on. Therefore the final expression for the n-soliton solution can be written in the form

$$ds^2 = f^{(ph)}(dz^2 - dt^2) + g_{ab}^{(ph)}dx^a dx^b, \tag{1.111}$$

where $f^{(ph)}$ is given by (1.110) and the matrix elements $g_{ab}^{(ph)}$ are given by (1.100) and (1.87).

1.5 Multidimensional spacetime

In this section we consider the application of the ISM described in sections 1.2, 1.3 and 1.4 to spacetimes with an arbitrary number of dimensions N, but which admit $N - 2$ commuting Killing vector fields ($N \geq 4$). There is some interest in this generalization in connection with ideas of using a compactified multidimensional world for the construction of a selfconsistent unified theory [162, 175, 132, 86]. Of course, the ansatz we are considering in this book is too poor for such purposes, given that it does not allow for a dependence of the metric in the extradimensional coordinates. All the multidimensional gravitational potentials should depend on two variables only, as in the four-dimensional case, otherwise the ISM described in the previous sections cannot be applied. Dependence in the extradimensional coordinates is important if one wishes to introduce into the theory nontrivial (non-Abelian) symmetry groups, which are related to the internal symmetries of the particles [86]. It may happen, however, that some of the multidimensional gravitational solitons and instantons that can be constructed by the method we are considering here could be used for the extradimensional unification approach. Bearing this in mind, we outline here the generalization of the ISM to multidimensional gravity, together with a simple example of its application in five dimensions. Further applications are given in section 5.4.3.

For definiteness, let us consider the case when all Killing vectors are spacelike and the metric is nonstationary. Thus we write the metric in the same form (1.36) with the only difference that now the indices a, b run over $N - 2$ values and $g_{ab}(t, z)$ is a $(N-2) \times (N-2)$ matrix. For the determinant of this matrix we adopt the same notation (1.38) and it is easy to show that the Einstein equations in vacuum $R_{AB}^{(N)} = 0$ ($A, B = 1, \ldots, N$) are equivalent to the system (1.39)–(1.42), where g, A and B are now $(N - 2) \times (N - 2)$ matrices. The function $\alpha(t, z)$ still satisfies the wave equation (1.44).

The basic fact is that the integration scheme of section 1.3 (excluding the last three formulas, (1.61)–(1.63)) and the procedure for the construction of the n-soliton solution described in section 1.4 remain unchanged. All the expressions in these sections are valid, with the exception of those which describe the transition from the metric coefficients g_{ab} and f to their physical values $g_{ab}^{(ph)}$ and $f^{(ph)}$. In fact, there is only one point where the structure of the equations differs from the $N = 4$ case: the explicit construction of the physical metric components $g_{ab}^{(ph)}$ and $f^{(ph)}$. It is obvious that instead of (1.61) we now have to write

$$g^{(ph)} = \alpha^{2/(N-2)} (\det g)^{1/(2-N)} g, \qquad (1.112)$$

then $\det g^{(ph)} = \alpha^2$, as it should be. Since for $\det g$ we have the same expression (1.99), we have

$$g^{(ph)} = \alpha^{-2n/(N-2)} \left(\prod_{k=1}^{n} \mu_k \right)^{2/(N-2)} g. \qquad (1.113)$$

From this it is easy to obtain the corresponding generalization of (1.62) and (1.63), and to write the physical matrices $A^{(ph)}$ and $B^{(ph)}$ in terms of the matrices A and B. By representing the physical component $f^{(ph)}$ as $f^{(ph)} = F_N f$, where f is given by (1.106), and substituting it together with $A^{(ph)}$ and $B^{(ph)}$ into (1.40)–(1.41), we get equations for the coefficient F_N. These equations can be integrated exactly: the final result for the metric component $f^{(ph)}$ is [285, 288]

$$f^{(ph)} = f_0 \alpha^{-n(n+4-N)/(N-2)} \det(\Gamma_{kl}) \prod_{k=1}^{n} \left[\mu_k^{2(n-3+N)/(N-2)} (\mu_k^2 - \alpha^2)^{(4-N)/(N-2)} \right]$$

$$\times \prod_{k,l=1;k>l}^{n} (\mu_k - \mu_l)^{4/(2-N)}. \qquad (1.114)$$

Of course, for $N = 4$, (1.113) and (1.114) coincide with (1.100) and (1.110), respectively. Also, we recall that $f^{(ph)}$ is only determined up to an arbitrary multiplicative constant. Thus, it is always possible to introduce a constant factor in (1.114) in order to correct the physical value of $f^{(ph)}$ if necessary.

Apart from the direct interpretation of the multidimensional metric components $g_{AB}^{(ph)}$ as gravitational potentials of the multidimensional spacetime, we can also use another well known interpretation of such a theory as ordinary four-dimensional general relativity with vector and scalar fields given by the extradimensional metric components. For this picture let us introduce here, and for the rest of this section, the following conventions:

1. The capital Latin indices $A, B, \ldots$ are the N-dimensional spacetime indices, and take N values.

2. The small Latin indices $i, k, l, \ldots$ from the second part of the alphabet correspond to the usual four-dimensional spacetime indices, and take four

values only. All fields depend on time and one space-like coordinate, which we denote as t and z, respectively.

3. The Greek indices $\alpha, \beta, \ldots$ label coordinates of the extra dimensions, and take $N - 4$ values.

4. The small Latin indices from the first part of the alphabet $a, b, \ldots, h$ label the ignorable coordinates (on which the fields have no dependence) and run over $N - 2$ values.

5. The small Latin indices from the first part of the alphabet with a bar, $\bar{a}, \bar{b}, \ldots, \bar{h}$ label the two ignorable coordinates of the four-dimensional spacetime (not the extradimensional sector), and take only two values.

First, let us look at the field contents of multidimensional gravity from the point of view of an effective four-dimensional theory. We start with the N-dimensional metric tensor, which we call 'physical' for convenience, and the N-dimensional Kaluza–Klein interval:

$$
\begin{aligned}
ds^2_{(N)} &= g^{(ph)}_{AB} dx^A dx^B \\
&= g^{(ph)}_{ik} dx^i dx^k + 2g^{(ph)}_{i\alpha} dx^i dx^\alpha + g^{(ph)}_{\alpha\beta} dx^\alpha dx^\beta,
\end{aligned} \tag{1.115}
$$

where $g^{(ph)}_{AB}$ depends only on the coordinates of the four-dimensional spacetime,

$$
g^{(ph)}_{AB} = g^{(ph)}_{AB}(x^i); \tag{1.116}
$$

this is the Kaluza–Klein ansatz. Let us define new fields G_{ik}, $A^{(\alpha)}_i$ and $\gamma_{\alpha\beta}$ by

$$
g^{(ph)}_{ik} = e^{2\sigma} G_{ik} + A^{(\alpha)}_i A^{(\beta)}_k \gamma_{\alpha\beta}, \tag{1.117}
$$

$$
g^{(ph)}_{i\alpha} = A^{(\beta)}_i \gamma_{\alpha\beta}, \tag{1.118}
$$

$$
g^{(ph)}_{\alpha\beta} = \gamma_{\alpha\beta}, \tag{1.119}
$$

where

$$
e^{2\sigma} = (\det \gamma_{\alpha\beta})^{-1/2}. \tag{1.120}
$$

Equations $R^{(N)}_{AB} = 0$ then describe a selfconsistent system for the interaction of the gravitational field, G_{ik}, with $N - 4$ vector fields $A^{(\alpha)}_i$ and $(N - 4)(N - 3)/2$ scalar fields $\gamma_{\alpha\beta}$. These vectors and scalars correspond to Abelian Yang–Mills fields and to some generalization of massless Klein–Gordon fields (presumably Higgs bosons); however, the interaction among these fields is unusual. Direct calculation shows that equations $R^{(N)}_{AB} = 0$ are equivalent to the system,

$$R_i^k - \frac{1}{2}\delta_i^k R = 2(\sigma_{;i}\sigma^{;k} - \frac{1}{2}\delta_i^k \sigma_{;m}\sigma^{;m})$$

$$+ \frac{1}{4}\left(\kappa_{\beta i}^{\alpha}\kappa_{\alpha}^{\beta k} - \frac{1}{2}\delta_i^k \kappa_{\beta m}^{\alpha}\kappa_{\alpha}^{\beta m}\right)$$

$$+ \frac{1}{2}e^{-2\sigma}\gamma_{\alpha\beta}\left(F_{mi}^{(\alpha)}F^{(\beta)mk} - \frac{1}{4}\delta_i^k F_{mn}^{(\alpha)}F^{(\beta)mn}\right), \quad (1.121)$$

$$F^{(\alpha)ik}_{\quad ;k} = 2\sigma_{;k}F^{(\alpha)ik} - \kappa_{\beta k}^{\alpha}F^{(\beta)ik}, \quad (1.122)$$

$$\kappa_{\alpha\,;k}^{\beta k} = \frac{1}{2}e^{-2\sigma}\gamma_{\alpha\mu}F_{mn}^{(\beta)}F^{(\mu)mn}, \quad (1.123)$$

where

$$F_{ik}^{(\alpha)} = A_{i,k}^{(\alpha)} - A_{k,i}^{(\alpha)}, \quad (1.124)$$

and

$$\kappa_{\beta i}^{\alpha} = \gamma^{\alpha\mu}\gamma_{\mu\beta,i}. \quad (1.125)$$

The raising and lowering of the four-dimensional indices, as well as the four-dimensional covariant differentiation in these equations are defined with respect to the metric G_{ik}, also R_i^k in (1.121) is the usual four-dimensional Ricci tensor for the metric G_{ik}. The matrix $\gamma^{\alpha\beta}$ in (1.125) is the inverse of $\gamma_{\alpha\beta}$ ($\gamma^{\alpha\mu}\gamma_{\mu\beta} = \delta_\beta^\alpha$). Covariant derivatives on the scalar σ are obviously $\sigma_{;k} = \sigma_{,k}$ and $\sigma^{;k} = G^{km}\sigma_{,m}$.

The conformal factor $\exp(2\sigma)$ in (1.117) was introduced in order to exclude second derivatives of the scalar fields in the stress-energy tensor on the right hand side of (1.121). Note that if we had interpreted $\exp(2\sigma)G_{ik}$ as the four-dimensional metric, instead of G_{ik}, then second derivatives of $\ln \det \gamma_{\alpha\beta}$ would appear in (1.121). With our choice we have a Klein–Gordon-like structure in the scalar sector. This choice is sometimes referred to as the Einstein frame. We should stress, however, that there are no sound theoretical reasons to single out a four-dimensional physical metric from a multidimensional spacetime.

Equations (1.121)–(1.125) were derived for a general metric of the form (1.115)–(1.116) in order to remind the reader of the possible four-dimensional physical interpretation of the coefficients $g_{AB}^{(ph)}$. But, if we wish to develop the integrable ansatz in such a theory, it is necessary to take the following particular case of metric (1.115):

$$ds_{(N)}^{2(ph)} = f^{(ph)}(t, z)(dz^2 - dt^2) + g_{\bar{a}\bar{b}}^{(ph)}(t, z)dx^{\bar{a}}dx^{\bar{b}}$$

$$+ 2g_{\bar{a}\alpha}^{(ph)}(t, z)dx^{\bar{a}}dx^{\alpha} + g_{\alpha\beta}^{(ph)}(t, z)dx^{\alpha}dx^{\beta}, \quad (1.126)$$

and to construct the n-soliton solution (1.113)–(1.114) for the coefficients $g_{AB}^{(ph)}(t, z)$ in this expression as explained in the beginning of this section. Then from (1.117)–(1.119) one can write this solution in terms of the potentials G_{ik}, $A_i^{(\alpha)}$ and $\gamma_{\alpha\beta}$, which will satisfy (1.121)–(1.125) automatically. The set of scalar fields $\gamma_{\alpha\beta}$ are just the components $g_{\alpha\beta}^{(ph)}$, according to (1.119). From (1.118) and

(1.126) it follows that the vectors $A_{\bar{a}}^{(\alpha)}$ are,

$$A_t^{(\alpha)} = 0, \quad A_z^{(\alpha)} = 0, \quad A_{\bar{a}}^{(\alpha)} = g_{\bar{a}\beta}^{(ph)} \gamma^{\beta\alpha}. \tag{1.127}$$

The corresponding four-dimensional spacetime interval is obtained from (1.117) and (1.126)

$$ds_{(4)}^2 = G_{zz}dz^2 + G_{tt}dt^2 + G_{\bar{a}\bar{b}}dx^{\bar{a}}dx^{\bar{b}}, \tag{1.128}$$

where

$$G_{zz} = -G_{tt} = e^{-2\sigma} f^{(ph)}, \quad G_{\bar{a}\bar{b}} = e^{-2\sigma}\left(g_{\bar{a}\bar{b}}^{(ph)} - A_{\bar{a}}^{(\alpha)}A_{\bar{b}}^{(\beta)}\gamma_{\alpha\beta}\right). \tag{1.129}$$

For illustrative purposes let us apply the described approach to five-dimensional spacetime and construct a five-dimensional 'black hole' solution which one can try to interpret as a four-dimensional black hole with some scalar and vector field sources. From the 'no hair' theorems we know that this is not possible. In fact, the explicit construction will show the problems with this type of solution.

In refs [23, 24] (see also section 8.3) it is proved that the Kerr black hole is just a double-soliton solution with two real-pole trajectories on a flat background. We can try to generalize this statement by saying that one may obtain an N-dimensional black hole as a double soliton with two real-pole trajectories on an N-dimensional flat background. The fact that we have presented the multidimensional approach in this section for the nonstationary metric only is not a problem because, as we will see, it is very simple to obtain the stationary version by a simple complex transformation of the time coordinate. Thus, let us start with the flat N-dimensional background

$$ds_{(N)}^2 = -dt^2 + dz^2 + (g_0)_{ab}\, dx^a dx^b, \tag{1.130}$$

where g_0 is the $(N-2) \times (N-2)$ diagonal matrix

$$g_0 = \text{diag}(t^2, 1, 1, \ldots, 1). \tag{1.131}$$

For such a background we have $\alpha = t$, $\beta = z$, i.e. canonical coordinates, and condition (1.38) is satisfied. The background generating matrix ψ_0 follows from (1.51) as

$$\psi_0 = \text{diag}(t^2 + 2z\lambda + \lambda^2, 1, 1, \ldots, 1), \tag{1.132}$$

which satisfies condition (1.52). We take two pole trajectories,

$$\left. \begin{aligned} \mu_1 &= w_1 - z + \sqrt{(w_1 - z)^2 - t^2}, \\ \mu_2 &= w_2 - z - \sqrt{(w_2 - z)^2 - t^2}, \end{aligned} \right\} \tag{1.133}$$

where w_1 and w_2 are two real arbitrary parameters, and we consider the spacetime region where the expressions under the square roots are both positive (the square roots are both defined as positive quantities). Now we perform the steps for the construction of the two-soliton solution as described in sections 1.3 and 1.4, but in N dimensions. We thus get the metric coefficients g_{ab} and the 2×2 matrix Γ defined in (1.83). Finally, from (1.113)–(1.114) we obtain the physical metric coefficients $f^{(ph)}$ and $g_{ab}^{(ph)}$. The main point is that this metric becomes stationary, remaining real and with the correct signature, after the complex transformation,

$$t = i\rho, \tag{1.134}$$

together with the replacement of $f^{(ph)}$ by (constant) $\times f^{(ph)}$, with some suitable constant. We then introduce instead of w_1 and w_2, two new real parameters z_1 and ζ defined by

$$w_1 = z_1 + \zeta, \qquad w_2 = z_1 - \zeta, \tag{1.135}$$

and instead of ρ and z, two new coordinates r and θ defined by

$$\rho = \left[(r - \mu)^2 - \zeta^2\right]^{1/2} \sin\theta, \qquad z = z_1 + (r - \mu)\cos\theta, \tag{1.136}$$

where μ is some new real arbitrary constant. The remaining two coordinates of the four-dimensional spacetime can be considered as the time t and the azimuthal angle φ. Finally, we need to set to zero some combination of the arbitrary parameters to ensure asymptotic flatness of the solution at space-like infinity $r \to \infty$. In this limit we then get the flat metric in standard spherical coordinates r, θ, φ. This is the way to obtain the N-dimensional generalization of the black hole metric. Its explicit N-dimensional expression is quite compact, but it does not seem to have enough physical interest for it to be worth giving its exact form here. Instead, we will write the explicit form of the five-dimensional version only, which contains the most relevant features of these solutions.

Five-dimensional Kerr solution. When $N = 5$ the spacetime interval (1.115) is

$$ds_{(5)}^2 = g_{ik}^{(ph)} dx^i dx^k + 2g_{i5}^{(ph)} dx^i dx^5 + g_{55}^{(ph)} (dx^5)^2. \tag{1.137}$$

Now we have only one scalar field σ and one vector field A_i and from (1.117)–(1.120) it follows that

$$g_{ik}^{(ph)} = e^{2\sigma} G_{ik} + A_i A_k e^{-4\sigma}, \qquad g_{i5}^{(ph)} = A_i e^{-4\sigma}, \qquad g_{55}^{(ph)} = e^{-4\sigma}. \tag{1.138}$$

Equations (1.121)–(1.125) of the four-dimensional theory now reduce to the system:

$$R_i^k - \frac{1}{2}\delta_i^k R = 6(\sigma_{;i}\sigma^{;k} - \frac{1}{2}\delta_i^k \sigma_{;m}\sigma^{;m})$$

$$+ \frac{1}{2}e^{-6\sigma}\left(F_{mi}F^{mk} - \frac{1}{4}\delta_i^k F_{mn}F^{mn}\right), \tag{1.139}$$

$$F_{;k}^{ik} = 6\sigma_{;k}F^{ik}, \tag{1.140}$$

$$\sigma_{;k}^{;k} = -\frac{1}{8}e^{-6\sigma}F_{mn}F^{mn}, \tag{1.141}$$

where

$$F_{ik} = A_{i,k} - A_{k,i}. \tag{1.142}$$

The double-solitonic solution for $g_{ik}^{(ph)}$, $g_{i5}^{(ph)}$ and $g_{55}^{(ph)}$, whose derivation has been described above, gives the following exact solution to (1.139)–(1.142):

$$ds_{(4)}^2 = G_{ik}dx^i dx^k$$

$$= \omega T \left[\frac{(r-\mu)^2 - v^2}{(r-\mu)^2 - v^2 \cos^2\theta} \right]^{1/3} \left(\frac{dr^2}{\Delta} + d\theta^2 \right)$$

$$+ \frac{1}{\omega T} \left\{ -(\omega - 2\mu r)dt^2 - 4\mu qar\sin^2\theta \, dt d\varphi \right.$$

$$\left. + \left[(r^2 + a^2)^2 q^2 - \Delta a^2 q^2 \sin^2\theta - \Delta \omega s^2 \frac{r-\mu-v}{r-\mu+v} \right] \sin^2\theta \, d\varphi^2 \right\} \tag{1.143}$$

$$e^{2\sigma} = \frac{1}{T} \left(\frac{r-\mu-v}{r-\mu+v} \right)^{1/3} \tag{1.144}$$

$$\left. \begin{array}{l} A_r = 0, \quad A_\theta = 0, \\ \\ A_\varphi = -\dfrac{2\mu sa}{\omega T^2} \left(\dfrac{r-\mu-v}{r-\mu+v} \right) r\sin^2\theta, \\ \\ A_t = \dfrac{qs}{T^2} \left[1 - \dfrac{r-\mu-v}{r-\mu+v} \left(1 - \dfrac{2\mu r}{\omega} \right) \right]. \end{array} \right\} \tag{1.145}$$

Here s, μ and a are arbitrary constants. The constant parameters v and q follow from

$$v^2 = \mu^2 - a^2, \quad q^2 = 1 + s^2, \tag{1.146}$$

and the functions T, ω and Δ are

$$T = \left[q^2 - \frac{r-\mu-v}{r-\mu+v} \left(1 - \frac{2\mu r}{\omega} \right) s^2 \right]^{1/2}, \tag{1.147}$$

$$\omega = r^2 + a^2 \cos^2\theta, \tag{1.148}$$

$$\Delta = r^2 - 2\mu r + a^2. \tag{1.149}$$

This solution was obtained in ref. [22] where its asymptotic structure at space-like infinity $r \to \infty$ was also studied. Note, however, that the four-dimensional physical metric used in ref. [22] is $\exp(2\sigma)G_{ik}$, instead of G_{ik}. From (1.144) and (1.147) it follows that in this limit $\exp(2\sigma) \to 1$, which means that the first nonvanishing terms in the asymptotic expansion of the 'electromagnetic' field and the rotational metric component (the $t\varphi$-metric component) are the same as in [22]. Thus, from the point of view of a distant observer the solution (1.143)–(1.149) describes an object with mechanical angular momentum $\mu q a$, 'electric' charge $2qs(\mu + v)$ and 'magnetic' moment $2\mu s a$. These statements can also be verified by direct analysis of the asymptotic expansion at $r \to \infty$ of the metric (1.143) and the 'electromagnetic' tensor F_{ik}. This last tensor is obtained from the potentials (1.145), which have the following first nonvanishing terms to first order:

$$A_\varphi = -\frac{1}{r} 2\mu s a \sin^2 \theta, \quad A_t = \frac{1}{r} 2qs(\mu + v). \tag{1.150}$$

To identify the mass we need the second order term in the expansion of the tt-metric component. Due to the effect of the conformal factor $\exp(2\sigma)$ it is slightly different from the value indicated in ref. [22]. In fact, it is easy to see from (1.143) that

$$G_{tt} = -1 + \frac{2}{r}\left[\mu + \frac{1}{2}s^2(\mu + v)\right] + \cdots, \tag{1.151}$$

therefore, the mass is $\mu + (1/2)s^2(\mu + v)$. Apart from this, this object is a source of the scalar field σ, whose expansion is

$$e^{2\sigma} = 1 - \frac{1}{r}\left[s^2(\mu + v) + \frac{2}{3}v\right] + \cdots. \tag{1.152}$$

This interpretation could make sense for a very distant observer, but the physical meaning of the solution as a whole is unclear. It is definitely not a black hole because of the physical singularity at the sphere radius $r = \mu + v$, where one would expect a regular event horizon for a black hole (we assume that both μ and v are positive). We could introduce a new radial coordinate $R = r - (\mu + v)$ and accept as physical only the region $R > 0$, and then we could interpret the solution as describing some exotic point-like source with a naked singularity.

As we have mentioned already the impossibility of having a solution that represents a regular perturbation of the black hole is a consequence of the well known 'no hair' theorems, and our example just illustrates this situation. However, it is worth mentioning that this example shows in some sense the deviation from these theorems for the particular case of the extreme black hole. Indeed, if we put $v = 0$ (i.e. $a = \mu$) then from (1.143)–(1.149) the additional singularities due to the scalar and vector fields disappear. The solution represents in this case a regular finite perturbation of the extreme Kerr metric in the whole spacetime region outside the central ring singularity $\omega = 0$, i.e. at points which are not too close to this singularity. The character of this singularity differs from the unperturbed case because of the factor T (this factor gives the perturbation) which is singular even for $v = 0$ at points $r = 0$ and $\theta = \pi/2$, if one approaches these points in some definite way. Nevertheless, in connection with the 'no hair' theorems it is more important that these scalar and vector fields do not destroy the qualitative structure of the spacetime outside the central singularity. In any case this particular example has no great significance because it is unstable with respect to the appearance of the nonzero quantity v, and because the physical meaning of the extreme Kerr metric is unclear.

One last remark on this solution is the following. The ratio between the 'magnetic' and mechanical angular momenta for this object is $2\mu s a/\mu q a$, which gives $2s(1 + s^2)^{-1/2}$. This can be made exactly equal to 2 in the limit $s \to \infty$. It is interesting that such a limit really exists for the solution (1.143)–(1.149). To obtain it one needs to make $s \to \infty$ and $a \to 0$ keeping the product sa finite and assuming that v is negative (but $\mu > 0$). In the limit $a/\mu \ll 1$ we have the expansion $v = -\sqrt{\mu^2 - a^2} = -\mu + a^2/(2\mu) + \cdots$. It is a simple exercise to obtain the final form of the solution after taking this limit.

2

General properties of gravitational solitons

In this chapter we give some general properties of the gravitational soliton solutions. The simplest soliton solutions, those with fewer poles, are studied in general and the pole fusion limit is described in section 2.1. In section 2.2 the case of a diagonal, but otherwise arbitrary, background metric is considered. It turns out that the integration of the spectral equations for the background solution in this case reduces to quadratures and the one- and two-soliton solutions can be given in general. Section 2.3 is devoted to the characterization of the gravitational solitons by some of the properties that solitons have in nongravitational physics. We see that the properties of the solitons do not always have a correspondence in the gravitational case. But under some restrictions some of these properties such as the topological charge can be identified. Thus, we can identify gravitational solitons and antisolitons, and, in particular, a remarkable solution that is the gravitational analogue of the sine-Gordon breather.

2.1 The simple and double solitons

Here we give a suitable form to the one- and two-soliton solutions, the simplest particular cases of the multisoliton solution described in section 1.4, and investigate some of their general properties. *Everywhere in this chapter we deal only with physical values of the metric coefficients which obey the full system of Einstein equations (1.38)–(1.42) and, for simplicity, we omit the label 'ph' in these coefficients.*

One-soliton solution. The one-soliton solution was explicitly obtained in section 1.4. The physical value of the matrix $g^{(1)}$ follows from (1.91), (1.80) and (1.83)–(1.87) for $n = 1$ ($k, l = 1$) and from (1.100) and is given by,

$$g_{ab}^{(1)} = \alpha^{-1}\mu_1(g_0)_{ab} + (\alpha\mu_1 Q_{11})^{-1}(\alpha^2 - \mu_1^2)L_a^{(1)}L_b^{(1)}, \qquad (2.1)$$

where Q_{11} is defined as

$$Q_{11} = m_a^{(1)} m_b^{(1)} (g_0)_{ab}. \tag{2.2}$$

The arbitrary constants w_1 in (1.67) and $m_{0a}^{(1)}$ in (1.80) are real. Since the constants $m_{0a}^{(1)}$ enter into the numerator, $L_a^{(1)} L_b^{(1)}$, and the denominator, Q_{11}, of (2.1) in an homogeneous quadratic manner, the components $g_{ab}^{(1)}$ depend only on the ratio $m_{01}^{(1)} / m_{02}^{(1)}$. Consequently, the one-soliton solution depends on two real arbitrary constants.

The physical value of the coefficient $f^{(1)}$ for the one-soliton solution comes from (1.110) for $n = 1$ ($k, l = 1$):

$$f^{(1)} = C_f^{(1)} f_0 \alpha^{-1/2} (\alpha^2 - \mu_1^2)^{-1} \mu_1^2 Q_{11}, \tag{2.3}$$

where $C_f^{(1)}$ is an arbitrary constant. We recall that for μ_1 we have two possible choices, μ_1^{in} and μ_1^{out}, which are given by (1.68) and (1.69). Thus we have actually two one-soliton solutions.

Two-soliton solution. In the case of two-solitons we have two pole trajectories, μ_1 and μ_2, which are the roots of the quadratic equation (1.67) for $k = 1, 2$. Now we have two arbitrary constants, w_1 and w_2, which can be either complex conjugate to one another ($w_2 = \overline{w}_1$ and then $\mu_2 = \overline{\mu}_1$) or both real (then μ_1 and μ_2 are real). In either case w_k brings into the solution two arbitrary real constants. The physical metric components $g_{ab}^{(2)}$ follow from (1.83)–(1.87), (1.80) and (1.100) for $n = 2$ ($k, l = 1, 2$), and after some algebraic transformations can be written in the form

$$g_{ab}^{(2)} = (g_0)_{ab} + D^{-1} (\mu_2 - \mu_1)(\alpha^2 - \mu_1 \mu_2)[\mu_1^{-1}(\alpha^2 - \mu_1^2) Q_{22} L_a^{(1)} L_b^{(1)}$$

$$- \mu_2^{-1}(\alpha^2 - \mu_2^2) Q_{11} L_a^{(2)} L_b^{(2)} - (\mu_2 - \mu_1) Q_{11} Q_{22} (g_0)_{ab}], \tag{2.4}$$

where we have introduced the notation

$$Q_{kl} = m_a^{(k)} m_b^{(l)} (g_0)_{ab} \qquad (k, l = 1, 2) \tag{2.5}$$

and

$$D = Q_{11} Q_{22} (\alpha^2 - \mu_1 \mu_2)^2 - Q_{12}^2 (\alpha^2 - \mu_1^2)(\alpha^2 - \mu_2^2). \tag{2.6}$$

The expression for $g_{ab}^{(2)}$ can be written in several convenient forms. We have chosen (2.4) as the most suitable for our purposes. To derive this form we have used the identity,

$$(Q_{11} Q_{22} - Q_{12}^2)(g_0)_{ab} = Q_{11} L_a^{(2)} L_b^{(2)} + Q_{22} L_a^{(1)} L_b^{(1)} - Q_{12}(L_a^{(1)} L_b^{(2)} + L_a^{(2)} L_b^{(1)}), \tag{2.7}$$

which follows from the definitions (1.86) and (2.5) for the vectors $L_a^{(k)}$ and the quantities Q_{kl}. We remark also that det Q_{kl} can be expressed through the vectors $m_a^{(k)}$ as

$$\det Q_{kl} = Q_{11}Q_{22} - Q_{12}^2 = \alpha^2 \left(m_1^{(1)}m_2^{(2)} - m_1^{(2)}m_2^{(1)} \right)^2, \qquad (2.8)$$

whereas D can be rewritten, from (2.6), as

$$D = \alpha^2 Q_{11}Q_{22}(\mu_1 - \mu_2)^2 + (\alpha^2 - \mu_1^2)(\alpha^2 - \mu_2^2) \det Q_{kl}. \qquad (2.9)$$

From (1.110) for $n = 2$ $(k, l = 1, 2)$ we get the physical value of the coefficient $f^{(2)}$ for the two-soliton solution:

$$f^{(2)} = C_f^{(2)} f_0 \alpha^{-2} (\mu_1\mu_2)^3 (\alpha^2 - \mu_1\mu_2)^{-2} [(\alpha^2 - \mu_1^2)(\alpha^2 - \mu_2^2)]^{-1} (\mu_1 - \mu_2)^{-2} D, \qquad (2.10)$$

where $C_f^{(2)}$ is some arbitrary constant. Using (2.5)–(2.9) it is easy to prove that D is always real and has the same sign throughout the spacetime. The same is true for Q_{11} in (2.3) and the remaining products in (2.3) and (2.10). Therefore the signature of the one- and two-soliton metrics is preserved throughout the spacetime.

Let us return to the one-soliton solution. As can be seen from (1.68)–(1.69), the solution is restricted to the coordinate patch in the (ζ, η)-plane, or the (α, β)-plane, where

$$(\beta - w_1)^2 \geq \alpha^2. \qquad (2.11)$$

Since the functions α and β have the structure (1.45)–(1.46), the boundary of region (2.11), i.e. the lines $(\beta - w_1)^2 = \alpha^2$, is the pair of null lines $a(\zeta) = w_1/2$ and $b(\eta) = -w_1/2$, which form a light cone (see fig. 2.1). The solution will be real only outside this cone, i.e. inside the two regions defined by (2.11). On the cone we have $\mu_1^2 = \alpha^2$ and the matrix $g^{(1)}$ coincides exactly with the background matrix g_0, as can be seen from (2.1). The solution for $g^{(1)}$ can also be defined in the region $(\beta - w_1)^2 < \alpha^2$ using the following considerations, which have a general character and refer to all soliton solutions with real poles of the dressing matrix $\chi(\lambda, \zeta, \eta)$. A real pole $\lambda = \mu_1$ is given by one of the expressions (1.68)–(1.69) for $k = 1$ with a real constant w_1. When we move along the coordinate plane and go from the region (2.11) into a region where $(\beta - w_1)^2 < \alpha^2$, the function μ_1 becomes complex. Obviously a continuation of the matrix $g^{(1)}$ into this region will be the solution corresponding to the two-pole situation with $\lambda = \mu_1$ and $\lambda = \mu_2 = \bar{\mu}_1$, where

$$\mu_1 = (w_1 - \beta)\left(1 \mp i\sqrt{\alpha^2(\beta - w_1)^{-2} - 1}\right).$$

However, for both of these functions we have $|\mu_1|^2 = \alpha^2$ and the poles are located on the circle $|\lambda|^2 = \alpha^2$ of the λ-plane. Using (1.83)–(1.85) it is easy to prove in general that the dressing matrix χ is identically equal to the unit matrix (i.e. the vectors $n_a^{(k)}$ vanish) if all the poles are located on this circle. In this case the soliton solution $g = \chi(0)g_0$ coincides with the background solution g_0.

For the two-soliton situation this can be seen from (2.4). If $\mu_2 = \bar{\mu}_1$, then the second term in (2.4) vanishes due to the factor $\alpha^2 - \mu_1\mu_2 = \alpha^2 - |\mu_1|^2$ and thus we have $g_{ab}^{(2)} = (g_0)_{ab}$. Consequently in the region $(\beta - w_1)^2 < \alpha^2$ the one-soliton solution $g^{(1)}$ remains unperturbed and coincides identically with the background solution g_0. The one-soliton matrix $g^{(1)}$ while remaining continuous on the entire (α, β)-plane, suffers discontinuities in its first derivatives on the light cone $(\beta - w_1)^2 = \alpha^2$. These discontinuities are then reflected into the metric coefficient f, which is determined by (1.40)–(1.41) through the first derivatives of g. Indeed, it follows from (2.3) that $f^{(1)}$ contains the factor $(\alpha^2 - \mu_1^2)^{-1}$, which becomes singular on the light cone $\alpha^2 - \mu_1^2 = 0$. The coefficient $f^{(1)}$ should be determined by (2.3) in the region (2.11) and by its background value $f^{(1)} = f_0$ in the region $(\beta - w_1)^2 < \alpha^2$.

With this definition, the metric coefficient f as well as its derivatives can suffer discontinuities on the boundary between these two regions. From the physical point of view this suggests that the solitary waves which are generated in this way might be considered, at least in some cases, as gravitational shock waves; we shall find some examples of this in section 4.5. This is one of the peculiar features of the gravitational solitary waves generated by real poles of the matrix χ. However, when real poles are absent and the dressing matrix χ contains pairs of complex conjugate poles there are no discontinuities in the metric and its derivatives along the corresponding light cones. In this case the perturbed gravitational field is present in the whole spacetime and the metric differs from the background metric everywhere.

We remind the reader that on the axis $\alpha = 0$ in fig. 2.1 we usually (but not always) have a physical singularity. If α is a time-like variable (in this case we can chose $\alpha = t$) such a singularity is of the usual cosmological type and the picture of the soliton metric is that of a big bang cosmology in which at $t = 0$ two solitary shock waves are generated. Their wavefronts travel towards space-like infinity at the speed of light, giving rise to the perturbed regions I and II, in such a way that the intermediate region III remains unperturbed.

Let us now consider the case of two real poles. This case will be discussed with the aid of fig. 2.2, which shows the behaviour of $g^{(2)}$ in the (α, β)-plane. In this case, (1.68)–(1.69) give two determinations for μ_1 and two determinations for μ_2. For any choice of the pair (μ_1, μ_2) the following is true. Two singular light cones arise with equations $(\beta - w_k)^2 = \alpha^2$ $(k = 1, 2)$. In regions I, V and VI,

$$(\beta - w_k)^2 \geq \alpha^2, \quad k = 1, 2, \tag{2.12}$$

and in these regions we have two-soliton solutions. If we cross the null line $\alpha = \beta - w_1$ from region I into region II the function μ_2 will remain real, but μ_1 will become complex with modulus $|\mu_1| = \alpha$. Then the solution in

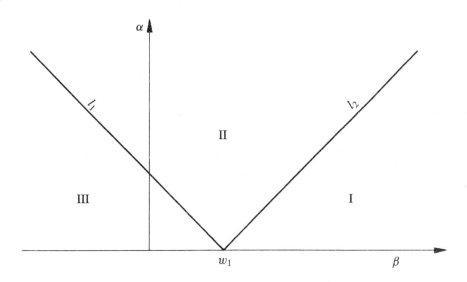

Fig. 2.1. Soliton solution with one real pole (w_1 real). Here the light cone $(\beta - w_1)^2 = \alpha^2$ divides the spacetime into three regions I, II and III, separated by the null lines l_1: $\alpha = -(\beta - w_1)$ ($\mu_1 = \alpha$), and l_2: $\alpha = \beta - w_1$ ($\mu_1 = -\alpha$). The one-soliton solution is defined only in regions I, where $\mu_1 < 0$, and III, where $\mu_1 > 0$. This solution may be matched throughout the light cone to the background solution in region II. The resulting solution, however, suffers discontinuities in its first derivatives on the light cone.

region II will be a three-soliton solution with one real pole, μ_2, and two complex conjugate poles μ_1 and $\overline{\mu}_1$ which are located on the circle $|\lambda|^2 = \alpha^2$ of the λ-plane. A simple inspection of (1.83)–(1.85) shows that two of the vectors $n_a^{(k)}$ corresponding to the poles μ_1 and $\overline{\mu}_1$ vanish, and the dressing matrix χ in region II will contain one pole only at $\lambda = \mu_2$. Hence, in region II we have a one-soliton solution and on the boundary $\alpha = \beta - w_1$ discontinuities arise.

If we now cross from region II into region III similar arguments show that we will have the background solution (g_0, f_0) in region III, with discontinuities also on the boundary $\alpha = \beta - w_2$. The whole schematic picture is shown in fig. 2.2. When α is a time-like variable we again may have some cosmological model with an initial big bang singularity on the axis $\alpha = 0$. Two pairs of gravitational shock waves in different space positions are now generated at the big bang and a new interaction phenomenon appears because two solitary waves collide head-on in region VI. In regions I and V we have also two interacting solitary waves but they run towards infinity one ahead of the other. In region IV (analogously in region II) the solution is one solitonic with one pole μ_1. The collision process of the waves in region VI takes a finite time; after this time the waves decouple and determine an inner unperturbed region III.

When α is a time-like variable the one- and two-soliton solutions have obvious cosmological interpretations. However, α can also be space-like and

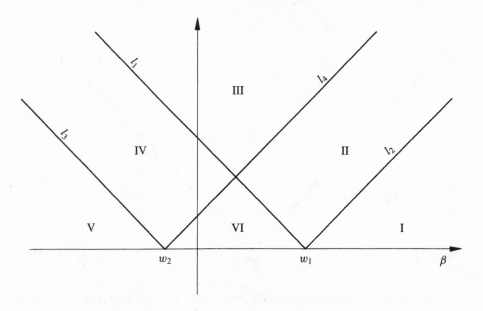

Fig. 2.2.   Soliton solution with two real poles (w_1 and w_2 real). Here the two light cones, for the two w_k ($k = 1, 2$), are defined by $(\beta - w_k)^2 = \alpha^2$, and divide the spacetime into six regions: I, II III, IV, V and VI. Now μ_1 is real in regions I, IV, V and VI, whereas μ_2 is real in regions I, II, V and VI. Thus the two-soliton solution is defined only in regions I (where $\mu_1 < 0$, $\mu_2 < 0$), V (where $\mu_1 > 0$, $\mu_2 > 0$) and VI (where $\mu_1 > 0$, $\mu_2 < 0$); region VI is the 'interaction' region. This solution may be matched with the one-soliton solution in region II throughout the null line l_2: $\alpha = \beta - w_1$ ($\mu_1 = -\alpha$) and with the corresponding one-soliton solution in region IV throughout the null line l_3: $\alpha = -(\beta - w_2)$ ($\mu_2 = \alpha$). The resulting solution may be matched with the background solution in region III throughout the null lines l_4: $\alpha = \beta - w_2$ ($\mu_2 = -\alpha$) and l_1: $\alpha = -(\beta - w_1)$ ($\mu_1 = \alpha$). The final solution, however, suffers discontinuities in its first derivatives on the light cones.

in this case we can choose $\alpha = z$. This corresponds to the cylindrical wave interpretation and α, in fact, becomes the radial coordinate. The coordinate β is now the time variable and x^1 and x^2 are the coordinates along the symmetry axis and the azimuthal angle. In this situation, fig. 2.1 shows the contraction of the cylindrical solitary wave (region III) to the axis of symmetry $\alpha = 0$. The back front of the wave $\alpha = -(\beta - w_1)$ contracts at the speed of light and behind the wave, in region II, it remains the unperturbed background spacetime. The wave collapses at the time $\beta = w_1$ and starts to reflect from the symmetry axis. In region I there is an expansion of the wave into the background space with a null wavefront $\alpha = \beta - w_1$. The analogous process for a double-solitonic cylindrical wave is shown in fig. 2.2. Now, there are two contracting waves in the past ($\beta < w_2$), the first one starts to reflect at the instant $\beta = w_2$ while the second (with wavefront $\alpha = -(\beta - w_1)$) continues its contraction until the time $\beta = w_1$. Between the times w_2 and w_1 an interaction region VI, the head-on collision region, is formed.

We have already mentioned that discontinuities arise only when real poles are present in the dressing matrix χ. When only pairs of complex conjugate poles appear the soliton metric is smooth everywhere (if the background metric is also smooth). In this case it is easy to evaluate the asymptotic behaviour of the solution in the limit $\alpha \to \infty$, when β is some fixed value. It follows from (1.68) and (1.69) that in this limit $|\mu_k| \to \alpha$ and all poles tend to be located on the circle $|\lambda|^2 = \alpha^2$, which means that the solution tends to the unperturbed background solution. In this asymptotic region the solitonic perturbation becomes small enough that all nonlinear effects are negligible. The solitary waves in this limit behave like gravitational waves in the linear approximation of general relativity.

2.1.1 Pole fusion

Let us now turn to discuss in some detail the relations between the perturbations $g^{(1)}$, $g^{(2)}$ and the background g_0. For the one-soliton case we can easily see that in general no choice of the arbitrary parameters $m_{0a}^{(1)}$ can lead from $(g^{(1)}, f^{(1)})$ to (g_0, f_0). This means that the one-soliton perturbation $g^{(1)}$ is, in general, a finite perturbation of g_0, i.e. by varying the parameters $m_{0a}^{(1)}$ one cannot, in general, find a solution $(g^{(1)}, f^{(1)})$ that is (in some suitable sense) arbitrarily close to the given background. The situation, however, is different in the case of two-soliton perturbations.

Let us now consider the limit $w_2 \to w_1$ and choose μ_1 and μ_2 to be the corresponding roots of (1.67) so that, also, $\mu_2 \to \mu_1$. Then, in general, when the arbitrary constants $m_{0a}^{(2)}$ and $m_{0a}^{(1)}$ are considered as independent of w_2 and w_1, the function $\det Q_{kl}$ from (2.8) will not vanish. Then (2.9) shows that D will not vanish when $w_2 \to w_1$. From (2.4) we can see that $g_{ab}^{(2)}$ tends to the background metric $(g_0)_{ab}$. By replacing $C_f^{(2)}$ in (2.10) with $C_f^{(2)\prime}(w_1 - w_2)^2$ we may arrange the terms in such a way that also $f^{(2)}$ will tend to f_0. Therefore in this case the background metric (g_0, f_0) is obtained as a limit of the two-soliton perturbation $(g^{(2)}, f^{(2)})$. This means that this family contains metrics that are arbitrarily close to the background.

On the other hand, when we choose the constants $m_{0a}^{(k)}$ in such a way that they depend on w_k and in the limit $w_2 \to w_1$ we have also $m_{0a}^{(2)} \to m_{0a}^{(1)}$ (and correspondingly $\mu_2 \to \mu_1$), the function D vanishes and, in general, (2.4) implies that $g^{(2)}$ does not tend to g_0, but to the new kind of one-soliton perturbation of g_0. This sort of 'pole fusion' procedure leads to a kind of one-soliton solution that corresponds to a second order pole of the dressing matrix χ. This is a new and nontrivial phenomenon in the framework of the ISM that we have described. It may seem that we are forced to use only the simple pole structure of the dressing matrix χ, but the existence of the 'pole fusion' limiting procedure shows that this is not the case. Indeed, if $w_2 \to w_1$, $\mu_2 \to \mu_1$ and we denote the relative small parameter by $\epsilon = w_2 - w_1$, then it is easy to show that for μ_1 and μ_2 we get the expansions

$$\mu_1 = \mu + \epsilon m_1 + O(\epsilon^2), \quad \mu_2 = \mu + \epsilon m_2 + O(\epsilon^2),$$

where μ, m_1 and m_2 are some finite regular quantities independent of ϵ.

If, in addition, we assume that in the limit $\epsilon \to 0$ the arbitrary constants $m_{0b}^{(2)}$ tend to the $m_{0b}^{(1)}$ according to the law $m_{0b}^{(2)} - m_{0b}^{(1)} \sim \epsilon$, we have the same behaviour for the vector $m_a^{(k)}$, i.e. $m_a^{(2)} - m_a^{(1)} \sim \epsilon$. Then it follows from (1.83) and (2.5)–(2.6) for $k, l = 1, 2$ that

$$\det \Gamma_{kl} = (\alpha^2 - \mu_1^2)^{-1}(\alpha^2 - \mu_2^2)^{-1}(\alpha^2 - \mu_1\mu_2)^{-2}D,$$

and it is obvious from (2.8)–(2.9) that $\det \Gamma_{kl} \sim \epsilon^2$. The evaluation of the vectors $n_a^{(k)}$ using (1.85) shows that the first terms of the expansion are singular: $n_a^{(k)} \sim \epsilon^{-1}$. On the other hand (1.82) shows that the sum $n_a^{(1)} + n_a^{(2)}$ has a regular expansion: $n_a^{(1)} + n_a^{(2)} = \text{finite} + O(\epsilon)$. This means that the matrices R_1 and R_2 in the dressing matrix χ, i.e.

$$\chi = I + R_1(\lambda - \mu_1)^{-1} + R_2(\lambda - \mu_2)^{-1},$$

have the following expansion:

$$R_1 = \epsilon^{-1}R + F_1 + O(\epsilon), \quad R_2 = -\epsilon^{-1}R + F_2 + O(\epsilon),$$

where the matrices R, F_1 and F_2 are independent of ϵ. Whereas each term $R_1(\lambda - \mu_1)^{-1}$ and $R_2(\lambda - \mu_2)^{-1}$ diverges separately at $\epsilon = 0$, the sum does not, i.e. the matrix χ has a well defined limit when $\epsilon \to 0$:

$$\lim_{\epsilon \to 0} \chi = I + \frac{(F_1 + F_2)(\lambda - \mu) + R(m_1 - m_2)}{(\lambda - \mu)^2},$$

and it is easy to see that $m_1 - m_2 \neq 0$. Consequently, this limit corresponds to the specific one-soliton solution which comes from a second order pole in the dressing matrix χ.

The same analysis can be performed for the general n-soliton case and as a result of the fusion of n poles into a single pole we can obtain the one-soliton solution when the matrix χ has one multiple pole of order n. If the number of poles n is even, there is the possibility of fusing them into two poles with multiplicity $n/2$. One example of such a fused two-soliton solution in the case of stationary gravitational fields is the well known Tomimatsu–Sato metric which we describe in chapter 8. In such a case the multiplicity $n/2$ is just the Tomimatsu–Sato distortion parameter δ.

2.2 Diagonal background metrics

We saw in section 1.4 that the main step in the construction of solitonic solutions in closed form is the computation of the set of vectors $m_a^{(k)}$. Once these vectors are known the rest can be performed by purely algebraic manipulations. However, to compute the $m_a^{(k)}$ one needs to integrate the system of differential equations (1.51), in order to find the background generating matrix $\psi_0(\lambda, \zeta, \eta)$. Fortunately in many cases we may choose a background metric g_0 which is simple enough that exact solutions for ψ_0 can be found. An important case corresponds to the election of a diagonal matrix g_0 since this produces a diagonal generating matrix ψ_0 as well [157, 192]. Of course, the concrete functional form of the diagonal components of $g_0(\zeta, \eta)$ is also of importance but in this and in the next sections we will only exploit the diagonality restriction of g_0. Even at this level one can extract many interesting results using the ISM approach to general relativity. Concrete functional dependences of g_0 will be used in the following chapters, where we describe several applications of the ISM. In this chapter only at the ends of this and the next section will we consider particular background metrics as examples.

First we note that for a diagonal g_0 it is more convenient to compute the vectors $m_a^{(k)}$ directly from (1.77), avoiding the problem of integrating ψ_0 from the differential system (1.51). Any diagonal solution g_0 of the Einstein equations (1.38)–(1.39) can be represented in the form

$$g_0 = \text{diag}(\alpha e^{u_0}, \ \alpha e^{-u_0}), \tag{2.13}$$

where $\alpha(\zeta, \eta)$ satisfies (1.44), and $u_0(\zeta, \eta)$ is a solution of the equation

$$(\alpha u_{0,\zeta})_{,\eta} + (\alpha u_{0,\eta})_{,\zeta} = 0. \tag{2.14}$$

The background metric in this case takes the form,

$$ds^2 = f_0(dz^2 - dt^2) + \alpha e^{u_0}(dx^1)^2 + \alpha e^{-u_0}(dx^2)^2, \tag{2.15}$$

and the coefficient f_0 can be calculated from (1.40)–(1.41) with the matrices A and B following from (1.42), where we should substitute g by g_0.

A simple analysis of (1.77) and (1.76) shows that in this case the general solution for the vectors $m_a^{(k)}$ is

$$m_1^{(k)} = A_k \mu_k^{-1/2} \exp(-\rho_k/2 - u_0/2), \quad m_2^{(k)} = A_k \mu_k^{-1/2} \exp(\rho_k/2 + u_0/2), \tag{2.16}$$

where each function ρ_k $(k = 1, 2, \ldots, n)$ can be found by quadratures from the following relations:

$$\rho_{k,\zeta} = \frac{\alpha + \mu_k}{\alpha - \mu_k} u_{0,\zeta}, \quad \rho_{k,\eta} = \frac{\alpha - \mu_k}{\alpha + \mu_k} u_{0,\eta}, \tag{2.17}$$

and where A_k are arbitrary constants. Although these constants do not appear in the final result we keep them in (2.16) in order to ensure the correct complex structure of the vectors $m_a^{(k)}$, which is necessary for the reality of the metric. The integrability condition for these two equations is satisfied automatically due to (1.66) and (2.14). Each function $\rho_k(\zeta, \eta)$ has an additive arbitrary constant which is complex in general; the values of these constants together with the parameters A_k, which are also complex in general, should be chosen in such a way that the reality conditions for the metric discussed in section 1.4 are satisfied: i.e. a real $m_a^{(k)}$ for a real μ_k and a complex conjugate pair of vectors $m_a^{(k)}$ for each complex conjugate pair of μ_k. Thus, we have reduced the problem of integration of $\psi_0(\lambda, \zeta, \eta)$ from (1.51) to the integration of the functions ρ_k from (2.17); but this is trivial because all the functions on the right hand side of (2.17) are given. The substitution of the vectors $m_a^{(k)}$ of (2.16) into the general formulas derived in section 1.4 will give us an n-soliton solution on the diagonal background.

The exact expressions for one and two gravisolitons on the diagonal background can be obtained from (2.1)–(2.3) and (2.4)–(2.10) after substitution of the vectors $m_a^{(k)}$ of (2.16) into $L_a^{(k)}$, Q_{11} and Q_{kl} of (1.86), (2.2) and (2.5), respectively. Let us give here the final results.

One-soliton solution. The one-soliton solution with the diagonal background metric (2.13)–(2.15) can be written in the form

$$g^{(1)} = \frac{1}{\mu_1 \cosh \rho_1} \left(\begin{matrix} (\mu_1^2 e^{\rho_1} + \alpha^2 e^{-\rho_1})e^{u_0} & \alpha^2 - \mu_1^2 \\ \alpha^2 - \mu_1^2 & (\alpha^2 e^{\rho_1} + \mu_1^2 e^{-\rho_1})e^{-u_0} \end{matrix} \right), \quad (2.18)$$

$$f^{(1)} = f_0 \alpha^{1/2} \mu_1 \cosh \rho_1 (\alpha^2 - \mu_1^2)^{-1}, \quad (2.19)$$

where the two possible choices for the function μ_1 are given by (1.68)–(1.69) for $k = 1$, and the function ρ_1 can be found from (2.17) also for $k = 1$.

We have to keep in mind that for any solution g of (1.38)–(1.39) the matrix $-g$ is also a solution, and that the same sign symmetry is valid for the metric component f. Due to this freedom one can always choose the correct signs in front of g and f in (2.18)–(2.19) in order to ensure positivity for f and the (++) signature for g.

Two-soliton solution. The two-soliton solution can be written in the following form:

$$g_{11}^{(2)} = \left\{ 1 + \frac{1}{D} \left[\cosh 2\tau + \frac{(\mu_1 + \mu_2)}{(\mu_1 - \mu_2)} \sinh 2\tau \right. \right.$$
$$\left. \left. + \cosh 2\sigma - \frac{(\alpha^2 + \mu_1 \mu_2)}{(\alpha^2 - \mu_1 \mu_2)} \sinh 2\sigma \right] \right\} \alpha e^{u_0}, \quad (2.20)$$

$$g_{22}^{(2)} = \left\{1 + \frac{1}{D}\left[\cosh 2\tau - \frac{(\mu_1 + \mu_2)}{(\mu_1 - \mu_2)}\sinh 2\tau\right.\right.$$
$$\left.\left. + \cosh 2\sigma + \frac{(\alpha^2 + \mu_1\mu_2)}{(\alpha^2 - \mu_1\mu_2)}\sinh 2\sigma\right]\right\}\alpha e^{-u_0}, \qquad (2.21)$$

$$g_{12}^{(2)} = \frac{2\alpha}{D}\left[\frac{(\mu_1 + \mu_2)}{(\mu_1 - \mu_2)}\sinh\sigma\sinh\tau + \frac{(\alpha^2 + \mu_1\mu_2)}{(\alpha^2 - \mu_1\mu_2)}\cosh\sigma\cosh\tau\right], \qquad (2.22)$$

where

$$D = 4\mu_1\mu_2\left[\alpha^2(\alpha^2 - \mu_1\mu_2)^{-2}\cosh^2\sigma + (\mu_1 - \mu_2)^{-2}\sinh^2\tau\right], \qquad (2.23)$$

and all possible values of the functions μ_1 and μ_2 are given by (1.68)–(1.69) for $k = 1$ and $k = 2$. To simplify the expressions we have introduced the notation

$$\sigma = \rho_1/2 + \rho_2/2, \quad \tau = \rho_1/2 - \rho_2/2, \qquad (2.24)$$

where the functions ρ_1 and ρ_2 follow from (2.17) for $k = 1$ and $k = 2$. For the metric coefficient $f^{(2)}$ we have,

$$f^{(2)} = f_0\mu_1\mu_2 D(\alpha^2 - \mu_1^2)^{-1}(\alpha^2 - \mu_2^2)^{-1}. \qquad (2.25)$$

Kasner background. A particularly interesting background metric is the Kasner solution which is introduced in section 4.2 and plays an important role in many applications. In this case the function u_0 in (2.13) is

$$u_0 = d\ln\alpha, \qquad (2.26)$$

where d is an arbitrary real parameter: the Kasner parameter. The integration of (2.17) is very easy if we take into account the expressions (1.66) which relate the derivatives with respect to the null coordinates ζ and η of the function α, with the analogous derivatives of the pole trajectories μ_k. The result is

$$\rho_k = d\ln\left(\frac{\mu_k}{\alpha}\right) + C_k, \qquad (2.27)$$

where C_k are the arbitrary constants, in general complex, that we have referred to above. Note that a diagonal limit of the solution (2.18)–(2.19) may be obtained by taking $C_1 \to \infty$, provided a constant proportional to $\exp(C_1)$ multiplies $f^{(1)}$ (the f coefficients are determined up to a multiplicative constant, see (1.110)); similarly diagonal limits of (2.20)–(2.25) may obtained by taking $C_1 = C_2 \equiv C \to \infty$, or $C_1 = -C_2 \equiv C \to \infty$, provided the multiplicative arbitrary constant in $f^{(2)}$ is proportional to $\exp(2C)$. Here we have assumed that the pole trajectories are real, thus the C_k are also real.

2.3 Topological properties

It was shown in the previous sections that from the mathematical point of view gravitational solitons indeed have the status of solitons. However, the place of these objects in the physical applications of general relativity is far from clear. This is because gravisolitons have a number of unusual properties with respect to their partners in nongravitational physics. Close examination shows that one finds difficulties in the physical interpretation of gravisolitons when trying to use an analogy with their nongravitational relatives. The reasons for this are the following:

1. The gravisolitons amplitudes and shapes are not preserved in time.

2. The gravisolitons velocities change in time and, furthermore, the definition of velocity is not clear.

3. In general, for real-pole trajectories, the field of the gravisolitons is not smooth in spacetime, it has discontinuities in the first derivatives at some null hypersurfaces as was explained in section 2.1.

4. There is no time evolution of the gravisolitons from a free (noninteracting) state at $t = -\infty$ to a free state again at $t \rightarrow \infty$ due to the unavoidable existence of cosmological singularities between these asymptotic regions. Therefore the description of the collision process needs special care.

5. There is no notion of energy (and consequently of its mass) for the gravisoliton. We remind the reader that any physical description of a soliton starts with the statement that this object represents some localized perturbation with finite energy. We also note that this is an important ingredient in soliton quantization [245]

6. It is not clear from our previous discussion whether a gravisoliton represents a topological object and whether some topological charge can be associated to it.

In this section we show that some of these problems can be solved at least up to a level where a qualitative understanding of them is possible. Our main interest now will concentrate on the last point referring to the topological properties of gravisolitons. In what follows we will see that despite the peculiarities in the physical properties of gravisolitons, there remain some analogies between them and the sine-Gordon kinks. By means of these analogies one can elucidate some of the physical aspects of the gravisoliton behaviour. It can be shown that for a wide class of cosmological and wave metrics the gravisolitons can be regarded as sine-Gordon kinks embedded in some (very special) external field. A number of the properties of sine-Gordon kinks can be generalized for their gravitational partners in an exact mathematical way. In this way one can show that the gravitational solitons are topological objects and that they can carry topological charge. There is attraction between a gravisoliton and an antigravisoliton and repulsion between two gravisolitons with the same charge. Also there is a bound state of a gravisoliton and an antigravisoliton which oscillates in time and which is the direct analogue of the sine-Gordon breather solution. Along with the clarification of these topological aspects we will gain some qualitative understanding of such notions as the velocity and head-on collision of gravisolitons. These results have been described in ref. [14]. However a number of problems, especially the exact formulation of the soliton energy, still remain for future development.

In order to have cosmological type solutions α should be time-like (i.e. $\alpha_{,\zeta}\alpha_{,\eta} < 0$) at least near the cosmological singularity. It has already been mentioned in section 1.2 that in this case the cosmological singularity corresponds to $\alpha = 0$ and, thus, we are forced to consider only half of the α axis ($\alpha \geq 0$). The variable β of (1.46) will be automatically space-like in the region where α is time-like. In this section, for simplicity, we restrict ourselves to the following topological structure of the spacetime. We assume that α is time-like and β is space-like everywhere, and that (α, β) form the single patch of the natural coordinates, which cover the whole two-dimensional section of the maximally extended physical spacetime. Furthermore, each pair of real numbers α, β in the ranges $0 < \alpha$, $-\infty < \beta < \infty$ represent one, and only one, point of this spacetime and vice versa. We consider as a second equivalent coordinate system in the same spacetime the variables (t, z): this means that the map between (t, z) and (α, β) is everywhere smooth and one to one in both directions. As a consequence we have to use only those functions $\alpha(t, z)$ and $\beta(t, z)$ for which the Jacobian $\alpha_{,t}\beta_{,z} - \alpha_{,z}\beta_{,t} = \frac{1}{2}(\alpha_{,\zeta}\beta_{,\eta} - \alpha_{,\eta}\beta_{,\zeta}) = -\alpha_{,\zeta}\alpha_{,\eta}$ has no zeros or infinities. It should be emphasized that all these restrictions are essential. Only for such types of gravisolitons can the close analogy between our integrable gravitational ansatz and the sine-Gordon theory be shown, and a sensible notion of topological charge be introduced. It may well happen that outside these additional conditions there is no way to characterize the solutions by some topological charge; for example, when each pair of coordinates (α, β) or (t, z) corresponds to two different points of the relevant two-dimensional section of the spacetime as was proposed in [122]. We do not consider these quite different (although possible) spacetime constructions in this section.

Let us turn again to the one-soliton (2.18)–(2.19) on the diagonal background (2.15). To ensure subluminal soliton speed in this solution it is necessary to choose the background metric function $u_0(t, z)$ to have space-like character ($u_{0,\zeta}u_{0,\eta} > 0$). In this case, as follows from (2.17), the variable $\rho_1(t, z)$ also has space-like character and the curve $\rho_1 = 0$ is time-like. Indeed, it can be seen from (2.18) that the field of the soliton perturbation is concentrated at the points where $\rho_1 = 0$. This is especially clear in the approximation in which α

and μ_1 can be considered as slowly varying functions with respect to ρ_1 and u_0, see below.

Thus in what follows we shall use only the background solutions in which u_0 is space-like and, also, only those functions $u_0(t, z)$ for which the pair (α, u_0) are acceptable time and space coordinates, i.e. for which the Jacobian $\alpha_{,\zeta} u_{0,\eta} - \alpha_{,\eta} u_{0,\zeta}$ nowhere vanishes or diverges.

Before we continue the analysis of simple soliton solutions let us look more carefully at the separation that we made by relations (1.68)–(1.69) of the poles μ_k into two classes: μ_k^{in} and μ_k^{out}. It is obvious that at each fixed spacetime point, μ_k belongs either to the class 'in' or to the class 'out'; and, more importantly, the same is true for the entire pole trajectory $\mu_k(\zeta, \eta)$. It can be seen that under the spacetime topological restrictions adopted earlier and as a consequence of general continuity requirements the property that the pole trajectory $\mu_k(\zeta, \eta)$ is μ_k^{in} or μ_k^{out} is global: if at some spacetime point $\mu_k = \mu_k^{in}$, then μ_k will remain μ_k^{in} everywhere, and likewise for μ_k^{out}. In other words, the pole trajectory can never cross the circle $|\lambda| = \alpha$. The simplest way to see this is to substitute into (1.67) the quantities μ_k and w_k in the form

$$\mu_k = |\mu_k| e^{i\varphi_k}, \quad w_k = \operatorname{Re} w_k + i \operatorname{Im} w_k. \tag{2.28}$$

Then (1.67) is equivalent to the following two relations:

$$\cos \varphi_k = |\mu_k|(|\mu_k|^2 + \alpha^2)^{-1}(2 \operatorname{Re} w_k - 2\beta), \quad \sin \varphi_k = 2|\mu_k|(|\mu_k|^2 - \alpha^2)^{-1} \operatorname{Im} w_k. \tag{2.29}$$

From the second relation it is obvious that no trajectory can cross from the region where $|\mu_k|^2 > \alpha^2$ to the region where $|\mu_k|^2 < \alpha^2$ and vice versa. One particular, but typical, example of the pole trajectories behaviour is shown in fig. 2.3. This figure corresponds to some fixed value of α in (2.29). In this case (2.29) represents a two-dimensional dynamical system in the phase space $(\operatorname{Re} \mu_k, \operatorname{Im} \mu_k)$ in which the function $\beta - \operatorname{Re} w_k$ serves as a dynamical parameter along the trajectories, and $\operatorname{Im} w_k$ corresponds to an arbitrary constant identifying each individual trajectory. This diagram shows very clearly the absolute separation of 'in' and 'out' pole trajectories.

The behaviour of the pole trajectories displayed on fig. 2.3 also confirms that in the case of a real pole μ_k the choice $\mu_k = \mu_k^{in}$ or $\mu_k = \mu_k^{out}$, solutions of the quadratic equation (1.67), has been made in the correct way. The problem is that in the case of real μ_k (real constant w_k) the entire spacetime consists of the two disconnected causal domains: $\beta - w_k < -\alpha$ and $\beta - w_k > \alpha$. There is no obvious physical reason why one cannot choose μ_k to be μ_k^{in} in one of these domains and μ_k^{out} in the other. The dynamical system we present in fig. 2.3 shows that such a choice would be against the natural continuity requirements. The correct choice of μ_k through the entire spacetime should result from the general complex picture by going continuously to the limit $\operatorname{Im} w_k \to 0$. In this way we obtain a unique rule for choosing the real solutions to (1.67). From fig.

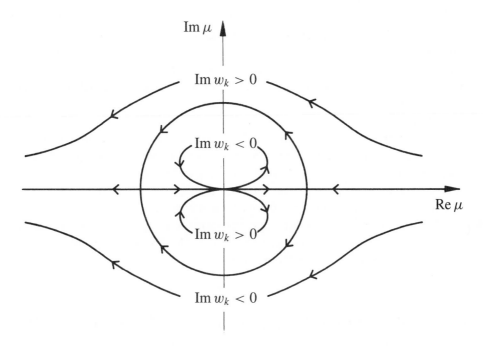

Fig. 2.3. The phase diagram for the pole trajectories $\mu_k(\alpha, \beta)$ for some fixed k, i.e. the solution of the quadratic equation (1.67) for arbitrary complex values of the constant w_k. The variable α is fixed, the dynamical parameter along the trajectories is $\beta - \mathrm{Re}\, w_k$ and the arrows indicate that this parameter increases. The different values of the parameters $\mathrm{Im}\, w_k$ correspond to different trajectories. For trajectories located on the axis and on the circle $|\mu_k| = \alpha$ we have $\mathrm{Im}\, w_k = 0$. Because the circle $|\mu_k| = \alpha$ is an invariant one-dimensional manifold for this dynamical system (it is also made of trajectories) no trajectory can cross this circle to go from region $|\mu_k| > \alpha$ to region $|\mu_k| < \alpha$ or vice versa.

2.3 it is easy to see that this rule requires that we should keep the 'in' or 'out' property of the trajectories globally, i.e. even through the causally disconnected regions. This rule is realized by (1.68)–(1.69) for the real μ_k when we take the prescription that the square root in these formulas should be understood everywhere as a positive quantity only.

We turn again to the one-soliton solution (2.18)–(2.19). It is quite clear from the above that there are, actually, two such solutions: one for $\mu_1 = \mu_1^{in}$ and another for $\mu_1 = \mu_1^{out}$. We will see in what follows that there is enough evidence to consider these two solutions as belonging to two different topological sectors and, consequently, as having different topological indices. In other words, it can be assumed that there is no homotopy between these two solutions. If such homotopy were to exist, it should manifest itself also in the approximation where the function α (together with β and μ_1) is slowly varying with respect to the functions ρ_1 and u_0. However, in this extremal case the theory based on (1.38)–(1.39) tends to coincide with the sine-Gordon theory, for which solutions with $\mu_1 = \mu_1^{in}$ and $\mu_1 = \mu_1^{out}$ emerge as well known topologically different solutions associated with topological charges 1 and -1. This correspondence helps us to find in the exact gravitational case the one-dimensional manifolds between which the one-soliton map acts, which is necessary for a sensible notion of homotopy.

To verify these assertions we need to change the metric components of matrix g for more suitable field variables which do not depend on arbitrary linear transformations of the dummy coordinates x^1 and x^2. It is possible to construct such invariants only from the matrices $g_{,\zeta}g^{-1}$ and $g_{,\eta}g^{-1}$. The first three nontrivial quantities of this kind are $\mathrm{Tr}[(g_{,\zeta}g^{-1})^2]$, $\mathrm{Tr}[(g_{,\eta}g^{-1})^2]$ and $\mathrm{Tr}(g_{,\zeta}g^{-1}g_{,\eta}g^{-1})$. The simplest invariants $\mathrm{Tr}(g_{,\zeta}g^{-1}) = 2\alpha_{,\zeta}\alpha^{-1}$ and $\mathrm{Tr}(g_{,\eta}g^{-1}) = 2\alpha_{,\eta}\alpha^{-1}$ are trivial, because they do not carry any information on the soliton behaviour. With the notation

$$[\ln(g_{11}\alpha^{-1})]_{,\zeta} = R_1 \cos(\gamma/2+\omega/2), \quad (g_{12}g_{11}^{-1})_{,\zeta}g_{11}\alpha^{-1} = R_1 \sin(\gamma/2+\omega/2),$$
$$(2.30)$$
$$[\ln(g_{11}\alpha^{-1})]_{,\eta} = R_2 \cos(\gamma/2-\omega/2), \quad (g_{12}g_{11}^{-1})_{,\eta}g_{11}\alpha^{-1} = R_2 \sin(\gamma/2-\omega/2),$$
$$(2.31)$$

we obtain

$$\mathrm{Tr}[(g_{,\zeta}g^{-1})^2] = 2R_1^2 + 2(\alpha_{,\zeta})^2\alpha^{-2}, \quad \mathrm{Tr}[(g_{,\eta}g^{-1})^2] = 2R_2^2 + 2(\alpha_{,\eta})^2\alpha^{-2},$$
$$(2.32)$$
$$\mathrm{Tr}(g_{,\zeta}g^{-1}g_{,\eta}g^{-1}) = 2R_1R_2 \cos\omega + 2\alpha_{,\zeta}\alpha_{,\eta}\alpha^{-2}.$$
$$(2.33)$$

Thus, the invariants we need are R_1, R_2 and ω. It is remarkable that (1.39), together with the selfconsistency conditions for (2.30)–(2.31), are reduced to the system:

$$\omega_{,\zeta\eta} + \left(\frac{R_{1,\eta}}{R_1} + \frac{\alpha_{,\eta}}{2\alpha}\right)\omega_{,\zeta} + \left(\frac{R_{2,\zeta}}{R_2} + \frac{\alpha_{,\zeta}}{2\alpha}\right)\omega_{,\eta}$$
$$= \left[\left(\frac{\alpha_{,\zeta}R_2}{2\alpha R_1}\right)_{,\zeta} + \left(\frac{\alpha_{,\eta}R_1}{2\alpha R_2}\right)_{,\eta} + R_1R_2\right]\sin\omega, \qquad (2.34)$$

$$2\alpha R_{1,\eta} + \alpha_{,\eta}R_1 + \alpha_{,\zeta}R_2 \cos\omega = 0, \quad 2\alpha R_{2,\zeta} + \alpha_{,\zeta}R_2 + \alpha_{,\eta}R_1 \cos\omega = 0, \quad (2.35)$$

which contains these invariants only, and the system:

$$(\gamma/2 + \omega/2)_{,\eta} = -R_2 \sin(\gamma/2 - \omega/2) + \alpha_{,\zeta}R_2(2\alpha R_1)^{-1} \sin\omega, \qquad (2.36)$$

$$(\gamma/2 - \omega/2)_{,\zeta} = -R_1 \sin(\gamma/2 + \omega/2) - \alpha_{,\eta}R_1(2\alpha R_2)^{-1} \sin\omega, \qquad (2.37)$$

from which the function γ can be found by quadratures, and which also involves the given invariants only. The integrability conditions of (2.36)–(2.37) is ensured by (2.34)–(2.35).

It is reasonable to consider the field ω (mod 2π) as the main gravisolitonic characteristic, because this field has qualitative features which we usually associate with solitons. Indeed, one can define the *solitonic vacuum states* as the exact solutions of the system (2.34)–(2.37), which correspond to the discrete set of constant values of ω, i.e. $\omega = 2\pi n$, where n is an integer. With these values of ω, (2.34)–(2.37) can be solved exactly. In fact, for the functions R_1 and R_2 we get $R_1 = \phi_{,\zeta}$, $R_2 = \phi_{,\eta}$, where ϕ is a solution of the equation $(\alpha\phi_{,\zeta})_{,\eta} + (\alpha\phi_{,\eta})_{,\zeta} = 0$. Equations (2.36)–(2.37) are now equivalent to the Bäcklund transformation and give for γ the kink solution: $\gamma = 4\arctan\exp[-(-1)^n\phi - C]$, where $C = $ constant. It turns out, however, that this is a fictitious or pure gauge kink because it can be removed by a linear transformation (with constant coefficients) of the dummy coordinates x^1 and x^2 (note that γ is not invariant with respect to such a transformation). Indeed, after computing the matrix g from (2.30)–(2.31) it is easy to see that the matrix g can be made diagonal $g = \text{diag}\,(\alpha\exp[(-1)^n\phi + C],\ \alpha\exp[-(-1)^n\phi - C])$, by this linear transformation. This means that $\gamma_{new} = 2\pi m$, where m is an integer. Since any diagonal solution for g has this form, we conclude that any diagonal matrix g represents one of the vacuum states with respect to the invariant gravisolitonic field ω. This picture conforms to our intuitive idea of considering the solutions of (1.39) with diagonal g as containing no true solitons because the Einstein field equations for g in this case are linear.

Because the function $\mu_1(\zeta, \eta)$ satisfies the differential equations (1.66), it is easy to show that for α time-like the variable $\mu_1\alpha^{-1}$ is also time-like. Then it can be seen that for the solution (2.18) the variables $\mu_1\alpha^{-1}$ and ρ_1 also form a pair of acceptable time and space coordinates, respectively. The analysis shows that for each fixed value of the new time $\mu_1\alpha^{-1}$ (i.e. on the straight lines $(\beta - w_1)\alpha^{-1} = $ constant in the (α, β)-plane) the function $\omega(\rho_1)$ acts as a regular map between the one-dimensional ρ_1-space and the ω-circle. The map is one to one in both directions and the angle ω covers exactly once the segment $[0, 2\pi]$ when ρ_1 runs between its natural boundaries (from $\rho_1(\alpha = 0)$ to $\rho_1(\alpha = \infty)$). It turns out that for this map the integer quantity $\text{sign}(\alpha^2 - \mu_1^2)$ corresponds to the Brouwer degree. This quantity is equal to $+1$ for $\mu_1 = \mu_1^{in}$ and to -1 for $\mu_1 = \mu_1^{out}$. The above arguments show that two related one-soliton solutions are associated with different topological indices and act like bridges between neighbouring vacua of the field ω. This is in agreement with general properties of topological solitons.

2.3.1 Gravisolitons and antigravisolitons

The previous analysis gives additional ground for considering the curve $\rho_1 = 0$ in the solution (2.18) as a soliton world line. A measure of the degree of nondiagonality of this solution, with respect to the diagonal background (2.13), can be evaluated as:

$$\frac{g_{12}^{(1)}}{\sqrt{(g_0)_{11}(g_0)_{22}}} = \frac{g_{12}^{(1)}}{\alpha} = \frac{1}{\cosh \rho_1} \left(\frac{\alpha}{\mu_1} - \frac{\mu_1}{\alpha} \right). \tag{2.38}$$

For a fixed value of the 'time' $\mu_1 \alpha^{-1}$ this measure tends to zero for large absolute values of ρ_1. Consequently, when ρ_1 is far from zero the metric (2.18) is effectively diagonal, but it does not tend to the background. However, we have shown already that for a diagonal metric the gravisoliton disappears, thus the soliton is mainly concentrated around the curve $\rho_1 = 0$.

Due to the close relation between gravisolitons and sine-Gordon kinks we can support qualitatively the topological nature of gravisolitons for the extreme case when α (or a and b in (1.45)) is a slowly varying variable with respect to ρ_1 and u_0. In this case the variables β and μ_1 are also slowly varying because they are expressed algebraically in terms of a and b. In the first approximation (2.35) gives $R_{1,\eta} = 0$, $R_{2,\zeta} = 0$ and, without loss of generality, one can take $R_1 = $ constant and $R_2 = $ constant. From (2.34) it follows that $\omega_{,\zeta\eta} = R_1 R_2 \sin \omega$ and we have for γ a Bäcklund transformation (2.36)–(2.37). The solution (2.18) has its own natural place in this approximation, (2.14) gives $u_{0,\zeta\eta} = 0$ and for u_0 space-like one can choose the coordinates in such a way that $u_0 = mz$, where $m > 0$ is an arbitrary constant. Integrating (2.17), which is trivial in the first order approximation (α and μ_1 are constants), and substituting (2.18) into (2.30)–(2.31) (at this step we differentiate only the rapidly varying functions ρ_1 and u_0) we obtain $R_1 = R_2 = m$ and

$$\omega = 4 \arctan e^{\rho_1}, \quad \rho_1 = m \frac{\alpha^2 + \mu_1^2}{\alpha^2 - \mu_1^2} \left(z + \frac{2\alpha\mu_1}{\alpha^2 + \mu_1^2} t + z_0 \right), \tag{2.39}$$

where z_0 depends only on the slowly varying functions α, β and μ_1, i.e. it is a constant in the first order approximation. Due to the obvious identity

$$\left(\frac{\alpha^2 + \mu_1^2}{\alpha^2 - \mu_1^2} \right)^{-2} \equiv 1 - \left(\frac{2\alpha\mu_1}{\alpha^2 + \mu_1^2} \right)^2, \tag{2.40}$$

we see that (2.39) describes a sine-Gordon soliton with mass m and local velocity $v = -2\alpha\mu_1(\alpha^2 + \mu_1^2)^{-1}$. It is clear now that the 'slow-α' approximation is valid when this soliton is heavy enough, i.e. when its mass is much bigger than the spacetime derivatives of α (and we are not too close to the singularity $\alpha = 0$). We see also that a topological charge $+1$ or -1, i.e. the sign($\alpha^2 - \mu_1^2$), can be associated with this soliton. This coincides with the result of our previous exact analysis.

The approximation of slowly varying α becomes exact if $\alpha = $ constant, exactly. In this case the equation for ω is the exact sine-Gordon equation and we have stationary poles ($\mu_k = $ constant) in the matrix χ in (1.64). Consequently, we can say that for the exact sine-Gordon theory the soliton topological charge can be defined by the location of the stationary pole $\lambda = \mu_1$ with respect to the circle $|\lambda| = \alpha$. The nontrivial and remarkable fact is that also for our gravitational ansatz (which represents a deformed sine-Gordon theory) the notion of topological charge can be maintained and described in terms of the same terminology. The generalization is straightforward: the stationary position of the pole $\lambda = \mu_1$ inside or outside the circle $|\lambda| = \alpha$ in the exact sine-Gordon theory implies the confinement of the entire pole trajectory $\lambda = \mu_1(\zeta, \eta)$ inside or outside the circle $|\lambda| = \alpha$ in general relativity.

Following this line of thought we may call the solution (2.18)–(2.19) for the case $\mu = \mu_1^{in}$ a gravisoliton (S) and for the case $\mu = \mu_1^{out}$ an antigravisoliton (A). However, the real physical manifestation of the topological charge can be seen only in the collision process of two such objects. If the notion of topological charge was introduced in the correct way attractive forces in the system SA and repulsive forces should be present in the systems SS and AA. It is somewhat problematic to see this in a direct way but we can use the same trick as in the sine-Gordon theory. First we need to show that there are three types of two-soliton solutions: the first describes the SS scattering, the second describes the SA scattering, and the third describes the time oscillating bound state of two solitons. The third solution if it exists can be called the *gravibreather*. If it turns out that the gravibreather can represent the SA bound state only, and that the combinations SS and AA do not have solutions of this kind, this would be a proof of the presence of attraction between gravisoliton and antigravisoliton and of repulsion between gravisolitons of the same charge. If so, a real metric of correct signature for the gravibreather should follow from the SA scattering solution by its analytic continuation to purely imaginary values of the relative collision velocity of the colliding gravisolitons. At the same time the analogous analytic continuation of the SS and AA type solutions should lead to an unphysical (complex) metric tensor. Let us show that this is really the case.

The two-soliton solution on the background metric (2.15) was given by (2.20)–(2.25). We know that μ_1 and μ_2 can be either both real or complex conjugate to each other. Because β can be replaced by $\beta +$ constant without any physical consequences we can set

$$w_1 = w, \quad w_2 = -w \tag{2.41}$$

in this solution. From (1.68)–(1.69) it follows that for real μ_1 and μ_2 we should take real w, whereas the complex conjugate pair of μ corresponds to purely imaginary values of w.

Let w be real and positive, and choose the real μ_1 and μ_2 to be $\mu_1 = \mu_1^{in}$ and $\mu_2 = \mu_2^{in}$. The region (in the α, β coordinate plane) of the gravisoliton head-on collision is now the triangle VI (see fig. 2.2) with vertices ($\alpha = w, \beta = 0$), ($\alpha = 0, \beta = w$) and ($\alpha = 0, \beta = -w$). Indeed because u_0 is space-like, from

(2.17) and (2.24) it follows that σ is space-like and τ is time-like. Let us look at the asymptotic form of the matrix $g^{(2)}$ in (2.20)–(2.23) in the region VI for large absolute values of σ and τ. It is not difficult to prove that at an early time, $\tau \ll -1$, and in the far region to the left, $\sigma \ll -1$ (but inside VI), the asymptotic form of the solution (2.20)–(2.23) coincides exactly with the functional form of the one-soliton solution for some diagonal background metric (vacuum) and pole trajectory μ_2. At $\tau \ll -1$ but far to the right, $\sigma \gg 1$, (2.20)–(2.23) gives the one-soliton solution, which corresponds again to the diagonal background metric and pole trajectory $\alpha^2 \mu_1^{-1}$. When τ increases these two solitons will collide (the σ distance between them decreases) and at $\tau \gg 1$ the state $g^{(2)}$ decays into two free solitons again: in the region $\sigma \ll -1$, (2.20)–(2.23) give the one-soliton metric with pole $\alpha^2 \mu_1^{-1}$ on the diagonal background, and in the region $\sigma \gg 1$, $g^{(2)}$ coincides with the one-soliton solution with pole μ_2 and diagonal background metric. Thus the picture is clear: the two-soliton solution (2.20)–(2.23) before and after the collision describes a pair of free gravisolitons on the vacuum backgrounds and the 'free poles' associated with them are $\alpha^2 \mu_1^{-1}$ and μ_2 instead of μ_1 and μ_2. This means that the two-soliton solution in the case $\mu_1 = \mu_1^{in}$ and $\mu_2 = \mu_2^{in}$ describes the collision process between a gravisoliton with an in-pole and an antigravisoliton with an out-pole, i.e. scattering in the SA system.

An analogous situation arises in the case in which $\mu_1 = \mu_1^{out}$ and $\mu_2 = \mu_2^{out}$. In this case we have scattering between a gravisoliton with an in-pole $\alpha^2 \mu_1^{-1}$ and an antigravisoliton associated with the out-pole μ_2. A similar analysis shows that in the case $\mu_1 = \mu_1^{out}$ and $\mu_2 = \mu_2^{in}$, or vice versa, the solution (2.20)–(2.23) describes the scattering of two gravisolitons of the same charge, i.e. scattering in the SS or the AA systems. In this case, as follows from (2.17)–(2.24), σ is time-like and τ is space-like. The asymptotics for large absolute values of σ and τ are the same as in the previous case, and at the initial ($\sigma \ll -1$) and final ($\sigma \gg 1$) stages of the collision we have again a pair of free gravisolitons with 'free poles' μ_2 and $\alpha^2 \mu_1^{-1}$, but now both of them are in-poles (or out-poles for $\mu_1 = \mu_1^{in}$ and $\mu_2 = \mu_2^{out}$).

There is also a third class of two-soliton solutions in (2.20)–(2.23). It corresponds to the case in which μ_1 and μ_2 form a complex conjugate pair. For u_0 space-like this solution becomes the gravitational analogue of the breather. In this case $\mu_2 = \bar{\mu}_1$, the variable τ and $\mu_1 - \mu_2$ become purely imaginary, but the metric (2.20)–(2.25) remains real with the correct physical signature. After the substitution $\tau = i\tau'$ we get the real variable τ' which is time-like, as can be seen from (2.17)–(2.24), and the two-soliton solution appears to be oscillating in time τ' (but not periodically). This solution corresponds to the purely imaginary values of the constant w in (2.41). Thus, the gravibreather can be considered as the analytic continuation in w of one of the two-soliton solutions with real μ_1 and μ_2. The main question now is of which type is it: SS, AA or SA?

2.3.2 The gravibreather solution

It is a simple task to prove that the gravibreather emerges from the SA state only. The analytic continuation we need comes from the general solutions to the equations

$$\mu_1^2 + 2(\beta - w)\mu_1 + \alpha^2 = 0, \quad \mu_2^2 + 2(\beta + w)\mu_2 + \alpha^2 = 0, \qquad (2.42)$$

for arbitrary complex values of the constant w. Let us define the function $F(s)$ by the equation $F^2 = s^2 - \alpha^2$, where s is complex valued and α is considered formally as a fixed real parameter. This function is analytic on the Riemann surface containing two sheets glued to each other at the cut between the points $s = \alpha$ and $s = -\alpha$. On the first sheet we have $F > 0$ for $(\mathrm{Im}\, s = 0, \mathrm{Re}\, s > \alpha)$, $F < 0$ for $(\mathrm{Im}\, s = 0, \mathrm{Re}\, s < -\alpha)$, and $\mathrm{Im}\, F > 0$ for $(\mathrm{Im}\, s > 0, \mathrm{Re}\, s = 0)$, $\mathrm{Im}\, F < 0$ for $(\mathrm{Im}\, s < 0, \mathrm{Re}\, s = 0)$. At the corresponding points of the second sheet, F has the opposite signs. Using the function $F(s)$ we obtain two distinct pairs of the w-analytical solutions to (2.42),

$$\mu_1 = w - \beta + F(w - \beta), \quad \mu_2 = -w - \beta - F(w + \beta), \qquad (2.43)$$

and

$$\mu_1 = w - \beta + F(w - \beta), \quad \mu_2 = -w - \beta + F(w + \beta). \qquad (2.44)$$

On the real w axis in the real regions of μ_1 and μ_2 the choice (2.43) gives $\mu_1 = \mu_1^{out}$ and $\mu_2 = \mu_2^{out}$ for the first w-sheet and $\mu_1 = \mu_1^{in}$ and $\mu_2 = \mu_2^{in}$ for the second. In contrast, in the same real regions the choice (2.44) corresponds to the 'out–in' or 'in–out' (depending on the sheet) pair μ_1 and μ_2.

Using the definition of F it is easy to prove that on the imaginary w axis this function has the property that

$$\overline{F}(w + \beta) = -F(w - \beta). \qquad (2.45)$$

Consequently the choice (2.43), and only this choice, gives the complex conjugate pair $\mu_2 = \overline{\mu}_1$ when w becomes purely imaginary. This means that the gravibreather follows from (2.20)–(2.23) by analytic continuation from the real to purely imaginary values of w, only in those cases in which for the real values of w the poles μ_1 and μ_2 form an 'in–in' or 'out–out' pair, i.e. (as has already been shown) just in those cases in which the two-soliton solution represents the collision between a gravisoliton and an antigravisoliton.

The choice (2.44) corresponds to collision of gravisolitons of the same charge. However, in these cases μ_1 and μ_2 for imaginary w cannot be complex conjugate to each other, i.e. we obtain in this w region an unphysical solution with a complex valued metric tensor.

The last thing we have to show is that the w continuation above is equivalent to the analytic continuation of the solution (2.20)–(2.23) from the real to the imaginary values of the relative collision velocity of the colliding solitons. It has been shown before that the measure of the local soliton velocity corresponding to some one-soliton pole trajectory μ is $-2\alpha\mu(\alpha^2 + \mu^2)^{-1}$. This expression is invariant under the interchange $\mu \rightarrow \alpha^2\mu^{-1}$, thus it is the same for a gravisoliton and an antigravisoliton. Due to this property and to the previous analysis we conclude that the solution (2.20)–(2.23) for any pair of real μ_1 and μ_2 describes the collision of two gravisolitons with initial velocities $v_1 = -2\alpha\mu_1(\alpha^2+\mu_1^2)^{-1} = \alpha(\beta - w)^{-1}$ and $v_2 = -2\alpha\mu_2(\alpha^2+\mu_2^2)^{-1} = \alpha(\beta+w)^{-1}$. Inside region VI, i.e. in the collision region, we have $v_1 < 0$ and $v_2 > 0$. The functions α and β in these formulas should be referred to that symbolic point in the interior of region VI where the world lines of the colliding solitons intersect. The relativistic formula for the relative velocity is $v_{rel} = (v_2 - v_1)(1 + v_1 v_2)^{-1} = 2\alpha w(w^2 - \alpha^2 - \beta^2)^{-1}$. Thus, the purely imaginary values of w indeed correspond to the purely imaginary values of the relative velocity of the colliding gravisolitons.

Independently of its topological properties, the gravibreather can be interesting in its own right. Let us write this solution here in a more suitable and simpler form than the general expressions (2.20)–(2.25). It is convenient to choose coordinates in such a way that $\alpha = q \sinh t \cosh z$, and $\beta = q \cosh t \sinh z$, where $q > 0$ is an arbitrary constant. We take the simplest space-like solution of u_0 in (2.14), i.e. $u_0 = 2k\beta$, where $k = $ constant. This background represents vacuum solutions for the Bianchi VI_0 homogeneous cosmological model. Then the two-soliton perturbation on the background (2.15) with the poles $\mu_1 = q(\sinh z - i)(1 - \cosh t)$ and $\mu_2 = q(\sinh z + i)(1 - \cosh t)$ (these are solutions of (2.42), where $w = -iq$) is the following exact solution to the Einstein equations:

$$ds^2 = \frac{D}{(q \sinh t \cosh z)^{1/2}} e^{k^2 q^2 \sinh^2 t \cosh^2 z}(dz^2 - dt^2) + \frac{q \sinh t \cosh z}{D}$$
$$\times \left\{ \left[(\cosh t \cosh \sigma - \sinh \sigma)^2 + (\sinh z \sin \tau - \cos \tau)^2 \right] \right.$$
$$\times e^{2kq \cosh t \sinh z}(dx^1)^2$$
$$+ \left[(\cosh t \cosh \sigma + \sinh \sigma)^2 + (\sinh z \sin \tau + \cos \tau)^2 \right]$$
$$\times e^{-2kq \cosh t \sinh z}(dx^2)^2$$
$$\left. + 4(\cosh t \cosh \sigma \cos \tau - \sinh z \sinh \sigma \sin \tau)dx^1 dx^2 \right\}, \qquad (2.46)$$

where

$$\left. \begin{array}{l} D = \sinh^2 t \cosh^2 \sigma + \cosh^2 z \sin^2 \tau, \\ \sigma = 2kq \sinh z + \sigma_0, \\ \tau = 2kq(\cosh t - 1) + q\tau_0. \end{array} \right\} \qquad (2.47)$$

Here σ_0 and τ_0 are arbitrary constants. The substitutions $t \rightarrow t/iw$, $z \rightarrow z/iw$ and $q \rightarrow iw$ replace (2.46) by the two-soliton solution related to the SA scattering process inside region VI.

The gravibreather (2.46) represents an inhomogeneous cosmological model which starts with an anisotropic Kasner-like behaviour near the singularity ($t = 0$), approaches the background solution (2.15) at $t \rightarrow \infty$, and oscillates in time in between. Some numerical results with details of the gravibreather behaviour can be found in refs [177, 110].

3

Einstein–Maxwell fields

The purpose of this chapter is to describe the integration scheme for Einstein–Maxwell equations. We begin in section 3.1 by writing the Einstein–Maxwell equations in a suitable form when the spacetime admits, as in chapter 1, an orthogonally transitive two-parameter group of isometries. We then formulate in section 3.2 the corresponding spectral equations which take in this case the form of 3×3 matrix equations. It turns out that one cannot simply generalize the procedure of chapter 1, since some extra constraints have to be imposed on the linear spectral equations to be able to reproduce the Einstein–Maxwell equations as integrability conditions of such linear equations. In sections 3.3 and 3.4 we show how these problems can be overcome and the n-soliton solution can be constructed. Because the procedure is rather involved we formulate the basic steps in a recipe of 11 points which should be useful for practical calculations. Finally in section 3.5, as an illustration of the procedure given, the analogue of the sine-Gordon breather in the Einstein–Maxwell context is deduced and briefly described.

3.1 The Einstein–Maxwell field equations

In sections 1.2–1.4 we established the complete integrability of Einstein equations in vacuum for the metric (1.36) by means of the ISM, and the same will be done for the stationary analogue of this metric in chapter 8. However, the inclusion of matter, i.e. the appearance of a nonzero right hand side in the Einstein equations, generally destroys the applicability of the ISM. This is because the stress-energy tensor produces a nonvanishing right hand side in the basic equation (1.39) that prevents the application of the ISM. However, there are some special cases in which the ISM works even when matter is included. A few of these cases will be described in sections 5.4.2 and 5.4.3 but these are more or less simple generalizations of the vacuum case. Here we will describe a new and really nontrivial situation: the complete integrability of the coupled Einstein–Maxwell field equations.

Some results for the Einstein–Maxwell system can be extracted from the five-dimensional vacuum gravitational equations. We saw in section 1.5 that such five-dimensional systems represent from the four-dimensional point of view the dynamics of coupled gravitational, electromagnetic and massless scalar fields; see (1.139)–(1.142). The scalar field corresponds to the g_{55} component of the five-dimensional metric tensor. The source for this field is the electromagnetic invariant $F_{ik}F^{ik}$ that appears on the right hand side of the wave equation for such a scalar field. If one wishes to avoid the presence of the scalar field it is necessary to impose on the electromagnetic field the constraint $F_{ik}F^{ik} = 0$. For this special case the Einstein–Maxwell equations are equivalent to the five-dimensional gravitational vacuum equations. In this special case it is a simple matter to generalize the ISM approach of sections 1.2–1.4 to the five-dimensional situation and construct soliton solutions. The 'L–A pair' for such a constrained gravitational–electromagnetic system has been described in ref. [13]. However, this is again a trivial generalization of the four-dimensional vacuum case and there is no way to extend this standard extradimensional approach to include the general electromagnetic field with $F_{ik}F^{ik} \neq 0$ avoiding, simultaneously, the appearance of scalar fields.

The main new step for the solution of the Einstein–Maxwell equations was made by Alekseev [4] in 1980, although the most detailed and comprehensive account of his approach was given in his 1988 paper [5]. The key idea of this method cannot be simply extracted from our sections 1.2–1.4 and it is no overstatement to say that it is again part of one of those miracles that pervade the field of integrable systems. Thus for the sake of the completeness of this book we will present in this chapter Alekseev's approach to the problem of integration of the Einstein–Maxwell equations. We follow closely ref. [5], although translated to our language. However, apart from the Einstein–Maxwell breather, which we will describe in section 3.5, we will not consider other applications to calculations of concrete solutions in closed form. Practically all known results in this field can be found in ref. [5], and refs [73, 74, 120, 110, 123].

It should be mentioned here that the integrability ansatz of the Einstein–Maxwell equations and the n-soliton solutions was also given by Neugebauer and Kramer in 1983 [226] in the framework of their development of the pseudopotential method [178]. In 1981 Cosgrove [65] obtained one-soliton solutions for this system in the framework of the Hauser–Ernst formalism [138]. However, we follow Alekseev's approach because it is closely related to the method we are using in this book.

First let us describe the integrable ansatz of the Einstein–Maxwell equations. It is convenient not to restrict ourselves from the beginning to the nonstationary metric (1.36), and include also the stationary case by writing the metric in the form,

$$ds^2 = f(x^\rho)\eta_{\mu\nu}dx^\mu dx^\nu + g_{ab}(x^\rho)dx^a dx^b. \tag{3.1}$$

In this chapter the Greek indices take only two values and correspond to the coordinates t, z for the nonstationary case and to ρ, z for the stationary metric (8.1). The two-dimensional matrix $\eta_{\mu\nu}$ is

$$\eta_{\mu\nu} = \begin{pmatrix} -e & 0 \\ 0 & 1 \end{pmatrix}, \tag{3.2}$$

where $e = 1$ and $e = -1$ for the nonstationary and stationary solutions, respectively. For the determinant of the two-dimensional matrix g (with components g_{ab}) we adopt the notation

$$\det g = e\alpha^2. \tag{3.3}$$

In order to make the integrable ansatz compatible with the metric (3.1) one should assume the following structure for the electromagnetic potentials:

$$A_\mu = 0, \quad A_a = A_a(x^\rho). \tag{3.4}$$

Then, the only nonvanishing components for the covariant and contravariant electromagnetic tensor field are

$$F_{\mu a} = A_{a,\mu}, \quad F^{\mu a} = \frac{1}{f}\eta^{\mu\nu}g^{ac}A_{c,\nu}. \tag{3.5}$$

Here, and in the following, g^{ab} are the components of the inverse matrix of g_{ab}, i.e. $g^{ac}g_{cb} = \delta_b^a$.

Einstein–Maxwell equations (with an appropriate choice of units) can be written in the form

$$R_a^b = 2\left(F_{\lambda a}F^{\lambda b} - \frac{1}{2}\delta_a^b F_{\lambda c}F^{\lambda c}\right), \tag{3.6}$$

$$R_\nu^\mu = 2\left(F_{\nu c}F^{\mu c} - \frac{1}{2}\delta_\nu^\mu F_{\lambda c}F^{\lambda c}\right), \tag{3.7}$$

$$(f\alpha F^{\mu a})_{,\mu} = 0. \tag{3.8}$$

Because the two-dimensional trace in the Greek indices on the right hand side of (3.7) vanishes identically, these equations can be written in the following equivalent form:

$$R_\mu^\mu = 0, \quad R_\nu^\mu - \frac{1}{2}\delta_\nu^\mu R_\lambda^\lambda = 2\left(F_{\nu c}F^{\mu c} - \frac{1}{2}\delta_\nu^\mu F_{\lambda c}F^{\lambda c}\right). \tag{3.9}$$

Direct calculation shows that the first of these equations is

$$\eta^{\mu\nu}(\ln f)_{,\mu\nu} + \eta^{\mu\nu}(\ln \alpha)_{,\mu\nu} + \frac{1}{4}g^{ab}g^{cd}\eta^{\mu\nu}g_{bc,\mu}g_{da,\nu} = 0, \qquad (3.10)$$

and that the second does not contain the second derivatives of the metric coefficient f:

$$\begin{aligned}
\eta^{\nu\rho}&\left[\frac{1}{2}(\ln f)_{,\mu}(\ln\alpha)_{,\rho} + \frac{1}{2}(\ln f)_{,\rho}(\ln\alpha)_{,\mu}\right] - \frac{1}{2}\delta^{\nu}_{\mu}\eta^{\rho\sigma}(\ln f)_{,\rho}(\ln\alpha)_{,\sigma} \\
&- \eta^{\nu\rho}(\ln\alpha)_{,\mu\rho} + \frac{1}{2}\delta^{\nu}_{\mu}\eta^{\rho\sigma}(\ln\alpha)_{,\rho\sigma} \\
&- \frac{1}{4}g^{ab}g^{cd}\left[\eta^{\nu\rho}g_{bc,\mu}g_{da,\rho} - \frac{1}{2}\delta^{\nu}_{\mu}\eta^{\rho\sigma}g_{bc,\rho}g_{da,\sigma}\right] \\
&= g^{cd}A_{d,\rho}\left[2\eta^{\nu\rho}A_{c,\mu} - \delta^{\nu}_{\mu}\eta^{\rho\sigma}A_{c,\sigma}\right].
\end{aligned} \qquad (3.11)$$

The trace of this last equation vanishes identically. Consequently it gives only two independent relations from which the metric coefficient f can be found by quadratures if the matrix g_{ab} and the potentials A_a are known. As in the vacuum case, (3.10) will then be satisfied due to the Bianchi identities, and we can forget about this equation from now on. Also, as in the vacuum case, the integration of the coefficient f from (3.11) does not present a major difficulty and will be carried out at the end of the procedure. Thus we turn our attention to the problem of the matrix g_{ab} and the potentials A_a from the system (3.6) and (3.8). It is easy to see that this system does not contain the coefficient f and that it forms a closed and selfconsistent system of equations for g_{ab} and A_a. Calculating R^b_a and using the definitions (3.5) we can write these equations in the form:

$$\eta^{\mu\nu}\frac{1}{\alpha}(\alpha g^{bc}g_{ac,\mu})_{,\nu} = -4\eta^{\mu\nu}g^{bc}A_{a,\mu}A_{c,\nu} + 2\delta^b_a\eta^{\mu\nu}g^{cd}A_{c,\mu}A_{d,\nu}, \qquad (3.12)$$

$$\eta^{\mu\nu}(\alpha g^{ac}A_{c,\mu})_{,\nu} = 0. \qquad (3.13)$$

It is important that the trace of the right hand side of (3.12) vanishes identically and that the function $\alpha(x^\mu)$, in accordance with its definition (3.3), should satisfy the vacuum 'wave' equation,

$$\eta^{\mu\nu}\alpha_{,\mu\nu} = 0. \qquad (3.14)$$

In this chapter we will often use a matrix notation. Thus, for definiteness in any matrix (M_{ik}, M^{ik}, $M^i_{\ k}$ or $M_i^{\ k}$) the first index, independent of its up or down position, will always enumerate the rows, and the second index will enumerate the columns. This rule, of course, also applies to the matrix (3.2) which has already been introduced. For the Kroneker delta, however, we will not distinguish the first and the second indices since it is irrelevant in this case and we will write δ_i^k. Let us introduce two-dimensional antisymmetric matrices,

$$\epsilon^{\mu\nu} = \epsilon_{\mu\nu} = \begin{pmatrix} 0 & 1 \\ -1 & 0 \end{pmatrix}, \tag{3.15}$$

and the same for the Latin indices,

$$\epsilon^{ab} = \epsilon_{ab} = \begin{pmatrix} 0 & 1 \\ -1 & 0 \end{pmatrix}. \tag{3.16}$$

Now we are in the position to start the description of the integration scheme. Following Alekseev's suggestion of exploiting the duality properties of the electromagnetic field, we introduce some auxiliary potentials B_a which will only play an intermediary role and will not be present in the final results. In terms of the original potentials A_a, these are defined by,

$$B_{a,\mu} = -\frac{1}{\alpha}\eta_{\mu\nu}\epsilon^{\nu\lambda}g_{ab}\epsilon^{bc}A_{c,\lambda}. \tag{3.17}$$

It is easy to verify that the integrability condition for this equation, $\epsilon^{\mu\nu}B_{a,\mu\nu} = 0$, coincides with the Maxwell equation (3.13). Relation (3.17) can also be written in its inverse form:

$$A_{a,\mu} = \frac{1}{\alpha}\eta_{\mu\nu}\epsilon^{\nu\lambda}g_{ab}\epsilon^{bc}B_{c,\lambda}. \tag{3.18}$$

Let us now combine A_a and B_a into a single complex electromagnetic potential Φ_a, defined by

$$\Phi_a = A_a + iB_a, \tag{3.19}$$

then (3.17)–(3.18) are, respectively, the imaginary and real parts of the equation for Φ_a:

$$\Phi_{a,\mu} = -\frac{i}{\alpha}\eta_{\mu\nu}\epsilon^{\nu\lambda}g_{ab}\epsilon^{bc}\Phi_{c,\lambda}, \tag{3.20}$$

from which the Maxwell equations for Φ_a trivially follow:

$$\eta^{\mu\nu}(\alpha g^{ac}\Phi_{c,\mu})_{,\nu} = 0. \tag{3.21}$$

By direct calculation one can show that Einstein equations (3.12) can be written as

$$\eta^{\mu\nu}\frac{1}{\alpha}(\alpha g^{bc}g_{ac,\mu})_{,\nu} = -2g^{bc}\eta^{\mu\nu}\overline{\Phi}_{a,\mu}\Phi_{c,\nu}. \tag{3.22}$$

The imaginary part on the right hand side of this equation vanishes because the left hand side is real. That this is indeed the case is a consequence of (3.17). Also due to this relation and the identity

$$\epsilon^{ad}\epsilon_{bc} = \delta_b^a\delta_c^d - \delta_c^a\delta_b^d, \tag{3.23}$$

the real part of the right hand side of (3.22) coincides exactly with the right hand side of (3.12). It is thus clear that any solution of (3.19)–(3.22), g_{ab} and $A_a = \text{Re}\,\Phi_a$, is also a solution of the Einstein–Maxwell equations (3.12)–(3.13). Note the auxiliary role played by the potential B_a.

3.2 The spectral problem for Einstein–Maxwell fields

In agreement with the general ideas of the ISM we should now find a way to represent the Einstein–Maxwell equations (3.19)–(3.22) as selfconsistency conditions of a linear spectral problem. Our experience in five-dimensional geometry suggests that we should look for the solution in the framework of the same spectral problem (1.51), but for three-dimensional matrices A, B and ψ. First let us rewrite the three-dimensional version of (1.51) in terms of the coordinates x^μ introduced in (3.1)–(3.2), in order to develop a universal approach both to the stationary and the nonstationary cases. This three-dimensional generalization is straightforward and can be written as

$$\Pi_\mu \psi = \left(\frac{e\lambda}{\lambda^2 - e\alpha^2} \eta_{\mu\rho} \epsilon^{\rho\sigma} K_\sigma - \frac{e\alpha}{\lambda^2 - e\alpha^2} K_\mu \right) \psi, \qquad (3.24)$$

where the operators Π_μ are

$$\Pi_\mu = \partial_\mu + \frac{2e(\lambda^2 \eta_{\mu\rho} \epsilon^{\rho\sigma} \alpha_{,\sigma} - \lambda\alpha\alpha_{,\mu})}{\lambda^2 - e\alpha^2} \partial_\lambda, \qquad (3.25)$$

and $\eta_{\mu\rho}$ and $\epsilon^{\rho\sigma}$ are given by the definitions (3.2) and (3.15). The matrices K_μ and ψ are now three-dimensional and the function $\alpha(x^\mu)$ is the same as before, i.e. it satisfies (3.14). For the nonstationary case we have $e = 1$ and $x^1 = t$, $x^2 = z$, and then the metric (3.1) coincides with (1.36) . If we use the null coordinates (1.14) it is easy to see that $\Pi_2 + \Pi_1 = D_1$, $\Pi_2 - \Pi_1 = D_2$, where D_1 and D_2 are the operators introduced in (1.49). Taking the sum and the difference of (3.24) for $\mu = 2$ and $\mu = 1$, we get exactly the three-dimensional version of the spectral problem (1.51), where $A = -(K_2 + K_1)$ and $B = K_2 - K_1$. For the stationary case one should take $e = -1$ and $x^1 = z$, $x^2 = \rho$, then when $\alpha = \rho$, it follows from (3.25) that $\Pi_1 = D_1$ and $\Pi_2 = D_2$ where D_1 and D_2 are now the operators (8.7) which are defined in chapter 8. Equations (3.24) for $\mu = 1$ and $\mu = 2$ now coincide directly with the three-dimensional spectral problem (8.6), where $V = K_1$ and $U = K_2$.

It is easy to see that due to the 'wave' equation (3.14) for the function α the operators Π_μ commute:

$$\Pi_\mu \Pi_\nu - \Pi_\nu \Pi_\mu = 0, \qquad (3.26)$$

and that the selfconsistency conditions for (3.24) are

$$\eta^{\mu\nu} K_{\mu,\nu} = 0, \qquad (3.27)$$

$$\epsilon^{\mu\nu} \left(K_{\mu,\nu} + \frac{1}{\alpha} K_\mu K_\nu - \frac{1}{\alpha} \alpha_{,\nu} K_\mu \right) = 0. \qquad (3.28)$$

The second of these equations implies that the two matrices K_μ can be written in terms of a single matrix X in the form

$$K_\mu = \alpha X_{,\mu} X^{-1}. \qquad (3.29)$$

Then (3.28) is just the integrability condition of (3.29) for X, and (3.27) gives a really nontrivial condition in the form of the following differential equation for the matrix X

$$\eta^{\mu\nu}(\alpha X_{,\mu}X^{-1})_{,\nu} = 0. \tag{3.30}$$

In general this equation does not reproduce the Einstein–Maxwell system, but it does so for some special class of matrices X. To single out this class we should impose on X some additional constraints that are not a consequence of the selfconsistency conditions of the spectral problem (3.24), but that are compatible with them. To formulate these constraints in the three-dimensional matrix form let us introduce first the following matrix Ω:

$$\Omega = \begin{pmatrix} 0 & 1 & 0 \\ -1 & 0 & 0 \\ 0 & 0 & 0 \end{pmatrix}. \tag{3.31}$$

In what follows we shall use the small Latin indices from the first part of the alphabet (i.e. from a to h) for the enumeration of the first and second rows and columns of the three-dimensional matrices, whereas for the third rows and columns we shall use the star symbol. With this convention the matrix Ω, for example, can be written in the following form:

$$\Omega = \begin{pmatrix} \Omega^{ab} & \Omega^{a*} \\ \Omega^{*b} & \Omega^{**} \end{pmatrix} = \begin{pmatrix} \epsilon^{ab} & 0 \\ 0 & 0 \end{pmatrix}. \tag{3.32}$$

It is now convenient to introduce two special combinations made up of the matrix X and its derivatives. Thus we define U_μ by

$$U_\mu = ie\alpha\eta_{\mu\rho}\epsilon^{\rho\sigma}X^{-1}X_{,\sigma} + 4e(\alpha^2 X^{-1})_{,\mu}\Omega. \tag{3.33}$$

The additional constraints we need to impose on the matrix X can now be written as

$$X = X^\dagger, \tag{3.34}$$

$$XU_\mu = -4ie\alpha\eta_{\mu\rho}\epsilon^{\rho\sigma}\Omega U_\sigma, \tag{3.35}$$

where † means Hermitian conjugation. These are the two fundamental constraints. However, we can still impose three new constraints on the matrix X. These new constraints are weaker, are easily imposed and do not represent a loss of generality. In fact, the first is

$$X^{**} = 2. \tag{3.36}$$

It is easy to see from (3.31)–(3.35) that $X^{**}_{,\mu} = 0$, so that $X^{**} =$ constant. Due to the invariance of the equations with respect to the rescaling $X \to cX$, $\alpha \to c\alpha$ (c is an arbitrary constant) the value of the constant X^{**} can be chosen at will. We chose 2 in (3.36) in order to make a more direct comparison with ref. [5].

The second of these new constraints is the requirement that the determinants of the two-dimensional blocks constructed from the first and second rows and columns of the matrices U_μ are not zero. We recall that we are using the following notation for the components of the matrices X and X^{-1}:

$$X = \begin{pmatrix} X^{ab} & X^{a*} \\ X^{*b} & X^{**} \end{pmatrix}, \quad X^{-1} = \begin{pmatrix} (X^{-1})_{ab} & (X^{-1})_{a*} \\ (X^{-1})_{*b} & (X^{-1})_{**} \end{pmatrix}. \tag{3.37}$$

Consequently the upper and left two-dimensional blocks of matrices U_μ, as follows from definition (3.33), are

$$(U_\mu)_a^{\ b} = ie\alpha\eta_{\mu\rho}\epsilon^{\rho\sigma}[(X^{-1})_{a*}X_{,\sigma}^{*b} + (X^{-1})_{ac}X_{,\sigma}^{cb}] + 4e[\alpha^2(X^{-1})_{ac}]_{,\mu}\epsilon^{cb}. \tag{3.38}$$

Because $\det X \neq 0$ and $\det \Omega = 0$, it follows from (3.35) that $\det U_\mu = 0$. Thus our second new constraint can be formulated as

$$\text{rank}(U_\mu) = 2, \quad \det[(U_\mu)_a^{\ b}] \neq 0. \tag{3.39}$$

These properties will be satisfied automatically by the construction of the solutions and, in practice, do not mean a loss of generality.

The third of these new constraints is that *at least one* of the diagonal elements of the two-dimensional matrix

$$X^{ab} - \frac{1}{2}X^{a*}X^{*b} \tag{3.40}$$

does not vanish. This condition can also be easily imposed and does not represent a loss of generality for the Einstein–Maxwell fields.

It is easy to show that the unique structure for the matrix X that follows from the constraints (3.34)–(3.40) is

$$\left. \begin{aligned} X &= \begin{pmatrix} -4\epsilon^{ac}g_{cd}\epsilon^{db} + 8\epsilon^{ac}\Phi_c\epsilon^{bd}\overline{\Phi}_d & 4\epsilon^{ac}\Phi_c \\ 4\epsilon^{bc}\overline{\Phi}_c & 2 \end{pmatrix}, \\ X^{-1} &= \begin{pmatrix} -\frac{1}{4}\epsilon_{ac}g^{cd}\epsilon_{db} & -\frac{1}{2}\epsilon_{ac}g^{cd}\Phi_d \\ -\frac{1}{2}\epsilon_{bc}g^{cd}\overline{\Phi}_d & \frac{1}{2} + g^{cd}\overline{\Phi}_c\Phi_d \end{pmatrix}, \end{aligned} \right\} \tag{3.41}$$

where g^{ab} are the components of the two-dimensional matrix g^{-1}, the inverse of g. We recall that the two-dimensional matrix g, with components g_{ab}, is real, symmetric and has determinant $\det g = e\alpha^2$. Equation (3.20), which is satisfied by the complex electromagnetic potentials Φ_a, is now a consequence of the $(a*)$-components of (3.35). Also, substitution of this form of the matrix X into the selfconsistency equation (3.30) exactly reproduces the Einstein–Maxwell equations (3.21)–(3.22) and nothing else.

At this point we have finished the first step of our procedure: the formulation of the basic additional constraints for the matrix X in three-dimensional form, i.e. constraints (3.34)–(3.35). However, we are still far from the solution of our problem. In fact, to find a way to construct a solution of the spectral equation (3.24) which satisfies the constraints is not a trivial task. To this end we introduce a new generating matrix φ, which is related to the generating matrix ψ of the vacuum case by

$$\psi = (X - 4i\lambda\Omega)\varphi. \tag{3.42}$$

Substitution of this expression into (3.24) shows that due to the additional constraint (3.35) this new generating matrix φ satisfies the following spectral equation:

$$\Pi_\mu\varphi = \left[\frac{i\lambda(\lambda^2 + e\alpha^2)}{(\lambda^2 - e\alpha^2)^2}U_\mu - \frac{2ie\alpha\lambda^2}{(\lambda^2 - e\alpha^2)^2}\eta_{\mu\rho}\epsilon^{\rho\sigma}U_\sigma\right]\varphi, \tag{3.43}$$

where U_μ are the matrices (3.33). The advantage of this representation of our spectral problem is that it consists of rational functions not only with respect to the original spectral parameter λ, but also with respect to a new parameter w defined by

$$w = -\frac{1}{2}\left(\lambda + 2\beta + \frac{e\alpha^2}{\lambda}\right), \tag{3.44}$$

where β is the second independent solution of the 'wave' equation (3.14), which has the following connection to the function α:

$$\beta_{,\mu} = -e\eta_{\mu\rho}\epsilon^{\rho\sigma}\alpha_{,\sigma}. \tag{3.45}$$

In nonstationary cases, when $e = 1$, the pair α and β is just (1.45) and (1.46). Due to this connection the parameter $w(x^\mu, \lambda)$ satisfies the identity

$$\Pi_\mu w = 0. \tag{3.46}$$

Relation (3.44) can be understood as a transformation $\lambda = \lambda(\alpha, \beta, w)$ from the parameter λ to the new spectral parameter w. After this transformation is applied to any generating matrix $\varphi(x^\mu, \lambda)$ it must be understood that such a matrix becomes a function of x^μ and w only (more precisely as $\varphi[x^\mu, \lambda(\alpha, \beta, w)]$). In this sense and due to identity (3.46), for any matrix φ we have

$$\Pi_\mu\varphi = (\partial_\mu\varphi)_w, \tag{3.47}$$

where the right hand side is the usual partial derivatives with respect to the coordinates x^μ performed under the assumption that w is some free parameter independent of x^μ. The key point now is that the application of this transformation to (3.43) shows its rational dependence on w together with a simple structure of differential operators:

$$\frac{\partial\varphi}{\partial x^\mu} = \frac{1}{2i}\left[\frac{w+\beta}{(w+\beta)^2 - e\alpha^2}U_\mu + \frac{e\alpha}{(w+\beta)^2 - e\alpha^2}\eta_{\mu\rho}\epsilon^{\rho\sigma}U_\sigma\right]\varphi. \tag{3.48}$$

The analyticity of this equation with respect to the spectral parameter w is important, because it allows us to apply to the construction of its solitonic solutions the dressing procedure used in the vacuum case, but with the meromorphic structure of the dressing matrices in the complex w-plane. At the same time the simplicity of the differential operators allows us to impose the additional constraints (3.34)–(3.35) in a simple way. It is worth emphasizing that it is not possible to have both of these properties simultaneously satisfied for the spectral equation in its original form (3.24). This does not mean that the original form is not appropriate at all, it just means that it needs a more sophisticated treatment than the standard approach described in sections 1.2–1.4.

3.3 The components g_{ab} and the potentials A_a

The construction of the n-soliton solution for the metric components g_{ab} and the electromagnetic potentials A_a can now proceed following steps similar to the vacuum case as described in sections 1.2–1.4. However, the different analytic structure and the complexity of the constraints in this case make the procedure a little more cumbersome. First we need to build the n-solitonic solution of the spectral problem (3.48) in general, i.e. without assuming any additional structure for the matrices U_μ (i.e. a structure that does not follow automatically from (3.48) itself). After that we will impose all the necessary additional constraints (i.e. the conditions that follow from (3.33)–(3.40)).

3.3.1 The n-soliton solution of the spectral problem

Let us start this first stage with the introduction of a new matrix $\Lambda_\mu^{\ \nu}$:

$$\Lambda_\mu^{\ \nu} = \frac{1}{2i} \frac{(w + \beta)\delta_\mu^\nu + e\alpha\eta_{\mu\rho}\epsilon^{\rho\nu}}{(w + \beta)^2 - e\alpha^2},$$ (3.49)

and then the spectral equation (3.48) takes the form

$$\varphi_{,\mu} = \Lambda_\mu^{\ \nu} U_\nu \varphi.$$ (3.50)

Let $\varphi^{(0)}$ and $U_\mu^{(0)}$ be some background solution of (3.50) with some given functions α and β. Then we search for the new 'dressed' solution, corresponding to the same functions α and β, of the form

$$\varphi = \chi\varphi^{(0)}.$$ (3.51)

Because $\varphi^{(0)}$ is a solution, from (3.50) we obtain the following equation for the dressing matrix χ:

$$\chi_{,\mu} = \Lambda_\mu^{\ \nu}(U_\nu\chi - \chi U_\nu^{(0)}).$$ (3.52)

Now we will use small Latin indices from the second part of the alphabet (i.e. indices i, j, k, l, ...) to enumerate quantities related to the poles of matrix χ. We assume that χ and χ^{-1} have n simple poles,

$$\chi = I + \sum_{k=1}^{n} \frac{R_k}{w - w_k}, \qquad \chi^{-1} = I + \sum_{k=1}^{n} \frac{S_k}{w - \widetilde{w}_k}. \tag{3.53}$$

Here and in what follows we do not assume summation over repeated indices i, k, l, ... Such a summation will be always indicated by the symbol $\sum$. At this stage w_k and $\widetilde{w}_k$ can be arbitrary functions of the coordinates x^μ. Also in what follows we consider that all the $2n$ functions w_k and $\widetilde{w}_k$ are different. From the identity $\chi \chi^{-1} = I$ we have the following conditions for the matrices $R_k(x^\mu)$ and $S_k(x^\mu)$:

$$R_k \chi^{-1}(w_k) = 0, \qquad \chi(\widetilde{w}_k) S_k = 0, \tag{3.54}$$

where expressions of the type $F(w_k)$ mean the values of the function $F(w, x^\mu)$ at $w = w_k$. The dependence on the coordinates x^μ is omitted for simplicity. Equations (3.54) imply that we can look for matrices R_k and S_k of the form

$$(R_k)_A^{\ B} = n_A^{(k)} m^{(k)B}, \qquad (S_k)_A^{\ B} = p_A^{(k)} q^{(k)B}. \tag{3.55}$$

Here and in the rest of this chapter we use capital Latin letters from the first part of the alphabet (i.e. from A to H) to enumerate the matrix components. We should keep in mind that the final results will be applied to the three-dimensional case, thus in agreement with our previous prescriptions the capital Latin indices will take the three values $A = (a, *)$, $B = (b, *)$, ..., $H = (h, *)$. Nevertheless, the construction of the solution of the spectral equation (3.48) that we are carrying out is valid for matrices of any dimension.

The substitution of (3.55) into (3.54) gives two systems of algebraic equations from which one can express all vectors $n_A^{(k)}$ and $q_A^{(k)}$ in terms of the vectors $m_A^{(k)}$ and $p_A^{(k)}$ as

$$\sum_{l=1}^{n} \frac{p_B^{(k)} m^{(l)B}}{w_l - \widetilde{w}_k} n_A^{(l)} = p_A^{(k)}, \tag{3.56}$$

$$\sum_{l=1}^{n} \frac{m^{(k)B} p_B^{(l)}}{w_k - \widetilde{w}_l} q^{(l)A} = -m^{(k)A}. \tag{3.57}$$

If we now introduce the $n \times n$ matrix T_{kl} (i.e. a matrix with respect to indices i, k, l, ...) and its inverse matrix $(T^{-1})_{kl}$,

$$T_{kl} = \frac{p_B^{(k)} m^{(l)B}}{w_l - \widetilde{w}_k}, \qquad \sum_{l=1}^{n} T_{il}(T^{-1})_{lk} = \delta_{ik}, \tag{3.58}$$

we obtain

$$q^{(k)A} = -\sum_{l=1}^{n}(T^{-1})_{lk}m^{(l)A}, \quad n_A^{(k)} = \sum_{l=1}^{n}(T^{-1})_{kl}p_A^{(l)} \tag{3.59}$$

for the vectors $q^{(k)A}$ and $n_A^{(k)}$.

To obtain the vectors $m^{(k)A}$ and $p_A^{(k)}$ we use (3.52), which can be written in the form

$$\Lambda_\mu^\nu U_\nu = \chi_{,\mu}\chi^{-1} + \Lambda_\mu^\nu \chi U_\nu^{(0)}\chi^{-1}, \tag{3.60}$$

or, equivalently, as

$$\Lambda_\mu^\nu U_\nu = -\chi(\chi^{-1})_{,\mu} + \Lambda_\mu^\nu \chi U_\nu^{(0)}\chi^{-1}. \tag{3.61}$$

All the terms in these equations are meromorphic functions of w that vanish at $w \to \infty$. Thus to satisfy these equations it suffices to eliminate the residues of all their poles. The left hand sides of both equations are regular at the points $w = w_k$ and $w = \widetilde{w}_k$. The first terms on the right hand sides generate the second order poles at these points if w_k and $\widetilde{w}_k$ depend on the coordinates x^μ. Consequently, the first result we have from (3.60) and (3.61) is that

$$w_k = \text{constant}, \quad \widetilde{w}_k = \text{constant}. \tag{3.62}$$

Note that due to the simplicity of the differential operators in the spectral equation (3.50) the poles and zeros of matrices χ and χ^{-1} in the w-plane are stationary points and not trajectories as in the vacuum case.

Now the right hand sides of (3.60)–(3.61) contain only simple poles at the points $w = w_k$ and $w = \widetilde{w}_k$, and the elimination of their residues gives the following equations for the matrices R_k and S_k:

$$R_{k,\mu}\chi^{-1}(w_k) + \Lambda_\mu^\nu(w_k)R_k U_\nu^{(0)}\chi^{-1}(w_k) = 0, \tag{3.63}$$

$$\chi(\widetilde{w}_k)S_{k,\mu} - \Lambda_\mu^\nu(\widetilde{w}_k)\chi(\widetilde{w}_k)U_\nu^{(0)}S_k = 0. \tag{3.64}$$

The solution of these equations can be expressed in terms of the background generating matrix $\varphi^{(0)}$ in the same way as was used in section 1.4 for the two-dimensional case. It is easy to check that if we substitute the matrices R_k and S_k into (3.63)–(3.64), take into account the conditions (3.54) and the fact that $\varphi^{(0)}$, $U_\mu^{(0)}$ is a solution of (3.50), then the system (3.63)–(3.64) is a set of differential equations for the vectors $m^{(k)A}$ and $p_A^{(k)}$. The general solution of which is[†]

$$m^{(k)A} = k_B^{(k)}[(\varphi^{(0)})^{-1}(w_k)]^{BA}, \tag{3.65}$$

[†] In (3.65) and (3.66) for the vectors $m^{(k)A}$ and $p_A^{(k)}$ there may also be arbitrary complex factors which can depend on the index k and the coordinates x^μ. However, such factors are not present in the final expressions for the matrices R_k and S_k, we therefore set them equal to 1.

$$p_A^{(k)} = l^{(k)B}[\varphi^{(0)}(\widetilde{w}_k)]_{AB}, \tag{3.66}$$

where $k_B^{(k)}$ and $l^{(k)B}$ are arbitrary constants (i.e. $2n$ arbitrary constant vectors), and we write the components of the matrix φ^{-1} (inverse of φ) with upper indices, i.e.

$$[\varphi^{-1}(w)]^{AD}[\varphi(w)]_{DB} = \delta_B^A.$$

The structure of the coefficients Λ_μ^ν, see (3.49), shows that in (3.60)–(3.61) we still have poles with nonzero residues at the two points where $(w + \beta)^2 - e\alpha^2 = 0$. If we define $\sqrt{e}$ as

$$\sqrt{e} = 1 \quad \text{if } e = 1, \qquad \sqrt{e} = i \quad \text{if } e = -1, \tag{3.67}$$

these poles can be written as $w = w_+$ and $w = w_-$, where

$$w_+ = -\beta + e\sqrt{e}\alpha, \qquad w_- = -\beta - e\sqrt{e}\alpha. \tag{3.68}$$

Elimination of the residues of (3.60), or (3.61), at these poles does not produce any new constraints on the matrices R_k and S_k, but gives the values of the matrices U_μ in terms of R_k, S_k and the background matrices $U_\mu^{(0)}$:

$$U_\mu = \frac{1}{2}\left[\chi(w_+)U_\mu^{(0)}\chi^{-1}(w_+) + \chi(w_-)U_\mu^{(0)}\chi^{-1}(w_-)\right]$$
$$+ \frac{1}{2}e\sqrt{e}\eta_{\mu\rho}\epsilon^{\rho\sigma}\left[\chi(w_+)U_\sigma^{(0)}\chi^{-1}(w_+) - \chi(w_-)U_\sigma^{(0)}\chi^{-1}(w_-)\right]. \tag{3.69}$$

With this formula we have finished the construction, in general, of the n-soliton solution of the spectral equation (3.50). This means that we can now express the matrices U_μ, φ and χ in terms of the background solution $U_\mu^{(0)}$, $\varphi^{(0)}$ up to the freedom of choosing arbitrary constants w_k, $\widetilde{w}_k$ representing the positions of the poles of the matrices χ and χ^{-1} in the w-plane, and the freedom of choosing the arbitrary constants $k_B^{(k)}$, $l^{(k)B}$ in the vectors $m^{(k)A}$ and $p_A^{(k)}$.

It is clear that one can use this freedom to further specify the solution when necessary. This, in fact, is necessary because the solution we have constructed for the matrices U_μ does not guarantee that these are the same matrices that can be expressed in terms of X by (3.33), and that such X satisfies (3.30) and the additional constraints (3.34)–(3.40). It is remarkable, and nontrivial, that all these additional requirements can indeed be satisfied due to the above freedom of parameters. This is a consequence of the fact that our spectral equations have 'conserved integrals' (some authors call them 'involutions'), i.e. some expressions quadratic in the generating matrix that give zero under the action of the operators Π_μ.

This property has already been used in section 1.3 and will be used again in section 8.1. In fact, the additional conditions (1.57) and (8.12) are consequences of this property. It is instructive to clarify this point more carefully. As we have seen the two-dimensional system (1.51), and also (8.6), can be represented by (3.24)–(3.25) with two-dimensional matrices ψ and K_μ. Let us imagine that we impose on the solutions of this two-dimensional problem the constraint

$$\Pi_\mu \left[\tilde{\psi} \left(\frac{e\alpha^2}{\lambda} \right) W \psi(\lambda) \right] = 0,$$

where W is some matrix function of λ and x^μ. Because of identity (3.46), this means that the expression inside the square brackets above can depend only on the parameter w:

$$\tilde{\psi} \left(\frac{e\alpha^2}{\lambda} \right) W \psi(\lambda) = Q(w),$$

where $Q(w)$ is some arbitrary matrix. If we choose this matrix to be symmetric then we can make it the same for all solutions ψ because we have the freedom of the transformation $\psi(\lambda) \rightarrow \psi(\lambda)\gamma(w)$, where $\gamma(w)$ is an arbitrary matrix. This transformation changes nothing in the spectral equation (3.24) but transforms the matrix Q according to $Q(w) \rightarrow \tilde{\gamma}(w)Q(w)\gamma(w)$. We recall that $w(e\alpha^2/\lambda) = w(\lambda)$, as follows from the definition of w, and that $\gamma(w)$ is insensitive to the replacement $\lambda \rightarrow e\alpha^2/\lambda$. By this transformation we can force the symmetric matrix $Q(w)$ to have the same fixed value $Q^{(0)}(w)$ for all solutions, including the background solution $\psi^{(0)}$. In other words, each solution can be normalized with the help of an appropriate matrix $\gamma(w)$ in such a way that for any pair of solutions $\psi^{(0)}$ and ψ we can write

$$\tilde{\psi} \left(\frac{e\alpha^2}{\lambda} \right) W \psi(\lambda) = \tilde{\psi}^{(0)} \left(\frac{e\alpha^2}{\lambda} \right) W^{(0)} \psi^{(0)}(\lambda),$$

where $W^{(0)}$ is the matrix W constructed with the background solution generated by $\psi^{(0)}$. If we apply the dressing formula $\psi = \chi \psi^{(0)}$ to this relation the result can be written as

$$W^{-1} = \chi(\lambda)(W^{(0)})^{-1} \tilde{\chi} \left(\frac{e\alpha^2}{\lambda} \right).$$

The choice of matrix W depends on the type of additional constraints we need for the final solution. In the two-dimensional pure gravitational case (the vacuum case) it was necessary to make the metric tensor g symmetric and this corresponds to the choice $W = g^{-1}$. In such a case we have

$$g = \chi(\lambda)g_0 \tilde{\chi} \left(\frac{e\alpha^2}{\lambda} \right),$$

which is exactly (1.57) for the nonstationary metric ($e = 1$), and (8.12) for the stationary field ($e = -1$). In both cases this condition ensures that g is symmetric.

3.3.2 The matrix X

Let us now return to the Einstein–Maxwell three-dimensional problem. Here the analogue of the two-dimensional metric tensor is the three-dimensional matrix X. However, this matrix needs to be Hermitian, not symmetric. Furthermore, the dressing matrix $\chi(w)$ is rational on w, which means that the replacement $\lambda \to e\alpha^2/\lambda$ is irrelevant in this case. This suggests that we impose the basic additional constraints (3.34) and (3.35) assuming the existence of a 'conserved integral' of the following form:

$$\partial_\mu \left[\varphi^\dagger(w, x^\mu) W(w, x^\mu) \varphi(w, x^\mu) \right] = 0 \qquad (3.70)$$

with some, as yet unknown, matrix W. The definition of the Hermitian conjugation of matrix functions that depend on the complex parameter w is the following: to obtain the Hermitian conjugation $M^\dagger(w, x^\mu)$ of any matrix $M(w, x^\mu)$ as a function of w one should first calculate the value of the matrix M at the complex conjugate point $\overline{w}$, i.e. the value $M(\overline{w}, x^\mu)$, and then take the usual Hermitian conjugate of this value.

The existence of the integral (3.70) means that $\varphi^\dagger W \varphi = Q(w)$, where Q depends only on w, not on the coordinates x^μ. We impose that matrix $Q(w)$ be Hermitian: $Q = Q^\dagger$. In this case the freedom of the transformation $\varphi(w, x^\mu) \to \varphi(w, x^\mu)\gamma(w)$, which obviously exists for the spectral equation (3.50), allows us to normalize each solution in such a way that the matrix Q will have the same canonical form for each solution. Indeed, under this transformation Q transforms as $Q \to \gamma^\dagger Q \gamma$. Since Q is Hermitian its transformed form can be made universal, i.e. the same for all solutions, by choosing an appropriate transformation matrix $\gamma(w)$ for each solution. Moreover, this universal form can be made diagonal, real and independent of w. Thus, without loss of generality, and within the class of Hermitian matrices Q, the integral (3.70) can be written as

$$\varphi^\dagger(w, x^\mu) W(w, x^\mu) \varphi(w, x^\mu) = C, \qquad (3.71)$$

where

$$C = \mathrm{diag}\,(C_1, C_2, C_3), \quad C_1, C_2, C_3 = \text{constant}, \qquad (3.72)$$

and where the three constants, C_1, C_2 and C_3, are real. This immediately implies that

$$C = C^\dagger. \qquad (3.73)$$

In fact, even the constants C_1, C_2 and C_3 can be eliminated from the solutions by making their modulus equal to 1. However, we will keep the matrix C in the more general form (3.72) in order to leave open the possibility for more convenient choices of arbitrary parameters in the final form of the solution.

Since (3.71) is universal it is also valid for the background solution

$$\varphi^{(0)\dagger} W^{(0)} \varphi^{(0)} = C, \qquad (3.74)$$

where $W^{(0)}$ is the matrix W calculated for the background solution. From (3.71) and (3.74) we have

$$W^{-1} = \chi (W^{(0)})^{-1} \chi^\dagger. \tag{3.75}$$

Now let us assume that the matrix W^{-1} has no singularities at the points where the matrices χ and $\chi^\dagger$ have poles. Then in order to satisfy (3.75) one needs first to eliminate the residues of these poles on the right hand side of this relation, i.e. at $w = w_k$ and $w = \overline{w}_k$. Let us consider first the set of points $w = \overline{w}_k$. The residues at these points vanish if

$$\left(I + \sum_{k=1}^{n} \frac{R_k}{\overline{w}_l - w_k} \right) (W^{(0)})^{-1} (\overline{w}_l) R_l^\dagger = 0, \tag{3.76}$$

or, in components,

$$\sum_{k=1}^{n} \frac{m^{(k)D} \left[(W^{(0)})^{-1} (\overline{w}_l) \right]_{DB} \overline{m}^{(l)B}}{w_k - \overline{w}_l} n_A^{(k)} = \left[(W^{(0)})^{-1} (\overline{w}_l) \right]_{AD} \overline{m}^{(l)D}. \tag{3.77}$$

It follows from (3.73)–(3.74) that we should construct any background solution in such a way that the matrix $W^{(0)}$ is Hermitian. Now it is easy to check that the equation eliminating the residues on the right hand side of (3.75) at the second set of poles, i.e. at $w = w_k$, coincides exactly with (3.77). Therefore this is the only equation we need in order to have regularity of (3.75) at the points where the matrices χ and $\chi^\dagger$ are singular.

Equation (3.77) is an algebraic system from which the vectors $n_A^{(k)}$ can be expressed in terms of the vectors $m^{(k)A}$. Note, however, that this can be achieved in another way, since from the algebraic system (3.56) the vectors $n^{(k)A}$ can also be found in terms of $m^{(k)A}$. Thus if we substitute into such a system (3.66) for the vectors $p_A^{(k)}$, (3.56) takes the form

$$\sum_{k=1}^{n} \frac{m^{(k)D} \left[\varphi^{(0)} (\tilde{w}_l) \right]_{DB} l^{(l)B}}{w_k - \tilde{w}_l} n_A^{(k)} = \left[\varphi^{(0)} (\tilde{w}_l) \right]_{AD} l^{(l)D}. \tag{3.78}$$

Of course, this equation should coincide with (3.77). The coincidence takes place when

$$\tilde{w}_k = \overline{w}_k, \tag{3.79}$$

and[†]

$$\left[(W^{(0)})^{-1} (\overline{w}_l) \right]_{DB} \overline{m}^{(l)B} = \left[\varphi^{(0)} (\tilde{w}_l) \right]_{DB} l^{(l)B}. \tag{3.80}$$

The first condition shows that the poles of the inverse matrix χ^{-1} should be located at the points which are complex conjugate to the poles of the matrix

---

[†] There may also be arbitrary factors that depend on the index l on the right hand side of (3.80). However, they are not essential because they do not appear in the final form of the solution and we can set them equal to 1.

χ. To discover the second condition we should substitute (3.65) into (3.80) for the vectors $m^{(k)A}$ and the expression for $(W^{(0)})^{-1}$ in terms of $\varphi^{(0)}$ and C which follows from (3.74):

$$(W^{(0)})^{-1} = \varphi^{(0)} C^{-1} \varphi^{(0)\dagger}. \tag{3.81}$$

After this substitution we should take into account the conditions (3.72)–(3.73) for the matrix C and the fact that now $\widetilde{w}_k = \overline{w}_k$. Then the resulting form of (3.80) is very simple:

$$k_A^{(k)} = C_{AB} \overline{l}^{(k)B}, \tag{3.82}$$

where C_{AB} are the components of the diagonal matrix C. This allows us to write all the constants $k_A^{(k)}$ in terms of the constants $l^{(k)A}$, or vice versa.

It is easy to check that the same results are obtained if we start our analysis from the 'conserved integral' (3.75) written in its inverse form:

$$W = (\chi^{-1})^{\dagger} W^{(0)} \chi^{-1}. \tag{3.83}$$

In this case, again under the conditions that the matrix W has no singularities at the points $w = \widetilde{w}_k$ and $w = \overline{\widetilde{w}}_k$, and that matrix $W^{(0)}$ is Hermitian, we get only one system of algebraic equations for which the residues of all the poles on the right hand side of (3.83) vanish:

$$\sum_{k=1}^{n} \frac{p_D^{(k)} \left[W^{(0)}(\overline{\widetilde{w}}_l) \right]^{BD} \overline{p}_B^{(l)}}{\widetilde{w}_k - \overline{\widetilde{w}}_l} q^{(k)A} = \left[W^{(0)}(\overline{\widetilde{w}}_l) \right]^{DA} \overline{p}_D^{(l)}. \tag{3.84}$$

In addition, we already had the algebraic equations (3.57) for the vectors $q^{(k)A}$. Substituting (3.65) for $m^{(k)A}$ into (3.57) we obtain

$$\sum_{k=1}^{n} \frac{p_D^{(k)} \left[(\varphi^{(0)})^{-1}(w_l) \right]^{BD} k_B^{(l)}}{\widetilde{w}_k - w_l} q^{(k)A} = \left[(\varphi^{(0)})^{-1}(w_l) \right]^{DA} k_D^{(l)}. \tag{3.85}$$

Imposing that (3.84) and (3.85) should coincide leads again to (3.79) and to the following constraint:

$$\left[(\varphi^{(0)})^{-1}(w_l) \right]^{BD} k_B^{(l)} = \left[W^{(0)}(\overline{\widetilde{w}}_l) \right]^{BD} \overline{p}_B^{(l)}. \tag{3.86}$$

After substitution of (3.66) for the vectors $p_B^{(l)}$ and of the expression for $W^{(0)}$ in terms of $\varphi^{(0)}$ and C into this last equation, we obtain that this equation coincides exactly with (3.82). This shows the selfconsistency of the procedure.

3.3.3 Verification of the constraints

Up to this point we do not need to know the explicit structure of the matrices W and W^{-1}, apart from their regularity at the points $w = w_k$ and $w = \overline{w}_k$ and the Hermiticity of their background values. Under these conditions the relations (3.79) and (3.82) between the free constant parameters ensure the absence of poles at the points $w = w_k$ and $w = \overline{w}_k$ on the right hand side of (3.75), or (3.83). It is now time to fix the exact structure of matrix W in such a way that (3.83) is satisfied not only at the poles but everywhere in the complex w-plane, and that it also satisfies the constraints (3.33)–(3.35). The analysis based on refs [4, 5] shows that this goal can be achieved if we choose the matrix W to be a linear function of w of the following form:

$$W = X - \frac{1}{4}XEX + 4i(w + \beta)\Omega, \tag{3.87}$$

where

$$E = \begin{pmatrix} 0 & 0 & 0 \\ 0 & 0 & 0 \\ 0 & 0 & 1 \end{pmatrix}. \tag{3.88}$$

With this choice the matrix

$$W - (\chi^{-1})^\dagger W^{(0)} \chi^{-1} \tag{3.89}$$

clearly has no singularities at finite values in the w-plane. It has also no singularities at infinity because the matrix χ^{-1} tends to unity as $w \to \infty$ and $W \to W^{(0)}$ since the fixed constant matrix Ω has the same value for the background and dressed solutions. This eliminates the poles in (3.89) at infinity in the w-plane. However, this expression still has nonzero finite values at $w \to \infty$, which should vanish if we wish to satisfy (3.83). Using (3.53) and (3.87) it is easy to calculate the first nonvanishing term of the matrix (3.89) at $w \to \infty$. Equating this term to zero we get

$$X - \frac{1}{4}XEX = X^{(0)} - \frac{1}{4}X^{(0)}EX^{(0)} + 4i(S^\dagger \Omega + \Omega S), \tag{3.90}$$

where $X^{(0)}$ is the background value of the matrix X and

$$S = \sum_{k=1}^{n} S_k. \tag{3.91}$$

The components of the matrix S follow from (3.55) and (3.59),

$$S_A{}^B = -\sum_{k,l=1}^{n} (T^{-1})_{kl} p_A^{(l)} m^{(k)B}. \tag{3.92}$$

Now (3.83) is completely satisfied because (3.89) represents an analytic function at each point on the w-plane which vanishes at infinity. Such a function is everywhere zero in the w-plane.

Due to the special structure of the matrix E it is easy to prove by direct computation the Hermiticity of the matrix X from the Hermiticity of the matrix $X - \frac{1}{4}XEX$. Then it is easy to see from (3.90) that the Hermiticity of the matrix $X^{(0)}$ implies the Hermiticity of the dressed matrix X. Another important property follows from (3.90), namely, that $X^{**} = 2$ if $X^{(0)**} = 2$. Because the background solution $X^{(0)}$ satisfies, by definition, all the additional constraints (including (3.34) and (3.36)) (3.90) guarantees that all these constraints are satisfied for the dressed solution X.

Since (3.90) gives the matrix $X - \frac{1}{4}XEX$, we need to know how to calculate the matrix X from this. Let us introduce a new matrix

$$G = X - \frac{1}{4}XEX. \tag{3.93}$$

Note that this matrix G coincides exactly with the matrix G that appears in refs [4, 5]. This expression can easily be inverted under the condition that $X^{**} = 2$. The result is

$$X = G + GEG, \quad \text{if } X^{**} = 2. \tag{3.94}$$

Thus after calculating G from (3.90) we can obtain X using the above equation.

Now we need to show that the matrix X constructed in this way and the matrices U_μ, which we found in (3.69), indeed satisfy the additional constraints (3.33) and (3.35) (recall that in (3.69) we should take into account the additional restrictions (3.79) and (3.82) for the arbitrary constants). Let us start with the second constraint. To prove its validity it suffices to show that the constraint (3.35) is conserved under the dressing procedure, because the background solution satisfies this condition by definition. First we should remark that if $X^{**} = 2$, then it follows from (3.93) that the inverse of matrix G is

$$G^{-1} = X^{-1} + \frac{1}{2}E, \quad \text{if } X^{**} = 2. \tag{3.95}$$

Due to this property and the trivial identity $E\Omega = 0$ it is easy to see that we have an equivalent form of condition (3.35), which is obtained by just replacing X on the left hand side by G:

$$GU_\mu = -4ie\alpha\eta_{\mu\rho}\epsilon^{\rho\sigma}\Omega U_\sigma. \tag{3.96}$$

Another equivalent equation can be obtained by multiplying (3.96) by $e\sqrt{e}\eta_{\nu\lambda}\epsilon^{\lambda\mu}$, and taking the sum and the difference of the new equation and the original one. The result is

$$\left(G \pm 4ie\sqrt{e\alpha}\Omega\right)\left(U_\mu \pm e\sqrt{e}\eta_{\mu\rho}\epsilon^{\rho\sigma}U_\sigma\right) = 0. \tag{3.97}$$

From (3.87), (3.93) and (3.68) we have,

$$G \pm 4ie\sqrt{e\alpha}\Omega = W(w_\pm). \tag{3.98}$$

Using (3.83) we see that the dressing formulas for the first factors in (3.97) are

$$G \pm 4ie\sqrt{ea}\Omega = (\chi^{-1})^{\dagger}(w_{\pm})\left(G^{(0)} \pm 4ie\sqrt{ea}\Omega\right)\chi^{-1}(w_{\pm}). \qquad (3.99)$$

The dressing formulas for the second factors in (3.97) can be obtained by multiplying (3.69) by $e\sqrt{e}\eta_{\nu\lambda}\epsilon^{\lambda\mu}$ and taking the sum and the difference of this new equation with (3.69) itself. We thus obtain

$$U_{\mu} \pm e\sqrt{e}\eta_{\mu\rho}\epsilon^{\rho\sigma}U_{\sigma} = \chi(w_{\pm})\left(U_{\mu}^{(0)} \pm e\sqrt{e}\eta_{\mu\rho}\epsilon^{\rho\sigma}U_{\sigma}^{(0)}\right)\chi^{-1}(w_{\pm}). \qquad (3.100)$$

The product of these two last equations clearly shows that if the left hand side of (3.97) is zero for the background solution, it is also zero for the dressed solution. Thus we conclude that condition (3.35) is valid because the background solution verifies it and because we have already ensured that (3.50) and (3.70) are satisfied.

Formula (3.100) shows that under the dressing procedure the invariants of the matrices $U_{\mu} \pm e\sqrt{e}\eta_{\mu\rho}\epsilon^{\rho\sigma}U_{\sigma}$ are conserved. If, for example, the rank of these matrices for the background solution is 1 (this is really the case as can easily be seen using the background version of equation (3.97)) then it is also 1 for the dressed solution. Since U_{μ} are defined by the sum of these two matrices the rank of U_{μ} is 2, and this ensures the additional constraint (3.39). As we have already mentioned this condition was automatically satisfied by the construction method of the solution. Now we see that this is also due to the structure of the background solution.

It follows from (3.100) that the traces of the matrices $U_{\mu} \pm e\sqrt{e}\eta_{\mu\rho}\epsilon^{\rho\sigma}U_{\sigma}$ are also conserved under the dressing procedure. However, it is more convenient to deal directly with the traces of the matrices U_{μ}, by taking the trace of (3.69). From this equation we have simply that $\mathrm{Tr}\, U_{\mu} = \mathrm{Tr}\, U_{\mu}^{(0)}$. Since (3.33) is trivially valid for the background solution, we have that $\mathrm{Re}\,\mathrm{Tr}\, U_{\mu}^{(0)} = 0$, which one can easily verify using (3.41) for the background matrix $(X^{(0)})^{-1}$ and the fact that the two-dimensional matrix $g^{(0)}$ is real and symmetric. As a consequence we have that for the dressed matrices U_{μ},

$$\mathrm{Re}\,\mathrm{Tr}\, U_{\mu} = 0. \qquad (3.101)$$

Finally we need to prove that our matrices U_{μ} and X are connected by (3.33). Again, for the background solution such a relation is trivially satisfied because we started with a given matrix $X^{(0)}$ and the new matrices $U_{\mu}^{(0)}$ were just defined using (3.33). However, when we start from the spectral equation (3.50) we first obtain some solution for φ and U_{μ} and we introduce the matrix X later, in a way that is completely independent of (3.33). In this case (3.33) represents an additional constraint connecting U_{μ} and X.

To prove the validity of the constraint (3.33) one can start from the conserved integral,

$$\left[\varphi^{\dagger}\left(G + 4i(w + \beta)\Omega\right)\varphi\right]_{,\mu} = 0. \qquad (3.102)$$

After differentiation and by substitution into this formula of the expression (3.48) for $\varphi_{,\mu}$ (and its Hermitian conjugate $\varphi^\dagger_{,\mu}$), we multiply the result by $(w+\beta)^2 - e\alpha^2$ and obtain on the left hand side of (3.102) a quadratic polynomial in the spectral parameter w, or more precisely in $w + \beta$. Since our solution already ensures that condition (3.102) is satisfied, all the coefficients in this polynomial vanish. The zero value of the coefficient of the quadratic term gives the identity

$$(G + 4i\beta\Omega)_{,\mu} = 2(U^\dagger_\mu \Omega - \Omega U_\mu). \tag{3.103}$$

The remaining coefficients of the polynomial give nothing new: the linear coefficient just leads again to (3.103), and the free coefficient leads to (3.96).

It would be nice to prove the validity of (3.33) by the method we used before, i.e. by proving that the dressing procedure preserves this relation. However, we have no suitable dressing formula for the right hand side of (3.33). Instead, we can calculate the exact structure of the matrices U_μ from those equations for which we have already proved the validity. Then we can check the correctness of (3.33) by direct substitution. At this stage we know that (3.34)–(3.40), (3.101) and (3.103) are valid and that matrix X has the structure of (3.41). Detailed analysis in which a key role is played by (3.35) in the form (3.96) and by (3.103) and (3.101), shows that this system leads to the following unique structure for the matrices U_μ:

$$\left.\begin{aligned}
(U_\mu)_a{}^b &= g_{ac,\mu}\epsilon^{cb} - \frac{i}{\alpha}\eta_{\mu\rho}\epsilon^{\rho\sigma}g_{ac}\epsilon^{cd}g_{df,\sigma}\epsilon^{fb} + 2\Phi_{a,\mu}\overline{\Phi}_c\epsilon^{cb}, \\
(U_\mu)_a{}^* &= -\Phi_{a,\mu}, \\
(U_\mu)_*{}^b &= 2\overline{\Phi}_c\epsilon^{cd}(U_\mu)_d{}^b, \\
(U_\mu)_*{}^* &= 2\epsilon^{ab}\Phi_{a,\mu}\overline{\Phi}_b.
\end{aligned}\right\} \tag{3.104}$$

Direct substitution of this result together with the matrix X, see (3.41), into (3.33) shows that this equation is identically verified. This is the final step in the proof that the matrix X which follows from (3.90) is the same matrix X which first appeared in (3.29), and that it has the structure (3.41) and satisfies (3.30). As a consequence, the functions g_{ab} and $\text{Re}\,\Phi_a$, which can be extracted from (3.90) using (3.93), (3.94) and (3.41), indeed represent a solution of the Einstein–Maxwell equations.

3.3.4 Summary of prescriptions

Let us summarize now, step by step, the set of practical prescriptions for constructing n-soliton solutions of the Einstein–Maxwell equations starting with a given background solution.

1. Take some background solution $g_{ab}^{(0)}$ and $A_a^{(0)}$ of the Einstein–Maxwell equations (3.12)–(3.13). Calculate the determinant of the matrix $g_{ab}^{(0)}$ and

find the function $\alpha(x^\mu)$ from the relation $\alpha^2 = e \det g_{ab}^{(0)}$, after choosing some definite root of this quadratic equation, for example $\alpha > 0$.

2. Take the previous $g_{ab}^{(0)}$, $A_a^{(0)}$ and α, and find, using (3.17), the auxiliary potentials $B_a^{(0)}$ (up to two arbitrary real additive constants), and write the background value of the complex electromagnetic potentials $\Phi_a^{(0)} = A_a^{(0)} + i B_a^{(0)}$.

3. Substitute the values $g_{ab}^{(0)}$ and $\Phi_a^{(0)}$ into (3.41). This gives the background value $X^{(0)}$ of the matrix X.

4. Calculate the background matrices $U_\mu^{(0)}$ by substituting into (3.33) the previous values of $X^{(0)}$ and α.

5. Use (3.45) to find the function $\beta(x^\mu)$, up to some arbitrary real additive constant.

6. From (3.87) compute the background matrix $W^{(0)}$ in terms of $X^{(0)}$ and β.

7. Substitute α, β and $U_\mu^{(0)}$ into the spectral equation (3.48) and find the normalized solution for the background generating matrix $\varphi^{(0)}(w, x^\mu)$, i.e. the solution that satisfies (3.74) with the matrix C defined by (3.72)–(3.73).

8. Using the previous $\varphi^{(0)}$ construct the vectors $m^{(k)A}$ and $p_A^{(k)}$ according to (3.65)–(3.66), where $\widetilde{w}_k = \overline{w}_k$, and where the constants $k_A^{(k)}$ and $l^{(k)A}$ are related by (3.82).

9. With these values for $m^{(k)A}$ and $p_A^{(k)}$ construct the matrix T_{kl} using (3.58), and again taking $\widetilde{w}_k = \overline{w}_k$.

10. Substitute the matrix T_{kl} and the vectors $m^{(k)A}$ and $p_A^{(k)}$ into (3.92) to obtain the matrix S.

11. Finally, from (3.90) with the help of (3.93) and (3.94) calculate the components of the matrix X in terms of the components of the matrices $X^{(0)}$ and S. The matrix X thus obtained when written in the form (3.41) gives the dressed solution g_{ab} and A_a of the Einstein–Maxwell equations in terms of the X^{ab} and X^{a*} components of X as

$$g_{ab} = \frac{1}{4}\epsilon_{ca}\left(X^{cd} - \frac{1}{2}X^{c*}\overline{X}^{d*}\right)\epsilon_{db}, \tag{3.105}$$

$$A_a = \frac{1}{4}\epsilon_{ca}\,\mathrm{Re}\,X^{c*}. \tag{3.106}$$

3 *Einstein–Maxwell fields*

3.4 The metric component f

To complete the construction of the n-soliton solution for the Einstein–Maxwell field we need to compute the metric coefficient f from (3.11). First, let us transform this equation into a more convenient form. It is easy to show that, as a consequence of (3.17) and (3.19), the right hand side of (3.11) can be written as

$$2\eta^{\nu\rho}g^{cd}A_{c,\mu}A_{d,\rho} - \delta_{\mu}^{\nu}\eta^{\rho\sigma}g^{cd}A_{c,\sigma}A_{d,\rho} = \frac{1}{2}\eta^{\nu\rho}\left(g^{cd}\Phi_{c,\mu}\overline{\Phi}_{d,\rho} + g^{cd}\overline{\Phi}_{c,\mu}\Phi_{d,\rho}\right).$$

(3.107)

From (3.41) one can compute the components of the matrices $X_{,\mu}X^{-1}$ in terms of g_{ab} and Φ_a, and then it is easy to see that

$$g^{ab}g^{cd}g_{bc,\mu}g_{da,\rho} + 2g^{cd}\Phi_{c,\mu}\overline{\Phi}_{d,\rho} + 2g^{cd}\overline{\Phi}_{c,\mu}\Phi_{d,\rho} = \text{Tr}\left(X_{,\mu}X^{-1}X_{,\rho}X^{-1}\right).$$

(3.108)

We know already (see the right hand sides of equations (3.12) and (3.22)) that

$$\eta^{\mu\rho}g^{cd}\Phi_{c,\mu}\overline{\Phi}_{d,\rho} = 0.$$

(3.109)

Thus it follows from (3.108) that

$$g^{ab}g^{cd}\eta^{\rho\sigma}g_{bc,\rho}g_{da,\sigma} = \eta^{\lambda\sigma}\,\text{Tr}\left(X_{,\lambda}X^{-1}X_{,\sigma}X^{-1}\right).$$

(3.110)

Taking into account (3.107), (3.108) and (3.110) we get the following form for (3.11)

$$(\ln f)_{,\mu}(\ln\alpha)_{,\nu} + (\ln f)_{,\nu}(\ln\alpha)_{,\mu} - \eta_{\mu\nu}\eta^{\lambda\sigma}(\ln f)_{,\lambda}(\ln\alpha)_{,\sigma} = P_{\mu\nu},$$

(3.111)

where

$$P_{\mu\nu} = 2(\ln\alpha)_{,\mu\nu} - \eta_{\mu\nu}\eta^{\lambda\sigma}(\ln\alpha)_{,\lambda\sigma} + \frac{1}{2\alpha^2}\,\text{Tr}(K_{\mu}K_{\nu}) - \frac{1}{4\alpha^2}\eta_{\mu\nu}\eta^{\lambda\sigma}\,\text{Tr}(K_{\lambda}K_{\sigma}),$$

(3.112)

where K_{μ} are the matrices (3.29).

It is easy to solve (3.111) with respect to the first derivatives of the metric coefficient f:

$$(\ln f)_{,\mu} = e\alpha\eta_{\mu\gamma}\epsilon^{\gamma\sigma}D^{-1}\epsilon^{\lambda\rho}\alpha_{,\lambda}P_{\rho\sigma},$$

(3.113)

where

$$D = \eta^{\mu\nu}\alpha_{,\mu}\alpha_{,\nu}.$$

(3.114)

After some simple transformations involving the first two terms on the right hand side of (3.112), and after exploiting the fact that function α satisfies (3.14), we obtain

$$(\ln f)_{,\mu} = (\ln|D|)_{,\mu} - (\ln\alpha)_{,\mu}$$

$$+ \frac{1}{4}e\eta_{\mu\gamma}\epsilon^{\gamma\sigma}D^{-1}\epsilon^{\lambda\rho}(\ln\alpha)_{,\lambda}\left[2\,\text{Tr}(K_{\rho}K_{\sigma}) - \eta_{\rho\sigma}\eta^{\kappa\nu}\,\text{Tr}(K_{\kappa}K_{\nu})\right].$$

(3.115)

We have already explained (see the text after (3.25)) how to choose the index e, the coordinates x^μ and matrices K_μ for the nonstationary and stationary cases. Following these rules one can check that (3.115) gives (1.40) and (1.41) exactly for the nonstationary fields and (8.4) for the stationary case.

In principle, (3.115) can be considered as a final version of (3.11), from which the metric coefficient f can be calculated after the matrix X is known. However, it is more convenient to express the right hand side of (3.115) in terms of the matrix G, using (3.94), because the basic equation (3.90) for the n-soliton solution directly gives the matrix G, not X. From (3.94) it is easy to check that

$$\mathrm{Tr}(K_\mu K_\nu) = \alpha^2 \, \mathrm{Tr}\left(G_{,\mu} G^{-1} G_{,\nu} G^{-1}\right) + \alpha^2 \, \mathrm{Tr}\left(E G_{,\mu} G^{-1} G_{,\nu} + E G_{,\nu} G^{-1} G_{,\mu}\right). \tag{3.116}$$

Now, substituting this equation for $\mathrm{Tr}(K_\mu K_\nu)$ into (3.115), and using (3.103) we can write $G_{,\nu}$ in terms of the matrices U_μ. Then a rather long calculation, for which all previous information about the matrices G, U_μ, the potentials Φ_a and, in particular, (3.17) and (3.96) is used, we obtain the final result for the differential equation for f:

$$(\ln f)_{,\mu} = (\ln |D|)_{,\mu} - (\ln \alpha)_{,\mu} + i D^{-1} \epsilon^{\lambda \nu} \alpha_{,\nu} \, \mathrm{Tr}\left[(G^{-1} + E) U_\lambda^\dagger \Omega U_\mu\right]. \tag{3.117}$$

Let us return to (3.61) and take its asymptotic form at $w \to \infty$. Keeping only the first nonvanishing terms which are of order w^{-1} we have

$$U_\mu = U_\mu^{(0)} - 2i \, S_{,\mu}, \tag{3.118}$$

where S is the matrix (3.92). Using this result, (3.90) and (3.93) we can write the last term of (3.117) in terms of the matrix T_{kl} (see (3.58)) the vectors $p_A^{(k)}$, $m^{(k)A}$ and its derivatives. After that we can compute the coefficient f from (3.117) by quadratures. The integration is rather long and will not be done here. It was performed[†] by Alekseev [5] the final result is very simple:

$$f = C_0 f^{(0)} T \overline{T}, \tag{3.119}$$

[†] Instead of matrix S Alekseev used matrix $R = \sum_{k=1}^n R_k$, which is essentially the same because $R = -S$, as follows from the identity $XX^{-1} = I$. The two equations $\alpha = \alpha(x^\mu)$ and $\beta = \beta(x^\mu)$ can be inverted and we can write $x^\mu = x^\mu(\alpha, \beta)$, thus one can introduce the derivatives $\partial x^\mu / \partial \alpha$ and $\partial x^\mu / \partial \beta$. It is easy to check that $\partial x^\mu / \partial \beta = D^{-1} \epsilon^{\mu\nu} \alpha_{,\nu}$. This shows that the last trace term in (3.117) coincides with the trace term in the corresponding differential equation for f in ref. [5] up to a sign. But this is correct because the function $\beta(x^\mu)$ that we are using in this book has the opposite sign with respect to that used in ref. [5].

It should also be mentioned that the absence of the term $(\ln |D|)_{,\mu}$ in the corresponding equation for the coefficient f in ref. [5] appears to be an error. In fact, since the function α is the same for the background and for the dressed solutions, the form of (3.119) cannot change due to this error. The discrepancy, however, is only present in the explicit structure of the background coefficient $f^{(0)}$. Of course, this background value of f should be calculated from the correct equation (3.117). It is worth mentioning also that our complex electromagnetic potential Φ_a has the opposite sign with respect to that used in ref. [5], and that our matrix T_{kl} is called Γ_{kl} in that reference.

where $C_0 = $ constant, $f^{(0)}$ is the background value of the metric coefficient f and T is the determinant of the $n \times n$ matrix T_{kl},

$$T = \det T_{kl}. \tag{3.120}$$

To end this section it is worth making some remarks on the relation between soliton solutions described here for the particular case when $\Phi_a = 0$ (vacuum) and the vacuum soliton solutions which can be constructed using the technique we have described in sections 1.2–1.4. There is not yet a comprehensive analysis of this relation. However, the results obtained in [72, 74, 110] show that to all appearances the n-soliton vacuum solution corresponding to n poles in the complex w-plane in Alekseev's approach is equivalent to the $2n$-soliton solution corresponding to n pairs of complex conjugate poles in the complex λ-plane in the framework described in sections 1.2–1.4. By equivalent we mean that two solutions can be transformed into each other by a coordinate transformation. Nevertheless, it is clear that in the vacuum case the ISM described in sections 1.2–1.4 which uses the complex structure in the λ-plane gives in some sense a richer set of soliton solutions, since it also includes solutions which correspond to an odd number of poles in the λ-plane. There are no analogues of such solutions in the framework that uses the complex structure in the w-plane.

Finally we remark that all the formulas in this chapter become essentially simpler if one uses the null coordinates ζ and η instead of coordinates x^μ (which are complex conjugate in the stationary case),

$$\zeta = \frac{1}{2}x^2 + \frac{\sqrt{e}}{2}x^1, \quad \eta = \frac{1}{2}x^2 - \frac{\sqrt{e}}{2}x^1. \tag{3.121}$$

However, in this case instead of one expression for the formulas in terms of the Greek indices one should write two different expressions which correspond to the ζ and η components. For $e = 1$, (3.121) give coordinates (1.14) because, as we have mentioned already, for the nonstationary case $x^2 = z$ and $x^1 = t$.

3.5 Einstein–Maxwell breathers

We end this chapter with the explicit example of an Einstein–Maxwell soliton solution which generalizes the gravibreather described in section 2.3.2. This solution, called an Einstein–Maxwell breather, was first derived and studied by Garate and Gleiser [110]. For a certain range of parameters, the solution represents two electrogravitational plane waves which originate at an initial cosmological singularity. Near the singularity the solution displays a breather-like behaviour, but at late times one finds two weak electrogravitational waves propagating on a homogeneous Bianchi VI_0 cosmological background.

The solution we describe here is the same as the solution in ref. [110], but since our purpose is also to illustrate how the procedure described in this chapter works, we will derive it using our language. As usual, the main problem is to obtain a normalized background generating matrix $\varphi^{(0)}(w, x^\mu)$ of the spectral equation (3.48). Fortunately, this problem was solved by Garate and Gleiser [110] and we can use their result. Let us now proceed to the construction of the one-soliton solution (with respect to the spectral parameter w) following the programme, summarized in 11 steps, of section 3.3.4.

1. As our background metric we take the solution used at the end of section 2.3.2, i.e. metric (2.15) with $u_0 = 2k\beta$,

$$ds^2 = f_0(dz^2 - dt^2) + \alpha e^{2k\beta}(dx^1)^2 + \alpha e^{-2k\beta}(dx^2)^2, \qquad (3.122)$$

where α and β are the functions (1.45) and (1.46):

$$\alpha = a(z+t) + b(z-t), \qquad \beta = a(z+t) - b(z-t), \qquad (3.123)$$

and

$$f_0 = |\alpha_{,t}^2 - \alpha_{,z}^2| \alpha^{-1/2} e^{k^2 \alpha^2}. \qquad (3.124)$$

Here k is an arbitrary constant (note that it has the opposite sign with respect to the same constant in ref. [110]). If α is a time-like variable ($\alpha_{,t}^2 - \alpha_{,z}^2 > 0$), the coordinates can be chosen so that $\alpha = t$ and $\beta = z$. In this case (3.122) represents the vacuum Bianchi VI_0 model. However, keeping α and β arbitrary does not complicate the calculations and we will use this more general form in what follows. The background matrix $g_{ab}^{(0)}$ is obviously

$$g_{ab}^{(0)} = \mathrm{diag}\left(\alpha e^{2k\beta}, \alpha e^{-2k\beta}\right). \qquad (3.125)$$

The background value of the electromagnetic potential is chosen as

$$A_a^{(0)} = 0. \qquad (3.126)$$

Since we are dealing with a nonstationary metric the indicator e which appeared first in (3.2) should be $e = 1$, and the Greek indices now correspond to the coordinates t, z. The components of matrices (3.2) and (3.15) are $\eta_{tt} = -1$, $\eta_{zz} = 1$, and $\epsilon^{tz} = \epsilon_{tz} = 1$, $\epsilon^{zt} = \epsilon_{zt} = -1$.

2. The second step of the programme is trivial. In accordance with (3.126) we choose $B_a^{(0)} = 0$ and for the background value of the complex electromagnetic potential we have

$$\Phi_a^{(0)} = 0. \qquad (3.127)$$

3. From (3.41) it follows that with this choice for $g_{ab}^{(0)}$ and $\Phi_a^{(0)}$

$$X^{(0)} = \text{diag}\left(4\alpha e^{-2k\beta},\ 4\alpha e^{2k\beta},\ 2\right). \tag{3.128}$$

4. From (3.33) we obtain the following structure for the matrices $U_\mu^{(0)}$,

$$U_\mu^{(0)} = \begin{pmatrix} i\eta_{\mu\rho}\epsilon^{\rho\sigma}(\alpha_{,\sigma} - 2k\alpha\beta_{,\sigma}) & \left(\alpha e^{2k\beta}\right)_{,\mu} & 0 \\ -\left(\alpha e^{-2k\beta}\right)_{,\mu} & i\eta_{\mu\rho}\epsilon^{\rho\sigma}(\alpha_{,\sigma} + 2k\alpha\beta_{,\sigma}) & 0 \\ 0 & 0 & 0 \end{pmatrix}. \tag{3.129}$$

5. The function β was defined in (3.123) and it is easy to check that (3.45) is automatically verified.

6. From (3.87) the matrix $W^{(0)}$ becomes

$$W^{(0)} = \begin{pmatrix} 4\alpha e^{-2k\beta} & 4i(w + \beta) & 0 \\ -4i(w + \beta) & 4\alpha e^{2k\beta} & 0 \\ 0 & 0 & 1 \end{pmatrix}. \tag{3.130}$$

7. Now we arrive at the crucial point, namely, the construction of the background generating matrix $\varphi^{(0)}(w, x^\mu)$ of the spectral equation (3.48) which satisfies the additional condition (3.74). For this we first need to choose some concrete values for the components of the diagonal and real matrix C. As follows from the analysis in ref. [5] one convenient choice is

$$C = \text{diag}(-4,\ 4,\ 1). \tag{3.131}$$

As we explained in section 3.3.2 this choice can be made without loss of generality. Now the Garate–Gleiser [110] solution for $\varphi^{(0)}$ may be written as[†]

$$\varphi^{(0)}(w, x^\mu) = \begin{pmatrix} \varphi_{11}^{(0)} & \varphi_{12}^{(0)} & 0 \\ \varphi_{21}^{(0)} & \varphi_{22}^{(0)} & 0 \\ 0 & 0 & 1 \end{pmatrix}, \tag{3.132}$$

[†] In ref. [110] this solution was written in the form $\varphi^{(0)}K$, where $\varphi^{(0)}$ is given by (3.132)–(3.134) and K is some constant matrix. However, this complication is not necessary at this stage. If $\varphi^{(0)}K$ is a solution of (3.48), so is $\varphi^{(0)}$. Thus K can be taken to be the unit matrix without loss of generality. The arbitrary constants that make K will reappear later among the constants $k_A^{(k)}$ and $l^{(k)A}$ in the vectors (3.65)–(3.66).

where

$$\varphi_{11}^{(0)} = i\frac{e^{k\beta}}{\sqrt{2}}\left(\frac{\cosh k\Sigma}{\sigma_-} - \frac{\sinh k\Sigma}{\sigma_+}\right),$$

$$\varphi_{12}^{(0)} = \frac{e^{k\beta}}{\sqrt{2}}\left(\frac{\cosh k\Sigma}{\sigma_+} - \frac{\sinh k\Sigma}{\sigma_-}\right),$$

$$\varphi_{21}^{(0)} = -\frac{e^{-k\beta}}{\sqrt{2}}\left(\frac{\cosh k\Sigma}{\sigma_-} + \frac{\sinh k\Sigma}{\sigma_+}\right),$$

$$\varphi_{22}^{(0)} = -i\frac{e^{-k\beta}}{\sqrt{2}}\left(\frac{\cosh k\Sigma}{\sigma_+} + \frac{\sinh k\Sigma}{\sigma_-}\right),$$

$$(3.133)$$

and

$$\sigma_\pm = \sqrt{w + \beta \pm \alpha}, \qquad \Sigma = \sigma_+\sigma_-. \tag{3.134}$$

Direct substitution of this matrix $\varphi^{(0)}(w, x^\mu)$ together with the previous matrices $U_\mu^{(0)}$, $W^{(0)}$, and C into (3.48) and (3.74) shows that these equations are indeed satisfied (this is a rather long calculation). The functions $\sigma_\pm$ are defined in such a way that their real parts are always positive, and they satisfy the equation

$$\sigma_\pm(\overline{w}, x^\mu) = \sigma_\pm(w, x^\mu). \tag{3.135}$$

It is important to use this last property of $\sigma_\pm$ when checking the validity of (3.74).

8. Now we consider the case in which the dressing matrix χ of (3.53) has only one pole in the complex w-plane, namely $w = w_1$, where w_1 is some fixed complex number. From (3.132)–(3.134) we can compute the matrix $\varphi^{(0)}(w, x^\mu)$ at the point $w = \overline{w}_1$, i.e. $\varphi^{(0)}(\overline{w}_1, x^\mu)$, and from (3.66) we obtain the vector p_A (since there is only one pole we omit the index (1) on the vectors $m^{(1)A}$, $p_A^{(1)}$, and the constants $k_A^{(1)}$, $l^{(1)A}$):

$$p_a = l^c\left[\varphi^{(0)}(\overline{w}_1, x^\mu)\right]_{ac}, \qquad p_* = l^*, \tag{3.136}$$

where l^1, l^2 and l^* are three arbitrary complex constants. We should recall, however, that one of these constants (if not zero) can be fixed without loss of generality: this is a consequence of the rescaling freedom $l^A \to \zeta l^A$ for an arbitrary complex factor ζ (see the footnote in section 3.3.1). We keep this in mind but will preserve the general form for the constant vector l^A for symmetry reasons and to avoid losing particular cases when some of its components are zero.

From (3.65) the vector m^A, under the condition (3.82) and with matrix C given by (3.131), can be expressed in terms of the vector p_A as

$$m^a = \overline{p}_b N^{ba}, \qquad m^* = \overline{p}_*, \tag{3.137}$$

where the 2×2 matrix N (with components N^{ab}) is

$$N = \begin{pmatrix} 4\alpha e^{-2k\beta} & 4i(w_1 + \beta) \\ -4i(w_1 + \beta) & 4\alpha e^{2k\beta} \end{pmatrix}. \qquad (3.138)$$

This last result allows us to express the final form of the solution in terms of the vector p_a only.

9. In the single-soliton case the matrix T_{kl} of (3.58) has only the 11-component, which can be written with the help of (3.137) as

$$T_{11} = \frac{1}{w_1 - \overline{w}_1} \left(l^* \overline{l}^* + p_a \overline{p}_b N^{ba} \right). \qquad (3.139)$$

10. Now, from (3.92) and again using (3.137), we obtain the matrix $S_A{}^B$:

$$\left. \begin{array}{ll} S_*{}^* = -\dfrac{1}{T_{11}} l^* \overline{l}^*, & S_a{}^b = -\dfrac{1}{T_{11}} p_a \overline{p}_c N^{cb}, \\[3mm] S_*{}^b = -\dfrac{1}{T_{11}} l^* \overline{p}_c N^{cb}, & S_a{}^* = -\dfrac{1}{T_{11}} p_a \overline{l}^*. \end{array} \right\} \qquad (3.140)$$

11. Finally, following step 11 of section 3.3.4 and using equation (3.119) for the metric component f, we have the solution

$$ds^2 = f(t, z)(dz^2 - dt^2) + g_{ab}(t, z)dx^a dx^b, \qquad (3.141)$$

where the metric coefficients are given in terms of the vector p_a and the matrix N by

$$g_{ab} = g_{ab}^{(0)} - i \left(\frac{1}{T_{11}} p_a \overline{p}_c N^{cd} \epsilon_{db} - \frac{1}{\overline{T}_{11}} \overline{p}_b p_c \overline{N}^{cd} \epsilon_{da} \right)$$

$$- \frac{4 l^* \overline{l}^*}{T_{11} \overline{T}_{11}} p_a \overline{p}_b, \qquad (3.142)$$

$$f = C_0 f_0 T_{11} \overline{T}_{11}, \qquad (3.143)$$

where $g_{ab}^{(0)}$ and f_0 are the coefficients of the background solution, (3.125) and (3.124), respectively, and ϵ_{ab} is the antisymmetric matrix (3.16). It is easy to see that the two-dimensional matrix g_{ab} of (3.142) is indeed real and symmetric. The Maxwell potential A_a is

$$A_a = -i \left(\frac{1}{T_{11}} \overline{l}^* p_a - \frac{1}{\overline{T}_{11}} l^* \overline{p}_a \right), \qquad (3.144)$$

which is obviously real.

Thus, we have finished the construction of the solution. However, it is worth writing it in a more closed and compact form using some special choice of coordinates, i.e. of α and β. First, it is clear that for the one-soliton solution we can choose the position of the pole w_1 on the imaginary axis

$$w_1 = iq_0, \tag{3.145}$$

where q_0 is some arbitrary real and positive constant (this choice is again possible due to the freedom of the transformation $\beta \to \beta + \text{constant}$). We can now choose functions α and β as we did in section 2.3.2 for the gravibreather solution:

$$\alpha = q_0 \sinh t \cosh z, \qquad \beta = q_0 \cosh t \sinh z, \quad t > 0. \tag{3.146}$$

With this choice we have for the functions $\sigma_\pm$

$$\left.\begin{aligned}
\sigma_+ &= \sqrt{\frac{q_0}{2}} \left[e^{(z+t)/2} + i e^{-(z+t)/2} \right], \\[2mm]
\sigma_- &= \sqrt{\frac{q_0}{2}} \left[e^{(z-t)/2} + i e^{-(z-t)/2} \right],
\end{aligned}\right\} \tag{3.147}$$

and for Σ

$$\Sigma = q_0(\sinh z + i \cosh t). \tag{3.148}$$

Let us also make a definite choice for the arbitrary parameters l^a, which can be fixed by the rescaling freedom $l^a \to \zeta l^a$. Since one of the parameters, l^1 or l^2, should be different from zero (otherwise we have the trivial solution) we impose the following constraint:

$$(l^1)^2 + (l^2)^2 = \frac{1}{4}. \tag{3.149}$$

This constraint fixes two of the four arbitrary real parameters contained in l^a. The general solution of this equation is

$$\left.\begin{aligned}
l^1 &= \frac{1}{2} \cosh s_0 \cos t_0 + \frac{i}{2} \sinh s_0 \sin t_0, \\[2mm]
l^2 &= \frac{1}{2} \cosh s_0 \sin t_0 - \frac{i}{2} \sinh s_0 \cos t_0,
\end{aligned}\right\} \tag{3.150}$$

where s_0 and t_0 are two arbitrary real parameters. The vector p_a follows now from (3.136), but it is more convenient to give its conjugate value $\bar{p}_a$:

$$\left.\begin{aligned}
\bar{p}_1 &= \frac{i}{2\sqrt{2}} e^{k\beta} \left(\frac{\cosh \Upsilon}{\sigma_-} - \frac{\sinh \Upsilon}{\sigma_+} \right), \\[2mm]
\bar{p}_2 &= \frac{i}{2\sqrt{2}} e^{-k\beta} \left(\frac{\cosh \Upsilon}{\sigma_-} + \frac{\sinh \Upsilon}{\sigma_+} \right),
\end{aligned}\right\} \tag{3.151}$$

where

$$\Upsilon = kq_0 \sinh z - s_0 + i(kq_0 \cosh t - t_0). \tag{3.152}$$

With these expressions for p_a and from (3.139), (3.142) and (3.143), it is a simple exercise to compute the metric coefficients g_{ab} and f. If we write the arbitrary complex constant l^* in the form

$$l^* = v_0 e^{i\delta_0}, \tag{3.153}$$

where v_0 and δ_0 are two real arbitrary parameters, we have the solution (3.141) with the following metric coefficients:

$$
\begin{aligned}
g_{11} &= \frac{q_0 \sinh t \cosh z}{D_1} \left[\left(\cosh t \cosh \sigma - \sinh \sigma + v_0^2 \sinh t \right)^2 \right. \\
&\quad \left. + \left(\sinh z \sin \tau - \cos \tau - v_0^2 \cosh z \right)^2 \right] e^{2kq_0 \cosh t \sinh z}, \\
g_{22} &= \frac{q_0 \sinh t \cosh z}{D_1} \left[\left(\cosh t \cosh \sigma + \sinh \sigma + v_0^2 \sinh t \right)^2 \right. \\
&\quad \left. + \left(\sinh z \sin \tau + \cos \tau - v_0^2 \cosh z \right)^2 \right] e^{-2kq_0 \cosh t \sinh z}, \\
g_{12} &= 2 \frac{q_0 \sinh t \cosh z}{D_1} \left[\sinh z \sinh \sigma \sin \tau \right. \\
&\quad \left. - \cosh t \cosh \sigma \cos \tau - v_0^2 \left(\cosh z \sinh \sigma + \sinh t \cos \tau \right) \right];
\end{aligned}
\tag{3.154}
$$

$$f = \frac{D_1}{\sqrt{q_0 \sinh t \cosh z}} e^{k^2 q_0^2 \sinh^2 t \cosh^2 z}; \tag{3.155}$$

where the functions D_1, σ and τ are defined by

$$
\begin{aligned}
D_1 &= \left(\sinh t \cosh \sigma + v_0^2 \cosh t \right)^2 + \left(\cosh z \sin \tau - v_0^2 \sinh z \right)^2, \\
\sigma &= 2kq_0 \sinh z - 2s_0, \\
\tau &= 2kq_0 \cosh t - 2t_0 + \frac{\pi}{2}.
\end{aligned}
\tag{3.156}
$$

The electromagnetic potential (3.144) can be written in the form

$$
\begin{aligned}
A_a &= 4 \frac{q_0 v_0}{D_1} \operatorname{Re} \left\{ e^{i\delta_0} \bar{p}_a \left[v_0^2 \left(\sinh^2 z + \cosh^2 t \right) + \sinh t \cosh t \cosh \sigma \right. \right. \\
&\quad \left. \left. - \sinh z \cosh z \sin \tau - i \left(\sinh t \sinh z \cosh \sigma + \cosh z \cosh t \sin \tau \right) \right] \right\}.
\end{aligned}
\tag{3.157}
$$

Substituting the vector $\bar{p}_a$ of (3.151) in the above result we can get the final expression in closed form for the potential A_a. The explicit form, however, is rather long and we do not display it here.

When $v_0 = 0$ (i.e. $l^* = 0$) we have $A_a = 0$ and the metric (3.154)–(3.156) represents a vacuum solution which coincides exactly with the gravibreather solution (2.46)–(2.47). To obtain exact coincidence one takes a reflection of one of the coordinates x^a (for example $x^1 \to -x^1$) and relates the arbitrary parameters q_0, s_0 and t_0 with the parameters q, σ_0 and τ_0 which appear in (2.46)–(2.47) as: $q_0 = q$, $2s_0 = -\sigma_0$ and $2t_0 = 2kq - q\tau_0 + \pi/2$.

The explicit expressions (3.154)–(3.157) are simple enough that the asymptotic behaviour of the solution in different spacetime regions ($t \to \pm\infty$, $z \to \pm\infty$ and $t \sim z$) can be studied analytically. Such an explicit representation was not used by Garate and Gleiser [110], nevertheless they were able to describe the main asymptotic properties of this solution using both analytical and numerical analysis. These authors also studied numerically the electromagnetic energy density, as well as a scalar invariant proportional to the square of the Riemann curvature tensor for this metric. They showed that $|w_1| = q_0$ can be interpreted as the wave width and that, when $kq_0 \sim 1$, the spacetime near $z = 0$ and at early times $t \sim 0$ has an oscillating behaviour typical of a gravitational breather [177]. At late times, however, the solution describes two weak electromagnetic plane waves propagating in a Bianchi VI$_0$ background. These waves which are due to the solitonic perturbation of this background start near $z = 0$ at the initial cosmological singularity $t = 0$.

4

Cosmology: diagonal metrics from Kasner

One context in which the ISM described in chapter 1 has been widely used is the cosmological context, especially for the generation of exact inhomogeneous cosmological models. The purpose of this and the next chapter is to review such applications and to provide an overview of the corresponding soliton solutions and their physical significance.

In this chapter we concentrate on spacetimes that can be described by a diagonal metric obtained as soliton solutions from a Kasner background. This background, which is a homogeneous but anisotropic cosmological model, is reviewed in section 4.2. Section 4.3 is devoted to the characterization of diagonal metrics. Several physical relevant quantities including the Riemann tensor in an appropriate frame, the optical scalars and the Bel–Robinson superenergy tensor which are useful for the interpretation of these diagonal metrics in the cosmological context, and also in the cylindrically symmetric and plane-wave contexts, are introduced. A brief review of some key equations of the ISM adapted to canonical coordinates is given in section 4.4. The ISM is then used to generate diagonal soliton solutions. Since the relevant field equations for diagonal metrics are linear, the soliton solutions can be generalized in several ways. The relation between these solutions and the known general solution of the linear problem is given, and the solutions are classified according to the type (real or complex) and the number of pole trajectories which define them. The solutions with real poles are discussed in section 4.5. Some interesting connections between different anisotropic but homogeneous models, and some inhomogeneous generalizations are found. We discuss the matching of solitonic regions with their backgrounds and find solutions representing pulse waves propagating on Kasner backgrounds, the cosmic broom, and some related solutions called cosolitons. In section 4.6 solutions with complex-pole trajectories are analysed. Generally these are more interesting as cosmological models since the metrics are regular except at the cosmological singularity. Solutions with one pole trajectory are interpreted as composite universes. We describe the collision of soliton-like perturbations on an homogeneous Kasner background. A comparison of this collision with the soliton collision of nonlinear physics is discussed. Only some features of the truly nonlinear nondiagonal soliton solutions are preserved when the diagonal limit is taken.

4.1 Anisotropic and inhomogeneous cosmologies

The large structure of the present Universe seems to be isotropic and spatially homogeneous. Physical cosmology is based on the relativistic Friedmann–Lemaître–Robertson–Walker (FLRW) models which describe the Universe as completely homogeneous and isotropic in all its evolution [237, 301, 314, 240].

Cosmological models which are not FLRW have also been studied in the past and, particularly, in recent years. For a comprehensive review of exact inhomogeneous cosmological models see the book by Krasiński [180]. This study was motivated partly by the need to explore the consequences of Einstein's theory of gravity (the theory in which all cosmological models are described) and partly because the FLRW cannot explain several features and enigmas of the present Universe. For instance, trying to explain why the Universe is FLRW today when it might have been anisotropic at early times motivated the 'chaotic cosmology' program [221]. Also one would like to explain the fact that if one perturbs the FLRW models [204, 223], one encounters decaying modes which would have been much more important in the past, suggesting a finite deviation from FLRW at early epochs, or that purely statistical fluctuations in FLRW models cannot collapse fast enough to form the observed galaxies [209]. The study of non-FLRW models started with the anisotropic and spatially homogeneous models in which a three-dimensional isometry group G_3 acts simply transitively on the hypersurfaces of homogeneity [91, 18, 92]. These models are included in the Bianchi classification and some of them have the interesting property that they evolve towards isotropy [208, 257]. Spatially inhomogeneous cosmologies have been limited mainly to the case of spacetimes admitting an Abelian two-parameter group, G_2, of isometries. They were initiated by Gowdy's spatially compact models [126, 125]. Later, there was increased interest in such cosmologies [293, 294, 295, 298, 45, 2]. See the reviews by Carmeli *et al.* [48], MacCallum [210, 211, 212], Verdaguer [288] and Bonnor *et al.* [39] for a summary of the main work.

Part of this interest has been motivated by the possibility that there could exist a background of gravitational waves [255, 50]. Present and proposed experiments may be close to detecting it [315, 274] and much effort has been devoted to find the means for doing so. If these waves have a cosmological origin they would generally have a longer period than the waves generated at the present epoch and could be observable by new detection techniques, for example, by the Doppler tracking of interplanetary spacecraft [146, 28, 216], by monitoring perturbations to planetary orbits [27, 279, 217], by the imprints they leave on the cosmic microwave background [181, 44], or by scrutinizing the timing noise in pulsars [215, 29, 147, 253]. The last two methods give the strongest constraints, so far, on the present amount of gravitational waves of cosmological origin. One way in which a background of gravitational waves could arise would be as a superposition of waves from many individual bursts generated astrophysically at some time in the past [255, 249, 273]. The origin of some waves may be oscillating loops of cosmic strings [290, 280, 281]. In the inflationary models [133, 203, 3, 144], the quantum fluctuations of gravitational modes may lead to a spectrum of long wavelength gravitational waves [256]. Another possibility is that the waves may be primordial in the sense that they derive from the initial conditions of the Universe at the Planck time in an FLRW universe [131, 314], as suggested by some quantum cosmology models [134, 282].

Another possibility is that the primordial waves may reflect a purely classical irregularity in some initial structure of the Universe. In this case the waves would not look like radiation at sufficiently early times because they would have wavelengths larger than the Universe's particle horizon. Also their amplitude would severely distort the background spacetime. Thus, classical primordial waves could also arise if the early Universe deviated considerably from the smooth structure we observe today. Depending on the type of initial irregularity, one might expect an isotropic stochastic background with a wide range of wavelengths if the Universe started off completely chaotic [221, 248].

Finding exact cosmological solutions which evolve towards homogeneous cosmological models with a background of cosmological gravitational waves is of interest as a classical mechanism for the creation of such a background. Inhomogeneous cosmological solutions admitting a G_2 provide examples of such a mechanism, although, of course, they produce a very correlated gravitational wave background (all waves travelling in the same direction) rather than a stochastic background like the one probably present. Such a correlated background might be generated by a less extreme form of irregularity [148] or as consequence of the dissipation of an initial inhomogeneous cosmological singularity [308, 309].

Most of the inhomogeneous solutions studied so far can be considered as generalizations of homogeneous Bianchi models in which the homogeneity is broken in one direction. In fact, all Bianchi models from type I to type VII and the locally rotationally symmetric (LRS) types VIII and IX admit an Abelian subgroup G_2 of isometries and can be written [257] as

$$ds^2 = f(t)(dz^2 - dt^2) + \eta_{cd}(t)e_a^c(z)e_b^d(z)dx^a dx^b \qquad (a, b, c, d = 1, 2), \quad (4.1)$$

where the functions $e_a^c(z)$ depend only on the particular Bianchi type. One may break the homogeneity in the z-direction by assuming that f and η_{cd} are also functions of z: $f(t, z)$ and $\eta_{cd}(t, z)$. Adams *et al.* [2, 1] have studied solutions which describe gravitational waves in Bianchi backgrounds, particularly of type I, by solving the Einstein equations when the above assumption is made.

Spacetime metrics admitting an orthogonally transitive two-parameter group of isometries, i.e. admitting an Abelian G_2 group of isometries with 2-surfaces orthogonal to the group orbits, can be written [179] in block diagonal form:

$$ds^2 = f(t, z)(dz^2 - dt^2) + g_{ab}(t, z)dx^a dx^b, \qquad (a, b = 1, 2), \qquad (4.2)$$

where we have assumed two space-like commuting Killing vector fields ∂_{x^1} and ∂_{x^2}. This class of metrics is often referred to as Gowdy models, even when no compact spatial sections are considered. They include (4.1) and their inhomogeneous generalizations and were first introduced in ref. [20]. As can be seen from chapter 1, this metric coincides with (1.36) and the ISM allows one to generate new solutions in a systematic way once a particular background or seed solution is given.

4.2 Kasner background

Most soliton solutions of cosmological interest have been obtained by the ISM from a background metric consisting of one of the homogeneous Bianchi models. This is only natural because Bianchi models are well understood and classified and, as we have seen in the last section, they can be easily generalized to spacetimes with G_2 isometries by breaking their homogeneity along one direction. Also this may provide a first approach for a classification of inhomogeneous cosmologies [294].

Among all Bianchi models the simplest one, Bianchi I, which is given in vacuum by the Kasner metric [164, 184], has been the most fruitful. The ISM and generalizations applied to the Kasner metric lead to a large class of inhomogeneous solutions and even to large families of homogeneous Bianchi models, thus relating solutions which were previously unrelated. The reason why the Kasner metric is of special interest in cosmology may be related to the fact that in some sense the 'generic' cosmological solution of the Einstein equations near the cosmological singularity is described by a succession of Kasner epochs, as has been indicated in the analysis of refs [20, 21].

More specifically in the case of the collision of parallel-polarized plane waves, whose interaction region is described by a metric (1.36), or (4.2), with a diagonal g_{ab} (as we shall see in chapter 7), Yurtsever [308, 309] has proved that as the singularity is approached ($\alpha = t \to 0$), which corresponds to the cosmological singularity in our context, the metric is asymptotic to a Kasner solution for each fixed $\beta = z$, in agreement with the analysis of refs [20, 21]. Since, as we shall see, the ISM does not change the singular character of a background metric, although it may introduce new singularities, the use of the Kasner metric as background seems reasonable for the generation of a quite general family of cosmological solutions.

In canonical coordinates, i.e. in coordinates such as (1.36) with the condition that $\alpha = t$ and $\beta = z$ (see (1.45)–(1.46)), the Kasner metric can be written as

$$ds^2 = t^{(d^2-1)/2}(dz^2 - dt^2) + t^{1+d}dx^2 + t^{1-d}dy^2, \tag{4.3}$$

where d is an arbitrary parameter, the Kasner parameter. Expression (4.3) is related to the standard Kasner form [20, 184, 179]:

$$ds^2 = -dT^2 + T^{2p_1}dx^2 + T^{2p_2}dy^2 + T^{2p_3}dz^2, \tag{4.4}$$

where $p_1 + p_2 + p_3 = p_1^2 + p_2^2 + p_3^2 = 1$, by the time transformations $T = t^{(d^2+3)/4}$ and

$$p_1 = \frac{2(1+d)}{d^2+3}, \qquad p_2 = \frac{2(1-d)}{d^2+3}, \qquad p_3 = \frac{d^2-1}{d^2+3}. \tag{4.5}$$

The parameter d may be chosen either positive ($d > 0$) or negative ($d < 0$) since one may obtain one solution from the other by interchanging the coordinates x and y.

The value $|d| = 1$ corresponds to a region of the Minkowski spacetime, and $d = 0$ is an LRS space with Petrov type D metric. Other values of d correspond to Petrov type I metrics. The z axis is expanding for $|d| > 1$ and contracting for $|d| < 1$; as we shall see this leads to quite different behaviours for the generated soliton metrics.

4.3 Geometrical characterization of diagonal metrics

From a diagonal background in canonical coordinates (1.36), the ISM generally leads to nondiagonal metrics, i.e. metrics with two polarizations when they admit a wave interpretation. But one may restrict the parameters involved in the ISM so that all new solutions are diagonal, i.e. one polarization. This is simply done by taking some of the parameters to be zero. Therefore the diagonal metrics thus obtained may be considered as a limit of the more general nondiagonal soliton solutions and as such they deserve attention since they preserve some of the features of the more general case. In particular, the pole trajectories are the same and, since these trajectories characterize the shape and the propagation of the solitons, this feature will be preserved in the nondiagonal case. In contrast, a feature that will obviously not be represented is the typical nonlinear interaction between the modes of polarization that follows from (1.39) in the general case.

Furthermore, the diagonal metrics characterized by having hypersurface orthogonal Killing vectors are also of interest on their own. Equations (1.39) become linear when the metric is written in the Einstein–Rosen form [90] and a superposition principle applies to their solutions. This will lead to several generalizations of the soliton solutions, some of which contain interesting spacetimes.

A great advantage of the diagonal metrics is that a fairly complete study of the spacetime they describe and of their physical properties can be made. The Riemann tensor becomes explicitly calculable in most cases and some of the Riemann components and optical scalars have direct physical meaning. In this section we give the main elements for the intrinsic characterization of these metrics, which will be of use later.

4.3.1 Riemann tensor and Petrov classification

Metric (4.2), when $g_{12} = 0$, takes the form

$$ds^2 = f(t, z)(dz^2 - dt^2) + g_{11}(t, z)dx^2 + g_{22}(t, z)dy^2 , \qquad (4.6)$$

and the Riemann tensor can be written as the 6×6 matrix

$$R^{(\alpha\beta)}_{(\gamma\delta)} = \begin{pmatrix} E & B \\ -B & E \end{pmatrix}, \qquad (4.7)$$

where the indices are (01), (02), (03), (23), (31) and (12). Here E and B are the 'electric' and 'magnetic' 3×3 matrices whose nonzero components are

$$E_{11} = e_1, \quad E_{22} = e_2, \quad E_{33} = e_3, \quad B_{12} = B_{21} = b,$$

with

$$\left.\begin{aligned}
e_1 &= \frac{1}{2f}\left[\frac{g_{22}''}{g_{22}} - \frac{1}{2}\left(\frac{g_{22}'}{g_{22}}\right)^2 - \frac{1}{2}\frac{\dot{g}_{22}}{g_{22}}\frac{\dot{f}}{f} - \frac{1}{2}\frac{g_{22}'}{g_{22}}\frac{f'}{f}\right], \\[2mm]
e_2 &= \frac{1}{2f}\left[\frac{g_{11}''}{g_{11}} - \frac{1}{2}\left(\frac{g_{11}'}{g_{11}}\right)^2 - \frac{1}{2}\frac{\dot{g}_{11}}{g_{11}}\frac{\dot{f}}{f} - \frac{1}{2}\frac{g_{11}'}{g_{11}}\frac{f'}{f}\right], \\[2mm]
e_3 &= \frac{1}{2f}\left(\frac{1}{2}\frac{g_{11}'}{g_{11}}\frac{g_{22}'}{g_{22}} - \frac{1}{2}\frac{\dot{g}_{11}}{g_{11}}\frac{\dot{g}_{22}}{g_{22}}\right), \\[2mm]
b &= \frac{1}{2f}\left(\frac{\dot{g}_{11}'}{g_{11}} - \frac{1}{2}\frac{\dot{g}_{11}}{g_{11}}\frac{g_{11}'}{g_{11}} - \frac{1}{2}\frac{\dot{g}_{11}}{g_{11}}\frac{f'}{f} - \frac{1}{2}\frac{g_{11}'}{g_{11}}\frac{\dot{f}}{f}\right).
\end{aligned}\right\} \qquad (4.8)$$

Here a dot denotes ∂_t and a prime ∂_z. All these quantities are easily calculated for soliton solutions. Since we are dealing with vacuum solutions $e_1 + e_2 + e_3 = 0$.

For some purposes, particularly when considering spacetimes which exhibit propagation of waves, it is best to express the Riemann tensor in terms of an appropriate null tetrad of vectors $(\vec{n}, \vec{l}, \vec{m}, \vec{m}^*)$, which satisfy:

$$l_\mu n^\mu = -1, \qquad m_\mu m^{*\mu} = 1 \qquad (4.9)$$

(all other scalar products among these vectors vanish), and in terms of which the spacetime metric can be written as

$$g_{\mu\nu} = m_\mu m_\nu^* + m_\mu^* m_\nu - n_\mu l_\nu - l_\mu n_\nu. \tag{4.10}$$

For the metric (4.6) we choose the following null tetrad:

$$\left.\begin{aligned}
\vec{n} &= (1/\sqrt{2f})(\partial_t + \partial_z), \quad \vec{l} = (1/\sqrt{2f})(\partial_t - \partial_z), \\
\vec{m} &= (1/\sqrt{2g_{11}})\partial_x + i(1/\sqrt{2g_{22}})\partial_y, \quad \vec{m}^* = (1/\sqrt{2g_{11}})\partial_x - i(1/\sqrt{2g_{22}})\partial_y.
\end{aligned}\right\} \tag{4.11}$$

The nonvanishing components of the Riemann tensor, which are also those of the Weyl tensor, $C_{\mu\nu\alpha\beta}$, since $C_{\mu\nu\alpha\beta} = R_{\mu\nu\alpha\beta}$ in vacuum, are given in this tetrad by the following three scalars:

$$\left.\begin{aligned}
\Psi_0 &= R_{\mu\nu\alpha\beta} n^\mu m^\nu n^\alpha m^\beta = \frac{1}{2}(e_2 - e_1) + b, \\
\Psi_2 &= \frac{1}{2} R_{\mu\nu\alpha\beta} n^\mu l^\nu (n^\alpha l^\beta - m^\alpha m^{*\beta}) = -\frac{1}{2}e_3, \\
\Psi_4 &= R_{\mu\nu\alpha\beta} l^\mu m^{*\nu} l^\alpha m^{*\beta} = \frac{1}{2}(e_2 - e_1) - b.
\end{aligned}\right\} \tag{4.12}$$

These components now have a direct physical interpretation [265]: Ψ_0 and Ψ_4 represent the radiative part of the gravitational field, whereas Ψ_2 contains the Coulomb part; Ψ_0 gives the radiative component along the left-directed waves and Ψ_4 along the right-directed waves.

There are only two independent scalar invariants [179]:

$$I = \Psi_0\Psi_4 + 3\Psi_2^2, \qquad J = \Psi_2(\Psi_0\Psi_4 - \Psi_2^2), \tag{4.13}$$

and the d'Inverno and Russell-Clark [81] algorithm for the algebraic classification of the Riemann tensor reduces to the following: if $I^3 = 27 J^2$ the metric is of Petrov type D, otherwise it is type I. The case of plane waves $I = J = 0$, i.e. type N, cannot be treated in canonical coordinates, but the ISM can also be applied in this case.

It may also be convenient to use a boosted tetrad, i.e. instead of $(\vec{n}, \vec{l}, \vec{m}, \vec{m}^*)$ to use

$$(A\vec{n}, A^{-1}\vec{l}, \vec{m}, \vec{m}^*), \tag{4.14}$$

where A is a positive function. Then the Riemann, or Weyl, scalars (4.12), (Ψ_0, Ψ_2, Ψ_4), are simply replaced by

$$(A^2\Psi_0, \Psi_2, A^{-2}\Psi_4). \tag{4.15}$$

This is especially useful for studying the asymptotic behaviour of the metric on the null directions $\vec{n}$ or $\vec{l}$. Thus, for instance, let us consider a null congruence defined by the null vector $k^0 = k^3$ and $k^1 = k^2 = 0$. The geodesic equation $k^\beta k_{\alpha;\beta} = 0$ is simply

$$(\partial_t + \partial_z)(k^0 f) = 0,\qquad(4.16)$$

and we can choose the normalization $k^0 = (1/\sqrt{2})/f$ so that $\vec{k} = (1/\sqrt{f})\vec{n}$ is the tangent to the affinely parametrized null congruence. Now if we take $A = 1/\sqrt{f}$ in the null tetrad (4.14), $A^{-1}\vec{l}$, $\vec{m}$ and $\vec{m}^*$ are parallel transported along the null congruence $\vec{k}$. This can be easily checked by computing the commutators $[\vec{k}, \vec{m}]$, $[\vec{k}, A^{-1}\vec{l}]$ and realizing that for our metric (4.6) $[\vec{m}, \vec{m}^*] = 0$ always. The Weyl scalars (4.15) in such a parallel propagated tetrad give the physical tidal forces, so that if one of them diverges at some spacetime point, the null congruence finds infinite tidal forces and therefore a curvature singularity. Thus there is no need to compute the scalar invariants I and J (4.13) to find such a singularity. Similarly, the singularities may be checked along the direction $\vec{l}$, by considering the tetrad (4.14) with $A = \sqrt{f}$. Then $A\vec{l}$ is tangent to a null congruence with affine parametrization and $A\vec{n}, \vec{m}$, and $\vec{m}^*$ are parallel transported along it.

4.3.2 Optical scalars

For a spacetime with G_2 symmetry, such as (4.2), there is always a preferred null congruence. The covariant derivative of a geodesic null vector field k^α is invariantly characterized by three optical scalars: the expansion, θ, the shear, σ, and the rotation or twist, ω. These are given by [89, 179, 299]

$$\theta = \frac{1}{2}k^\alpha{}_{;\alpha}, \qquad \omega^2 = \frac{1}{2}k_{[\alpha;\beta]}k^{\alpha;\beta}, \qquad \sigma\sigma^* = \frac{1}{2}k_{(\alpha;\beta)}k^{(\alpha;\beta)} - \frac{1}{4}(k^\alpha{}_{;\alpha})^2.$$
$$(4.17)$$

Note also that $\theta + i\omega = k_{\alpha;\beta}m^\alpha m^{*\beta}$ and $\sigma = -k_{\alpha;\beta}m^\alpha m^\beta$ in terms of a null tetrad, but those quantities are invariant and independent of the choice of such a tetrad.

For a diagonal metric such as (4.6), $\omega = 0$ (the null congruence is hypersurface orthogonal) and the expansion and shear become

$$\theta = \frac{1}{2\sqrt{2}}\frac{1}{ft}, \qquad \sigma = \frac{1}{2\sqrt{2}}\frac{1}{f}\left(\frac{1}{t} - \frac{\dot{g}_{11}}{g_{11}} - \frac{\dot{g}_{22}}{g_{22}}\right).\qquad(4.18)$$

When a metric defines a preferred vector field a classification of the vector field is also a characterization of the metric. On the other hand, a geodesic null vector field can be interpreted as the tangent vector to optical rays. Therefore the expansion and shear are important quantities for studying metrics (4.6): they give physical information on the spacetime in which the rays propagate.

4.3.3 Superenergy tensor

In general relativity we do not have a local definition for the energy of the gravitational field. Several quantities have been proposed for the energy density or energy flux of the gravitational field. For instance, for spacetimes with cylindrical symmetry, Thorne [272] was able to define a C-energy flux vector that obeys a conservation law (we shall introduce this quantity in the cylindrical context in chapter 6). But in the general context one of the most interesting quantities is the Bel–Robinson superenergy tensor $T^{\mu\nu\alpha\beta}$, which is defined as the gravitational analogue of the electromagnetic stress-energy tensor [10, 11]. It therefore represents the energy density of local relative acceleration. However, the Bel–Robinson tensor does not have the dimensions of an energy density, but of the square of an energy density (superenergy). It is given by

$$T^{\mu\nu\alpha\beta} = R^{\mu\rho\alpha\sigma} R^{\nu\ \beta}_{\ \rho\ \sigma} + {}^{*}R^{\mu\rho\alpha\sigma*} R^{\nu\ \beta}_{\ \rho\ \sigma}, \tag{4.19}$$

where ${}^{*}R^{\mu\nu\alpha\beta}$ is the dual tensor of $R^{\mu\nu\alpha\beta}$, ${}^{*}R^{\mu\nu\alpha\beta} = \eta^{\mu\nu\rho\sigma} R_{\rho\sigma}^{\ \ \alpha\beta}$ ($\eta^{\mu\nu\rho\sigma}$ is the totally antisymmetric covariant Levi-Civita tensor), $\epsilon = T^{\mu\nu\alpha\beta} u_{\mu} u_{\nu} u_{\alpha} u_{\beta}$ represents the superenergy density as seen by an observer with velocity u^{μ} ($u^{\mu}u_{\mu} = -1$) and

$$P^{\alpha} = -(\delta^{\alpha}_{\mu} + u^{\alpha}u_{\mu})T^{\mu\nu\rho\sigma} u_{\nu} u_{\rho} u_{\sigma} \tag{4.20}$$

is the corresponding Poynting vector.

Now for metric (4.6) we may define an orthonormal tetrad given by

$$\vec{e}_0 = (1/\sqrt{f})\partial_t, \quad \vec{e}_1 = (1/\sqrt{g_{11}})\partial_x, \quad \vec{e}_2 = (1/\sqrt{g_{22}})\partial_y, \quad \vec{e}_3 = (1/\sqrt{f})\partial_z. \tag{4.21}$$

By projecting the Riemann tensor and its dual onto this tetrad, we can calculate the superenergy density and the Poynting vector, as seen by an observer at rest in this frame:

$$\epsilon = e_1^2 + e_2^2 + e_3^2 + 2b^2, \quad P^0 = P^1 = P^2 = 0, \quad P^3 = 2(e_1 - e_2)b. \tag{4.22}$$

We have a clear analogy with electromagnetism. There is only a superenergy flux along the z-direction. Note, for instance, that for the Kasner metric (4.3), $P^3 = 0$, since $b = 0$. Associated with these quantities we may define a three-velocity field as the ratio of the superenergy tensor flux (Poynting vector) to the superenergy density $v(t, z) = P^3/\epsilon$. Such a quantity is not covariant since it is a velocity measured with respect to the orthonormal frame (4.21). However, it may be significant when we analyse inhomogeneities propagating on a background, such as the Kasner background, in which such a field is zero.

We note that a similar three-velocity may be defined in Minkowski space with the electromagnetic field using the Poynting flux 'vector' $T^{0\mu}$, which satisfies $T^{0\mu}_{\ ,\mu} = 0$ ($T^{\mu\nu}$ is the electromagnetic stress-energy tensor). The three-velocity v^i may be written in this case as [222] $v^i = (\tanh\alpha)n^i$, where

$$[\tanh(2\alpha)]n^i = T^{0i}/T^{00} = 2(\mathbf{E} \times \mathbf{B})^i (\mathbf{E}^2 + \mathbf{B}^2)^{-1} \tag{4.23}$$

(the factor 2 in the tanh is due to the tensorial character of $T^{\mu\nu}$), n^i is unitary and **E** and **B** are the electric and magnetic fields. Now an observer propagating at such a velocity measures no flux of electromagnetic energy. For a null electromagnetic field ($\mathbf{E}^2 = \mathbf{B}^2$, $\mathbf{E} \cdot \mathbf{B} = 0$) such a speed is the speed of light. In analogy we shall see later that, for the gravitational case, in the regions where gravitational radiation dominates, the velocity $v(t, z) = P^3/\epsilon$ also approaches unity.

4.4 Soliton solutions in canonical coordinates

Given that the most used background solution in the cosmological context is the Kasner metric, which depends on t only and has $\det g_0 = t^2$, it is useful, for practical purposes, to rewrite some of the equations for the n-soliton solutions obtained by the ISM of sections 1.3 and 1.4 in terms of canonical coordinates. Thus, we impose that $t = \alpha$ and $z = \beta$. This determines the arbitrary functions that define α and β in (1.45)–(1.46). We recall from (1.14) that $z = \zeta + \eta$ and $t = \zeta - \eta$ and that $\partial_\zeta = \partial_z + \partial_t$ and $\partial_\eta = \partial_z - \partial_t$. Equation (1.38) is now

$$\det g = t^2, \tag{4.24}$$

where we assume that $t \geq 0$. The Einstein equations (1.39)–(1.42) become

$$U_{,t} - V_{,z} = 0, \tag{4.25}$$

$$(\ln f)_{,t} = -\frac{1}{t} + \frac{1}{2t} \operatorname{Tr}(U^2 + V^2), \tag{4.26}$$

$$(\ln f)_{,z} = \frac{1}{2t} \operatorname{Tr}(U \cdot V), \tag{4.27}$$

where U and V are the 2×2 matrices $U = tg_{,t}g^{-1}$ and $V = tg_{,z}g^{-1}$, which are related to the matrices A and B of (1.42) by: $2U = -(B + A)$ and $2V = B - A$. The linear system for the generating matrix $\psi(\lambda, t, z)$ associated with system (4.25) (i.e. the spectral equations, or the 'L–A pair') is

$$D_1\psi = -\frac{tV + \lambda U}{\lambda^2 - t^2}\psi, \qquad D_2\psi = -\frac{tU + \lambda V}{\lambda^2 - t^2}\psi, \tag{4.28}$$

where D_1 and D_2 are the operators

$$D_1 = \partial_z - \frac{2\lambda^2}{\lambda^2 - t^2}\partial_\lambda, \qquad D_2 = \partial_t - \frac{2\lambda t}{\lambda^2 - t^2}\partial_\lambda, \tag{4.29}$$

where λ is a complex spectral parameter. Equations (4.28) are trivially obtained by writing (1.51) in the new variables. Also, as before, the matrix $g(t, z)$ solution of (4.25) follows from the generating matrix when $\lambda = 0$:

$$g(t, z) = \psi(0, t, z). \tag{4.30}$$

We recall from sections 1.3 and 1.4 that the procedure for constructing the n-soliton solution is to start with a given particular background solution g_0 and to integrate the linear system (4.28) to find a solution $\psi_0(\lambda, t, z)$ for the background generating matrix. One then looks for solutions of ψ of the type $\psi = \chi\psi_0$, where χ is the dressing matrix, which is assumed to have the form (1.64), where the n pole trajectories $\mu_k(t, z)$ $(k = 1, 2, \ldots, n)$ satisfy equations (1.66) which in the new variables are

$$\mu_{k,z} = \frac{2\mu_k^2}{t^2 - \mu_k^2}, \qquad \mu_{k,t} = \frac{2t\mu_k}{t^2 - \mu_k^2}. \tag{4.31}$$

The solutions of these equations are the roots of the quadratic equation (1.67), namely

$$\mu_k^2 + 2(z - w_k)\mu_k + t^2 = 0, \tag{4.32}$$

where w_k are arbitrary complex constants. The solutions may be written as

$$\mu_k^\pm = w_k - z \pm [(w_k - z)^2 - t^2]^{1/2}, \tag{4.33}$$

where the superscripts $-$ and $+$ stand for the labels *in* and *out*, respectively, in (1.68)–(1.69) with the prescriptions that the branches of the square roots should be chosen in such a way that $|\mu_k^-| < t$ and $|\mu_k^+| > t$ (see section 1.4 for a detailed discussion of this prescription). Next one constructs the vectors $m_a^{(k)}$ according to (1.80), which now reads

$$m_a^{(k)} = m_{0b}^{(k)}[\psi_0^{-1}(\mu_k, t, z)]_{ba}, \tag{4.34}$$

where $m_{0b}^{(k)}$ are arbitrary complex constant vectors. With these vectors one may construct the symmetric matrix Γ_{kl} defined in (1.83). This is almost the final step in the construction of the n-soliton solution: next one inverts this matrix according to (1.84) and constructs the matrix g using (1.87). This matrix does not satisfy the condition (4.24), but this is easily overcome by the definition (1.100) of the physical matrix $g^{(ph)}$, which is a solution of (4.25) and satisfies (4.24). The physical coefficient $f^{(ph)}$ which solves (4.26)–(4.27) is then constructed using (1.110). Everywhere in this chapter we will use only the physical values of the metric components defined by (1.100) and (1.110) and, for simplicity, we will not use the label (ph) in the physical n-soliton solutions.

Kasner background. In this chapter we shall deal mostly with the Kasner metric as the background solution, thus we need to compute the corresponding background generating matrix ψ_0. We recall that this is the only nonalgebraic step in the construction of the n-soliton solution. Thus we start with the matrix $g_0 = \mathrm{diag}(t^{1+d}, t^{1-d})$, where d is the Kasner parameter, then the corresponding U and V matrices are, respectively, $U_0 = \mathrm{diag}(1 + d, 1 - d)$ and $V_0 = 0$. The linear equation (4.28) is now easily integrated and leads to

$$\psi_0(\lambda, t, z) = \mathrm{diag}\left((t^2 + 2z\lambda + \lambda^2)^{(1+d)/2}, (t^2 + 2z\lambda + \lambda^2)^{(1-d)/2}\right), \tag{4.35}$$

which satisfies (4.30), i.e. $g_0 = \psi(\lambda = 0)$. This background generating matrix is the starting point for most solutions in this chapter. The vectors $m_a^{(k)}$ of (4.34) are now written, using the definition of the pole trajectories (4.32), as

$$m_a^{(k)} = \left(m_{01}^{(k)}(2w_k\mu_k)^{-(d+1)/2}, m_{02}^{(k)}(2w_k\mu_k)^{(d-1)/2} \right). \qquad (4.36)$$

The constant vectors $m_{0a}^{(k)}$ have the normalization freedom $m_{0a}^{(k)} \rightarrow \zeta^k m_{0a}^{(k)}$, where ζ^k are some constants. This transformation does not change the matrix g of (1.87) and changes the coefficient f of (1.110) by a constant only. This result may be compared with the analogous result for $m_a^{(k)}$ obtained in (2.16) with the solutions (2.26)–(2.27) for the Kasner metric. The relation of the parameters used here $(m_{0a}^{(k)}, w_k)$ with the previous parameters (A_k, C_k) is simply

$$A_k^2 = m_{01}^{(k)} m_{02}^{(k)}(2w_k)^{-1}, \qquad e^{-C_k} = \frac{m_{01}^{(k)}}{m_{02}^{(k)}}(2w_k)^{-d}, \qquad (4.37)$$

where we still have the normalization freedom.

4.4.1 Generalized soliton solutions

Starting with a diagonal metric as the background solution we can easily generate diagonal soliton solutions by taking some of the arbitrary parameters $m_{0b}^{(k)}$ in (4.34) to be zero; here we shall take $m_{01}^{(k)} = 0$.

The general expression for the metric coefficients g_{11} and g_{22} can be obtained for n solitons by adding solitons one at a time. For any diagonal background metric the result is

$$g_{11} = \prod_{k=1}^{n}(\mu_k/t)(g_0)_{11}, \qquad g_{22} = t^2/g_{11}. \qquad (4.38)$$

In fact, from (1.87) we have that $g_{11} = (g_0)_{11}$ since the vectors $L_a^{(k)}$ defined in (1.86) satisfy $L_1^{(k)} = 0$, then (4.38) follows trivially from (1.100); recall that we now use the canonical coordinates $t = \alpha$ and $z = \beta$. Note that this result is general and independent of the background metric. When g_0 depends on t only, the coefficient f is best found by integrating (4.26)–(4.27) directly, rather than using (1.110). For the Kasner background we get [51],

$$f = f_0 t^{n(n-2d)/2} \prod_{k=1}^{n} \left[\mu_k^{2+d-n}(\mu_k^2 - t^2)^{-1} \right] \prod_{k,l=1;k>l}^{n} (\mu_k - \mu_l)^2. \qquad (4.39)$$

Here and below, $t \geq 0$ and we write all formulas containing the functions μ_k in a formal way, although we understand that each time a choice of sign is made that ensures the well defined sense of our expressions and the physical signature of the metric. Since the metric coefficient f is determined (up to a multiplicative constant) once the 'transversal' metric coefficients $g_{ab}(t, z)$ are known, some essential features of the metric can be seen from g_{ab}. However, we should bear in mind that the spacetime geometry is also determined by f and that this coefficient is related in a nonlinear way to g_{ab}. Thus a complete solution is not known until all metric coefficients f and g_{ab} are given.

We can now proceed to the generalization of the n-soliton solutions (4.38) and (4.39). The key to such a generalization is that the metric coefficients g_{11} and g_{22} can be written in the Einstein–Rosen form in terms of a single function $\Phi(t, z)$ that satisfies an hyperbolic linear equation. That is, by writing

$$g = \text{diag}(te^{\Phi}, te^{-\Phi}), \tag{4.40}$$

(4.25) are reduced to the familiar 'cylindrical' wave equation

$$\Phi_{,tt} + \frac{1}{t}\Phi_{,t} - \Phi_{,zz} = 0, \tag{4.41}$$

and (4.26)–(4.27) become

$$(\ln f)_{,t} = -\frac{1}{2t} + \frac{t}{2}\left[(\Phi_{,t})^2 + (\Phi_{,z})^2\right], \quad (\ln f)_{,z} = t\Phi_{,z}\Phi_{,t}. \tag{4.42}$$

Then (4.38) can be written as

$$\Phi = \Phi_0 + \Phi_s, \quad \Phi_s = \sum_{k=1}^{n} \ln(\mu_k/t), \quad \Phi_0 = d\ln t, \tag{4.43}$$

where Φ_s stands for the soliton part and Φ_0 for the background part, and the expression $d\ln t$ is applicable in the case of the Kasner background. Since Φ_s satisfies a linear equation one may easily generalize the above solutions. For instance, for complex-pole trajectories μ_k, the real and imaginary parts of Φ_s lead both to independent solutions. Also, any solution can be multiplied by any real or complex number to get another solution. In particular, this is equivalent in some cases to considering degenerate poles. These new solutions are called generalized soliton solutions. In the next sections we shall consider such generalized solutions for both real and complex poles, giving also the f coefficient. In each case we shall consider in detail solutions related to one and two poles since they are shown to give the generic behaviour for all cases.

General solution of the linear equation. For further reference and to allow comparison with the soliton solutions, it is worth reviewing here some of the well known solutions of the wave equation (4.41). Under the condition that Φ be bounded in z the general solution of (4.41) (apart from the last equation of

(4.43)) can be written in Fourier Bessel integrals [300, 2] as

$$\Phi_{FB} = \int_0^\infty \left\{ [A_k \sin(kz) + B_k \cos(kz)] J_0(kt) \right.$$
$$\left. + [C_k \sin(kz) + D_k \cos(kz)] N_0(kt) \right\} dk, \qquad (4.44)$$

where J_0 and N_0 are the Bessel and Neuman functions of order zero, and A_k, B_k, C_k and D_k are k-dependent coefficients. Note that if we impose compact spatial sections, as required in Gowdy cosmologies, the above integral is substituted by a sum (Fourier series) of terms periodic in z. Since $N_0(kt)$ is singular at $t = 0$ whereas $J_0(kt)$ is regular there, the parameters (C_k, D_k) and (A_k, B_k) play a very different role near $t = 0$. Adams *et al.* [2] call 'chaotic' the solutions with nonzero parameters (C_k, D_k), and nonchaotic those which have (A_k, B_k) as the only nonzero parameters. In fact, in the limit $t \to 0$:

$$\Phi_{FB} \approx \int_0^\infty [A_k \sin(kz) + B_k \cos(kz)] dk$$
$$+ \frac{2}{\pi} (0.577 - \ln 2 + \ln t) \int_0^\infty [C_k \sin(kz) + D_k \cos(kz)] dk, \quad (4.45)$$

so that the chaotic part diverges logarithmically and may be seen as contributing to the Kasner solution where it introduces z-dependent Kasner parameters. This term leads to the Kasner-like cosmological singularity [20].

At large t, however, both the chaotic and nonchaotic parts behave in a similar fashion, and the solutions Φ_{FB} can be considered as a superposition of travelling waves along the z axis with the amplitude decreasing as $t^{-1/2}$. As $t \to \infty$, (4.44) can be written as

$$\Phi_{FB} \approx \frac{1}{\sqrt{2\pi t}} \int_0^\infty \left[E_k^+ \cos(kt - kz + \phi_k^+) + E_k^- \cos(kt + kz + \phi_k^-) \right] \frac{dk}{\sqrt{k}}, \qquad (4.46)$$

where, following Adams *et al.* [2], the amplitude and phase of the waves are defined in terms of (A_k, B_k, C_k, D_k) by

$$(E_k^\pm)^2 = (B_k \pm C_k)^2 + (D_k \mp A_k)^2, \quad \tan\phi_k^\pm = (D_k \mp A_k)/(B_k \pm C_k). \quad (4.47)$$

We shall see later that many soliton solutions are not bounded at $|z| \to \infty$ and thus cannot be expressed in terms of the Fourier Bessel integrals (4.44). Instead the solutions in (4.43) may be expressed as superpositions of the well known solution of the linear wave equation (4.41) given by Lamb [183, 254],

$$\Phi \equiv \int_0^{s_k - z - t} \frac{g_k(v) dv}{[(s_k - z - v)^2 - t^2]^{1/2}}, \qquad (4.48)$$

where $g_k(v)$ are arbitrary bounded functions and s_k are arbitrary constants. The values that these functions, and parameters, take in the soliton case will be specified later.

There are some other soliton solutions, namely those resulting of the super-position of two opposite poles ('soliton–antisoliton' pairs) that are bounded at $|z| \rightarrow \infty$ and can be written in terms of the nonchaotic part of (4.44). This has been used by Feinstein [97] to define an effective energy for the corresponding soliton solutions.

4.5 Solutions with real poles

Real-pole trajectories are obtained from (4.33) with $w_k = z_k^0$ (real), the poles are then

$$\mu_k^{\pm} = z_k \pm (z_k^2 - t^2)^{1/2}, \quad z_k \equiv z_k^0 - z \qquad (4.49)$$

with either sign being allowed. In (4.49) the square root is positive for positive values of z_k and negative for negative values of z_k, then the minus and plus signs stand for the *in* and *out* labels, respectively, in (1.68)–(1.69). The two functions are not independent since, as a consequence of $\mu_k^- = \alpha^2/\mu_k^+$, we have $\mu_k^- = t^2/\mu_k^+$. The parameters z_k^0 are sometimes called the soliton 'origins' since they mark the origins of the light cones $z_k^2 = t^2$. It is obvious that these pole trajectories are only defined in terms of canonical coordinates outside such light cones $z_k^2 \geq t^2$. On the light cones $(\mu_k/t)^2 = 1$, from (4.43) it is obvious that Φ continuously matches a solution that does not include the k-pole (if we consider just one pole trajectory, then this matches the background metric). The first derivatives of the metric coefficients $g_{11}(t, z)$ and $g_{22}(t, z)$ are, however, discontinuous along such a light cone. This has usually been interpreted as implying that it leads to shock wave solutions [16]. For the true solitons (i.e. solutions corresponding to a collection of simple poles) such an interpretation is indeed possible. However, we will see in subsection 4.5.2 that for the so-called generalized soliton solutions (which correspond to the degenerate poles already mentioned in section 2.1) such a statement is not generally true: sometimes this discontinuity leads to a curvature singularity on the light cones and to continue the solution across the light cone is meaningless, and sometimes it simply reflects the fact that canonical coordinates behave badly there. It is thus clear that in terms of the canonical coordinate patch the spacetime is divided by a set of intersecting light cones with origins at z_k^0 as in fig. 4.1.

It is now convenient to write down the main asymptotic values for μ_k/t, which is the main ingredient of the solutions (4.43). We may distinguish four asymptotic regions in canonical coordinates:

(i) The future time-like infinity or 'causal region' ($|z_k| \ll t \rightarrow \infty$) for all k, which is contained in the intersection of the set of light cones with origins at $z = (z_1^0, z_2^0, \ldots, z_n^0)$.

(ii) The future null infinity ($|z_k| \sim |z| = t \rightarrow \infty$).

(iii) The space-like infinity ($t \ll |z_k| \sim |z| \rightarrow \infty$).

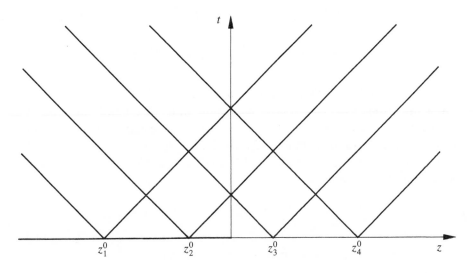

Fig. 4.1. This represents the light cones for four real-pole soliton solutions with equally spaced origins.

(iv) The 'initial region' ($|z_k|$, $|z_k| \gg t \to 0$), i.e. the region near $t = 0$ which is generally a cosmological singularity if the Kasner metric is used as the background.

In the regions $z_k^2 \geq t^2$ the limiting values of $\mu_k^\pm$ are $\mu_k^\pm/t \to 1$ if $z \to -\infty$ and $\mu_k^\pm/t \to -1$ if $z \to +\infty$ at future null infinity, and

$$\frac{\mu_k^-}{t} = \frac{t}{2z_k}\left[1 + 0\left(\frac{t^2}{z_k^2}\right)\right], \quad \frac{\mu_k^+}{t} = \frac{2z_k}{t}\left[1 + 0\left(\frac{t^2}{z_k^2}\right)\right], \tag{4.50}$$

both at space-like infinity and in the initial region.

The first set of generalized soliton solutions is obtained from (4.43) by simply multiplying each term in the sum by an arbitrary real parameter h_i [53]. Since this is equivalent to considering degenerate poles, the metric coefficient f may be found from (4.39) by taking appropriate limits. The final result is:

$$\Phi = d \ln t + \sum_{k=1}^{s} h_k \ln\left(\frac{\mu_k}{t}\right), \tag{4.51}$$

$$f = t^{[(d-g)^2 - 1]/2} \prod_{k=1}^{s}\left[\mu_k^{h_k(h_k + d - g)}(z_k^2 - t^2)^{-h_k^2/2}\right] \prod_{k,l=1;k>l}^{s}(\mu_k - \mu_l)^{2h_k h_l}, \tag{4.52}$$

where we have taken s pole trajectories and $g \equiv \sum_{k=1}^{2} h_k$.

Each term with index k (which we designate as Φ_k) in (4.51) is simply related to the solution (4.48). In fact, the superposition (4.51) corresponds to the case

$g_k(\nu) = h_k$ for $0 < \nu < \infty$ and $s_k = z_k^0$; then the integration of (4.48) leads to

$$\Phi_k = h_k \ln\left(\frac{\mu_k^+}{t}\right) \equiv h_k \cosh^{-1}\left(\frac{z_k}{t}\right). \tag{4.53}$$

Note that if we take μ_k^-, this is equivalent to changing h_k to $-h_k$.

It is interesting to point out how the soliton limits of solutions (4.51)–(4.52) for two poles, i.e. $h_1 = h_2 = 1$ or $h_1 = -h_2 = 1$, can be obtained from the nondiagonal two-soliton solutions (2.20)–(2.25). Thus the case $h_1 = h_2 = 1$ is recovered when the parameters C_1 and C_2 in (2.27) are $C_1 = C_2 \to \infty$, and one makes use of the identity

$$(t^2 - \mu_1\mu_2)(\mu_1 - \mu_2) = 2(w_2 - w_1)\mu_1\mu_2. \tag{4.54}$$

The case $h_1 = -h_2 = 1$ is recovered when the parameters C_1 and C_2 in (2.27) are $C_1 = -C_2 \to \infty$, and one makes use of the identity

$$(\mu_k^\pm)^2 - t^2 = \pm 2\mu_k^\pm\sqrt{z_k^2 - t^2}. \tag{4.55}$$

Another type of generalized soliton solution may be obtained by changing h_k to ih_k in (4.51) [53, 173]. In fact, according to (4.53) the above change leads to the solutions $h_k \cos^{-1}(z_k/t)$. Such solutions are valid inside the light cones $|z_k| \le t$, so that in some sense they are complementary to solution (4.51). For this reason they are also called *cosoliton* solutions. Note that this is equivalent to extending the solution for real poles $\ln(\mu_k/t)$ inside the above light cone and taking into account that the imaginary part of a complex solution of a linear equation is also a real solution of the same equation. The evaluation of the coefficient f is in this case a bit more involved.

The final result is

$$\Phi = d \ln t + \sum_{k=1}^{s} h_k \cos^{-1}\left(\frac{z_k}{t}\right), \tag{4.56}$$

$$f = t^{(d^2-1-g_s)/2} \exp\left[d \sum_{k=1}^{s} h_k \cos^{-1}\left(\frac{z_k}{t}\right)\right] \prod_{k=1}^{s}\left[(t^2 - z_k^2)^{h_k^2/2}\right]$$

$$\times \prod_{k,l=1;k>l}^{s}\left(\frac{z_k z_l - (t^2 - z_k^2)^{1/2}(t^2 - z_l^2)^{1/2}}{z_k z_l + (t^2 - z_k^2)^{1/2}(t^2 - z_l^2)^{1/2}}\right)^{(h_k h_l)/2}, \tag{4.57}$$

where $g_s \equiv \sum_{k=1}^{s} h_k^2$. Of course, by superposing solutions of the type (4.51) and (4.56), one may obtain new solutions. In the next subsections we shall consider solutions (4.51)–(4.52) and (4.56)–(4.57) by studying in detail solutions with one and two (opposite) poles. This will suffice for understanding the general case.

4.5.1 Generation of Bianchi models from Kasner

Here we consider the generalized soliton solution with one pole. That is, by taking $s = 1$ and denoting h_1 by h, solution (4.51)–(4.52), for μ_k^+, reads

$$\Phi = d \ln t + h \cosh^{-1} \left(\frac{z_1}{t} \right), \quad |z_1| \geq t, \tag{4.58}$$

$$f = t^{[(d-h)^2-1]/2} \mu_1^{hd} (z_1^2 - t^2)^{-h^2/2}. \tag{4.59}$$

This solution was first given by Wainwright *et al.* [297]. Its connection to the Kasner metric via the ISM was noted by Kitchingham [172] and Verdaguer [285]. Generally, it does not have a cosmological interpretation since, besides the singularity at $t = 0$, it is also singular at space-like infinity ($t \ll |z_k| \rightarrow \infty$) and on the light cone $z_1^2 = t^2$, with the exception of some particular values of the parameters h and d. This may be easily checked from the Riemann tensor, whose components using tetrad (4.11) are:

$$\Psi_0 = \Psi_+, \quad \Psi_4 = \Psi_-, \quad \Psi_2 = \frac{1}{2f} \left(\frac{(1 - d^2 - h^2)}{4t^2} + \frac{dhz_1}{2t^2(z_1^2 - t^2)^{\frac{1}{2}}} \right), \tag{4.60}$$

where

$$\Psi_{\mp} = \frac{1}{2f} \left(\frac{-d(d^2 - 1)}{4t^2} + \frac{3hd^2(z_1 \pm t)^{\frac{1}{2}}}{4t^2(z_1 \mp t)^{\frac{1}{2}}} \right.$$
$$\left. - \frac{3dh^2(z_1 \pm t)}{4t^2(z_1 \mp t)} + \frac{h(h^2 - 1)(z_1 \pm t)^{\frac{3}{2}}}{4t^2(z_1 \mp t)^{\frac{3}{2}}} \right),$$

with f given in (4.59). From this it follows that the hypersurface $z_1^2 = t^2$ is singular except when $h^2 = 1$ and $h^2 \geq 3/2$. Note that this includes the true soliton case, i.e. $h = \pm 1$ (the plus or minus sign depends on the pole trajectory election μ_1^+ or μ_1^-) in which case the solution may be obtained from the corresponding nondiagonal one-soliton solution. The singularity may be seen by approaching the hypersurface $-z_1 = t$ by the geodesic null vector $A^{-1}\vec{l}$ with $A = \sqrt{f}$, as was indicated in subsection 4.3.1. Then in the boosted tetrad (4.14) $A\vec{n}, \vec{m}$ and $\vec{m}^*$ are parallel propagated along $A^{-1}\vec{l}$ and the Riemann components become (4.15). When the null hypersurface is not singular one may match this solution to the Kasner background inside the light cone $z_1^2 \leq t^2$ and then the light cone does contain a shock wave. However, the singularity at space-like infinity prevents giving to (4.58)–(4.59) a reasonable meaningful physical interpretation. We shall discuss shock waves in the next subsection.

By far the most interesting case of (4.58)–(4.59) is when $h^2 = d^2 + 3$, since in this case the metric has only the cosmological singularity at $t = 0$ and the solution is the Ellis and MacCallum family of vacuum Bianchi models. This can be seen more clearly by introducing new coordinates (T, Z) related to (t, z) by the coordinate change, adapted to spatial homogeneity [48],

$$t = e^{-2aZ}\sinh(2aT), \quad z_1 = e^{-2aZ}\cosh(2aT), \tag{4.61}$$

where a is an arbitrary positive parameter. Then $\mu_1/t = [\tanh(aT)]^{-1}$ and the solution (4.58)–(4.59) for this special case becomes the Ellis and MacCallum [91] metric

$$\begin{aligned}
ds^2 &= [\sinh(2aT)]^{1+d^2}[\tanh(aT)]^{d\sqrt{3+d^2}}(dZ^2 - dT^2) \\
&\quad + [\sinh(2aT)]^{1+d}[\tanh(aT)]^{\sqrt{3+d^2}}e^{-2a(1+d)Z}dx^2 \\
&\quad + [\sinh(2aT)]^{1-d}[\tanh(aT)]^{-\sqrt{3+d^2}}e^{-2a(1-d)Z}dy^2.
\end{aligned} \tag{4.62}$$

The particular case $d = 0$ (axisymmetric background) corresponds to the Bianchi V model, whereas $d^2 = 1$ (Minkowski background) leads to the Kantowski–Sachs solutions of Bianchi type III. Other values of d correspond to Bianchi type VI_h (Bbi case). All these solutions have the cosmological singularity at $T = 0$ [207] and are of type I in the Petrov classification with the exception of $d^2 = 1$, which is of type D. In this way we see that some Bianchi models of types III, V and VI are related to Bianchi I models via generalized soliton solutions.

4.5.2 Pulse waves

Next we consider generalized soliton solutions with two opposite poles. The results of the last subsection suggest that when we have superposition of poles (μ_k^+ only or μ_k^- only) with the same signs of the constants h_k in (4.51) the metric will generally be singular at space-like infinity. This is due essentially to the asymptotic behaviour of the pole trajectories at space-like infinity as described in (4.50). It is obvious that if we superpose poles μ_k^+ (or μ_k^-) with opposite signs of the constants h_k we may get solutions which behave asymptotically like the Kasner background at space-like infinity. For instance, if we take $s = 2$, $h_1 = h = -h_2$, then according to (4.50) $\mu_1^\pm/\mu_2^\pm \to 1$ at space-like infinity and $\Phi \to d \ln t$ in (4.51). From (4.51)–(4.52) such a generalized soliton solution with opposite poles is

$$\Phi = d \ln t + h \left[\cosh^{-1}\left(\frac{z_1}{t}\right) - \cosh^{-1}\left(\frac{z_2}{t}\right) \right], \quad \min(|z_1|, |z_2|) \geq t, \tag{4.63}$$

$$f = t^{(d^2-1)/2}(\mu_1 - \mu_2)^{-2h^2}\mu_1^{h(h+d)}\mu_2^{h(h-d)}[(z_1^2 - t^2)(z_2^2 - t^2)]^{-h^2/2}, \tag{4.64}$$

where μ_1 and μ_2 both belong to type μ^+ or are both of type μ^-.

Solution (4.63)–(4.64) is known as the Carmeli and Charach pulse-wave solution [45, 48] and is supposed to represent pulse waves propagating on a

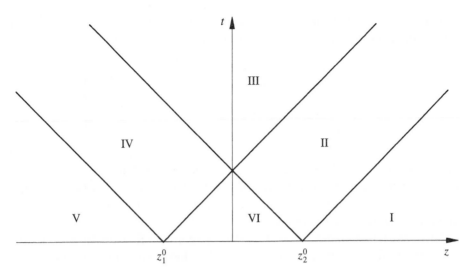

Fig. 4.2.   The two light cones with origins at z_1^0 and z_2^0 divide the canonical coordinate patch into six regions.

Kasner background. This is achieved as follows. The two light cones $z_1^2 = t^2$ and $z_2^2 = t^2$ divide the spacetime in canonical coordinates into the six regions shown in fig. 4.2.   Solution (4.63)–(4.64) applies to regions I, V and VI. This solution may be matched through the light-cone hypersurface to metric (4.58)–(4.59) for one pole in region II and with the soliton solution obtained from (4.58)–(4.59) changing z_1 to z_2 in region IV. Both these solutions may be matched to the Kasner background in region III. On the other hand we know that at space-like infinity, which is contained in regions I and V, the solution (4.51)–(4.52) will tend to the Kasner background also. Therefore one is tempted to interpret the resulting solution as an inhomogeneous cosmological model representing two pulse waves propagating at the speed of light on a Kasner background.

However, this interpretation cannot always be maintained, because for some values of the parameters the metric is singular on the null hypersurfaces defined by the light cones. Furthermore, the matching of soliton solutions with real poles to the background metric described in section 2.1 does not contain all necessary details and a more exact prescription for the matching must be given. Gleiser [118] was the first to consider such a detailed prescription for the matching of soliton solutions with real poles to the Minkowski background. His analysis based on Taub's study of spacetimes with distribution valued curvature tensors [269] can be generalized to the present case [70, 25, 130].   Other possible extensions of the soliton solutions beyond the null hypersurfaces have also been considered in ref. [130].

Before discussing the matching procedures, it is worth noting that (4.63) can be written in terms of the Fourier Bessel integrals (4.44) [97] as

$$
\left.
\begin{aligned}
\Phi &= d \ln t + \int_0^\infty A_k \sin\left[k\left(z - \frac{z_1^0 + z_2^0}{2}\right)\right] J_0(kt)dk, \\[2mm]
A_k &= \frac{2h}{k} \sin\left[\frac{k(z_2^0 - z_1^0)}{2}\right],
\end{aligned}
\right\} \tag{4.65}
$$

i.e. in the case $z_1^0 + z_2^0 = 0$ (as in fig. 4.2) the second term in Φ corresponds to the nonchaotic part of (4.44) with $B_k = 0$. This means that at late times the solution may be interpreted as a superposition of travelling waves (4.46) with amplitude and phase given by (4.47).

For the matching of the different regions, we consider first the Riemann tensor for metric (4.63)–(4.64). With the tetrad (4.11) this has the components

$$
\left.
\begin{aligned}
\Psi_0 &= \Psi_+, \quad \Psi_4 = \Psi_-, \\[2mm]
\Psi_2 &= \frac{1}{2f}\left[\frac{1 - d^2 - 2h^2}{4t^2} + \frac{hd}{2t^2}\left(\frac{z_1}{(z_1^2 - t^2)^{\frac{1}{2}}} - \frac{z_2}{(z_2^2 - t^2)^{\frac{1}{2}}}\right)\right. \\[2mm]
&\quad \left. + \frac{h^2}{2t^2}\frac{z_1 z_2 - t^2}{(z_1^2 - t^2)^{\frac{1}{2}}(z_2^2 - t^2)^{\frac{1}{2}}}\right],
\end{aligned}
\right\} \tag{4.66}
$$

where

$$
\begin{aligned}
\Psi_\mp &= \frac{1}{2f}\left[\frac{-d(d^2 - 1)}{4t^2} + \frac{3hd^2}{4t^2}\left(\frac{(z_1 \pm t)^{\frac{1}{2}}}{(z_1 \mp t)^{\frac{1}{2}}} - \frac{(z_2 \pm t)^{\frac{1}{2}}}{(z_2 \mp t)^{\frac{1}{2}}}\right)\right. \\[2mm]
&\quad - \frac{3h^2 d}{4t^2}\left(\frac{(z_1 \pm t)^{1/2}}{(z_1 \mp t)^{1/2}} - \frac{(z_2 \pm t)^{1/2}}{(z_2 \mp t)^{1/2}}\right)^2 \\[2mm]
&\quad + \frac{3h^3}{4t^2}\left(\frac{(z_1 \pm t)(z_2 \pm t)^{\frac{1}{2}}}{(z_1 \mp t)(z_2 \mp t)^{\frac{1}{2}}} - \frac{(z_1 \pm t)^{\frac{1}{2}}(z_2 \pm t)}{(z_1 \mp t)^{\frac{1}{2}}(z_2 \mp t)}\right) \\[2mm]
&\quad \left. + \frac{h(h^2 - 1)}{4t^2}\left(\frac{(z_1 \pm t)^{\frac{3}{2}}}{(z_1 \mp t)^{\frac{3}{2}}} - \frac{(z_2 \pm t)^{\frac{3}{2}}}{(z_2 \mp t)^{\frac{3}{2}}}\right)\right],
\end{aligned}
$$

with f given by (4.64). This must be compared with the curvature tensor (4.60) in the regions II and III. In region III (Kasner) the curvature tensor is simply deduced from any of the previous ones taking $h = 0$.

For simplicity, we shall only consider the matching between regions I and II through the light cone $-z_2 = t$ and between regions II and III through $-z_1 = t$. The singularities on the latter were discussed in the previous subsection and, as was the case for the one-pole solution, one can see from (4.66) that there is also a curvature singularity on $-z_2 = t$ unless $h^2 = 1$ or $h^2 \geq 3/2$. Again this can be seen by approaching the hypersurface $-z_2 = t$ by the geodesic null vector

$A^{-1}\vec{l}$ with $A = 1/\sqrt{f}$. Notice that there is no divergence on the light cone for the true two-soliton solutions, i.e. when $h \equiv \pm 1$.

To analyse the matching surfaces one must use regular coordinates near them. Note that canonical coordinates are not appropriate since the metric coefficient f diverges near the light cones so that such coordinates are ill defined there. It is useful to introduce null coordinates [70] and then take an affine parametrization of one of them near the hypersurface. For example in region II we define

$$u_1 = (-z_1 + t)/2, \quad v_1 = (-z_1 - t)/2. \tag{4.67}$$

Now $f(u_1, v_1)$ diverges on the light-cone hypersurface $v_1 = 0$. Thus we may define a new (affine) coordinate V_1 by $V_1 = v_1^{(2-h^2)/2}$ (if $h^2 \neq 2$) and the new f coefficient is smooth at $v_1 = 0$. We see that if $h^2 > 2$ the hypersurface $v_1 = 0$ is at infinity in the affine coordinate, therefore the global interpretation of that spacetime is not clear. However, if $h^2 < 2$ then the light cone $v_1 = 0$ corresponds also to $V_1 = 0$. Therefore the only reasonable values for a cosmological interpretation of the matched spacetime are

$$h^2 = 1, \quad 3/2 \leq h^2 < 2. \tag{4.68}$$

The matching between regions II and III is done as follows. We introduce null coordinates $\tilde{u}_1 = (-z_1 + t)/2$, $\tilde{v}_1 = (-z_1 - t)/2$ in region III. Such coordinates are regular at the null surface $-z_1 = t$ for the Kasner background. With coordinates (u_1, V_1) in region II and $(\tilde{u}_1, \tilde{v}_1)$ in region III all metric coefficients match continuously across the hypersurface $v_1 = 0$, although they have discontinuous first derivatives with respect to v. A similar analysis can be done for the matching between regions I and II, i.e. one introduces null coordinates on I, $u_2 = (-z_2 + t)/2$, $v_2 = (-z_2 - t)/2$, which are not regular at $v_2 = 0$, and then $V_2 = v_2^{(2-h^2)/2}$, which is regular. The matching with region II is done as before.

Then following Taub [269] one may compute the Ricci tensor components $R_{V_1 V_1}$ and $R_{V_2 V_2}$ on the matching hypersurfaces $v_1 = 0$ and $v_2 = 0$, respectively. These components depend on the jump across the hypersurface of the first derivative with respect to V_1, and V_2, of $\sqrt{\det g_{ab}}(= t)$. These components are

$$R_{V_1 V_1} = -\frac{1}{u_1} \delta(V_1), \quad R_{V_2 V_2} = -\frac{1}{u_2} \delta(V_2). \tag{4.69}$$

The delta functions signal the presence of a null fluid with negative energy density along the matching hypersurfaces. Note also that our spacetime has only a cosmological singularity. Therefore, provided (4.68) is satisfied, the spacetime resulting from this matching has a physical interpretation as a pair of gravitational shock waves which start as inhomogeneities on the z axis (region VI) and propagate in opposite directions on a Kasner background (regions V and

I). The shock waves are inhomogeneous regions (regions IV and II) with shock fronts formed by null fluids of negative energy density, i.e. no ordinary matter. These waves sweep the space in such a way that they leave the region in between (region III) in the exact homogeneous Kasner background. Some authors call this a 'cosmic broom'. By superposing solutions of this type one gets an example of a cosmological model in which some inhomogeneities near the initial cosmological singularity evolve towards a superposition of gravitational shock waves propagating on a homogeneous (Kasner) background.

4.5.3 Cosolitons

Let us now consider the family of generalized soliton solutions (4.56)–(4.57) that are valid in the intersection of the light cones $|z_k| \leq t$ ($k = 1, \ldots, s$) and which are known as cosolitons. To analyse this superposition of s pole trajectories we consider first, as always, the solution with one pole, i.e. $s = 1$:

$$\Phi = d \ln t + h \cos^{-1} \left(\frac{z_1}{t} \right), \quad |z_1| \leq t, \tag{4.70}$$

$$f = t^{(d^2 - h^2 - 1)/2} (t^2 - z_1^2)^{h^2/2} \exp \left[dh \cos^{-1} \left(\frac{z_1}{t} \right) \right], \tag{4.71}$$

where we have written $h \equiv h_1$. Using the null tetrad (4.11) the nonnull Riemann, or Weyl, scalars for this metric are

$$\left. \begin{aligned} \Psi_0 &= -(2f)^{-1} X^+, \quad \Psi_4 = -2(f)^{-1} X^-, \\ \Psi_2 &= -(8f)^{-1} [(1 + h^2 - d^2) t^{-2} - 2h z_1 d t^{-2} (t^2 - z_1^2)^{-1/2}], \end{aligned} \right\} \tag{4.72}$$

where f is defined in (4.71) and

$$\begin{aligned} X^\pm = {}& (t^2 - z_1^2)^{-1/2} \left[h z_1 t^{-2} (3d^2 - h^2 - 1) \pm h t^{-1} (3d^2 - h^2 - 3) \right] / 4 \\ &+ (t^2 - z_1^2)^{-1} h^2 d (2 + z_1^2 t^{-2} \pm 3 z_1 t^{-1}) / 2 \\ &+ (t^2 - z_1^2)^{-3/2} h \left[z_1 + z_1^3 \pm t (2 + h^2 z_1^2 t^{-2}) / 2 \right] \\ &+ d t^{-2} (d^2 - h^2 - 1) / 4. \end{aligned}$$

The algebraic classification of this tensor is easily performed by using the Russell–Clark algorithm as described after (4.13). One sees that for $h \neq 1$ the metrics are of Petrov type I. For $h = 0$ the metric, of course, reduces to the Kasner background which includes a metric representing a region of flat spacetime ($d^2 = 1$) and a type D metric ($d = 0$).

Unlike the generalized soliton solutions with one pole, (4.58)–(4.59), which are defined in the complementary region $|z_1| \geq t$ and which include a subfamily of spatially homogeneous metrics ($h^2 = d^2 + 3$), there are no values of the parameter h ($h \neq 1$) for which metric (4.70)–(4.71) is spatially homogeneous [234]. Such a metric has curvature singularities at $t = 0$ and on the light cone

$|z_1| = t$ for any value of the parameter h $(h \neq 1)$. This means that they have no clear cosmological interpretation and that, unlike the case of metric (4.58)–(4.59), it is not possible to match the spacetime they describe with another metric outside the light cone, in spite of the fact that the transversal metric coefficients match continuously with those of the Kasner metric or with those of (4.58)–(4.59). Therefore one cannot interpret the cosoliton solutions as representing shock waves on a Kasner background.

As we have done in the previous section one might now consider the solution with two opposite poles, i.e. take $s = 2$ and $h = h_1 = -h_2$ in (4.56)–(4.57). This solution is only valid in region III of fig. 4.2, and, due to the singularities on the light cones bounding such a region, it does not seem to lead to any meaningful cosmological model unless one considers that it represents some limit of complex-pole solutions. Such solutions will be the subject of the next section.

4.6 Solutions with complex poles

Complex pole trajectories are obtained from (4.33) when

$$w_k = z_k^0 - ic_k, \quad c_k \neq 0, \tag{4.73}$$

where z_k^0 and c_k are real. The main difference with respect to the real-pole trajectories used in section 4.5 is that these are now defined over all the canonical coordinate patch and the metrics that they induce are regular on the light cones $z_k^2 = t^2$. It is useful to write (4.33) as

$$\mu_k/t = \sqrt{\sigma_k}e^{i\gamma_k}, \tag{4.74}$$

where $\sqrt{\sigma_k}$ is understood as a positive quantity and the explicit forms of the functions $\sigma_k(t, z)$ and $\gamma_k(t, z)$ are

$$\sigma_k^\pm = L_k \pm (L_k^2 - 1)^{1/2}, \quad L_k \equiv (z_k^2 + c_k^2)t^{-2} + [1 - 2(z_k^2 - c_k^2)t^{-2} + (z_k^2 + c_k^2)^2 t^{-4}]^{1/2}, \tag{4.75}$$

$$\gamma_k = \cos^{-1}\left(\frac{2z_k\sqrt{\sigma_k}}{t(1 + \sigma_k)}\right), \tag{4.76}$$

where, again, the minus and plus signs stand for the labels *in* and *out*, respectively, of (1.68)–(1.69). The square roots in (4.75) are also understood to have only positive values. Note that as a consequence of $\mu_k^- = \alpha^2/\mu_k^+$, we have $\sigma_k^+ = (\sigma_k^-)^{-1}$ and that $0 < \sigma_k^- < 1$ and $1 < \sigma_k^+ < \infty$.

Some of the properties of the resulting metrics can be foreseen from the asymptotic values of the pole trajectories, since they are the main ingredient of the solution. For the four asymptotic regions defined in section 4.5, in the canonical coordinate patch, we have the values

$$\sigma_k^- = 1 - \frac{2|c_k|}{t} + O\left(\frac{c_k^2}{t^2}, \frac{c_k z_k^2}{t^3}\right), \quad \text{future time-like infinity,}$$

$$\sigma_k^- = 1 - 2\left[\frac{[c_k^2 + (z_k^0)^2]^{1/2} - \text{sign}(z)z_k^0}{t}\right]^{1/2} + O\left(\frac{z_k^0}{t}, \frac{c_k}{t}\right),$$

future null infinity,

$$\sigma_k^- = \frac{t^2}{4z_k^2}\left[1 + O\left(\frac{t^2}{z_k^2}, \frac{c_k^2}{z_k^2}\right)\right], \quad \text{space-like infinity,}$$

$$\sigma_k^- = \frac{t^2}{4(c_k^2 + z_k^2)}\left[1 + O\left(\frac{t^2}{z_k^2 + c_k^2}\right)\right], \quad \text{initial region.} \tag{4.77}$$

The time evolution of $\sigma_k^-(t, z)$ is represented in fig. 4.3 for $c_k = 0.2$. The slope of the curves between small and large values of z is governed by the parameter c_k: the smaller c_k the steeper the slope. Since the 'soliton-like' waves can be associated with the z-derivatives of σ_k the parameter c_k reflects the 'width' of the solitons. All the metric-dependent equations can be obtained from the pole equations (1.66). In terms of σ_k, these become

$$\sigma_{k,z} = \frac{8z_k \sigma_k^2(1 - \sigma_k)}{H_k(1 + \sigma_k)t^2}, \quad \sigma_{k,t} = \frac{2\sigma_k(1 - \sigma_i^2)}{H_k t}, \quad H_k \equiv (1 - \sigma_k)^2 + \frac{16c_k^2 \sigma_k^2}{(1 - \sigma_k)^2 t^2}. \tag{4.78}$$

After one derivation these equations enable one to find the Riemann tensor for the n-soliton with complex poles in terms of $\sigma_k(t, z)$. It is easy to see, using (4.77), that $\sigma_{k,z}$ has a maximum at null infinity, indicating that the corresponding soliton solutions contain inhomogeneities propagating at the speed of light as $t \to \infty$.

By using the linearity of (4.41) for Φ the soliton solutions (4.43) can be generalized as follows. First, any complex solution induces two real solutions corresponding to its real and imaginary parts. Second, any solution can be multiplied by an arbitrary real parameter. The generalized soliton solutions will essentially be of two classes: the superposition of the real parts of $\ln(\mu_k/t)$ (which correspond to the modulus of the pole trajectories), and the superposition of the imaginary parts of $\ln(\mu_k/t)$ (which correspond to the pole phases). In the first class we have from (4.43),

$$\Phi = d \ln t + \sum_{k=1}^{s} h_k \ln \sqrt{\sigma_k}, \tag{4.79}$$

where d, h_k are real parameters and s, the number of poles, is an integer. The parameter h_k indicates the k-pole degeneracy. The metric coefficient f can be

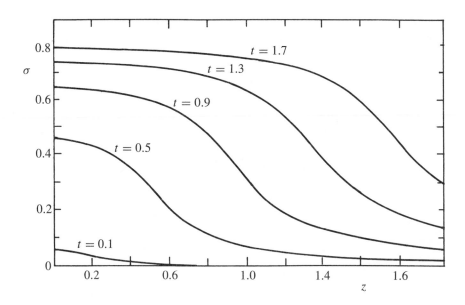

Fig. 4.3. This shows the function $\sigma_k^-(t, z)$ for different values of t, as defined by (4.75). The origin is $z_k^0 = 0$ and the width parameter is $c_k = 0.2$. The function is unchanged if one reverses the sign of z. When σ_k^- approaches 1, the corresponding soliton solution approaches the background metric.

found by either directly integrating (4.42) or, more easily, by taking appropriate limits in the corresponding soliton solution (4.39). The result, up to an arbitrary multiplicative parameter, is

$$f = t^{(d^2-1-g^2)/2} \prod_{k=1}^{s} \sigma_k^{h_k(2h_k+d-g)/2} (1 - \sigma_k)^{-h_k^2/2} H_k^{-h_k^2/4}$$

$$\times \prod_{k,l=1;k>l}^{s} \left\{ \left[(\sigma_k + \sigma_l)t^2 - \frac{8z_k z_l \sigma_k \sigma_l}{(1+\sigma_l)(1+\sigma_k)} \right]^2 - \frac{64c_k^2 c_l^2 \sigma_k^2 \sigma_l^2}{(1-\sigma_k)^2(1-\sigma_l)^2} \right\}^{h_k h_l/2},$$

$$(4.80)$$

where $g \equiv \sum_{k=1}^{s} h_k$, and H_k is given in (4.78). Note that the real-pole solution (4.51)–(4.52) may be obtained from this in the limit $c_k \to 0$.

The second class of generalized soliton solutions can be obtained from the imaginary part of the complex-pole trajectories (4.74), i.e.

$$\Phi = d \ln t + \sum_{k=1}^{s} h_k \gamma_k. \qquad (4.81)$$

The corresponding metric coefficient f can be obtained again by making appropriate use of (4.42). The computation here is more complex than in the previous case, however [113]. The final result, up to an arbitrary multiplicative constant, is

$$f = t^{(d^2+2g-1)/2} \exp\left(d \sum_{k=1}^{s} h_k \gamma_k\right)$$

$$\times \prod_{k=1}^{s} \sigma_k^{h_k(h_k-1)/2} \left[(z_k^2 - c_k^2 - t^2)^2 + 4c_k^2 z_k^2\right]^{h_k(h_k-1)/4}$$

$$\times (1 - \sigma_k)^{-h_k^2/2} H_k^{h_k(2-h_k)/4} t^{-h_k^2}$$

$$\times \prod_{k,l=1;k>l}^{s} (A_{kl}^-/A_{kl}^+)^{h_k h_l/2}, \tag{4.82}$$

where

$$A_{kl}^{\pm} = (\sigma_k + \sigma_l)(1 - \sigma_k^2)(1 - \sigma_l^2)t^2$$
$$- 8\sigma_k \sigma_l \left[z_k z_l(1 - \sigma_k)(1 - \sigma_l) \pm c_k c_l(1 + \sigma_l)\right].$$

Note that the solution (4.56)–(4.57) may be obtained from this in the limit $c_k \to 0$. Solutions (4.81)–(4.82) are also called cosoliton solutions.

These complex-pole solutions are also related to the solution (4.48) of the linear equation (4.41). In fact, each term with index k (which we designate by Φ_k) can be obtained from the integration of (4.48) when $g_k(\nu) = h_k$ and $s_k = z_k^0 - ic_k$ (i.e. s_k is now a complex parameter) which leads to

$$\Phi_k = h_k \ln\left(\frac{\mu_k^+}{t}\right) \equiv h_k \cosh^{-1}\left(\frac{z_k^0 - ic_k - z}{t}\right). \tag{4.83}$$

This solution is now complex, but its real and imaginary parts are the two real solutions (4.79) and (4.81), respectively. This is equivalent to Synge's method of complexification of the real wave solution (4.53), which amounts to making the parameter z_k^0 complex [263, 99, 100].

In the following three subsections we shall study these metrics in detail. As for the case of real poles, we shall consider the case for one and two opposite poles. The general case, as well as solutions obtained by superposition of (4.79) and (4.81), can be understood in terms of these two simple cases. As we shall see both lead to physically relevant cosmological solutions.

4.6.1 Composite universes

These are the generalized soliton solutions corresponding to the real part of $\ln(\mu_1/t)$ for one complex pole, $s = 1$, $h_1 = h$ and $c_1 = w$

$$\Phi = d \ln t + h \ln \sqrt{\sigma_1}. \tag{4.84}$$

The spacetimes they describe have been called composite universes [46, 99, 100], because when $w \to 0$, the wave solution (4.84) goes to the solution with real poles (4.58) in the region outside the light cone $z_1^2 \geq t^2$, whereas it goes to the background metric inside the light cone, i.e.

$$\sigma_1^{\pm} \to [z_1/t \pm (z_1^2/t^2 - 1)^{1/2}]^2, \quad z_1^2 \geq t^2; \quad \sigma_1^{\pm} \to 1, \quad z_1^2 \leq t^2. \qquad (4.85)$$

The first solution corresponds to the Wainwright *et al.* solution discussed in subsection 4.5.1 and the second to the Kasner metric. It would seem that it corresponds to the matching of spacetimes discussed in that subsection. However, this is not the case because when $w \to 0$, the f coefficient diverges inside the light cone as $1/w^2$. This does not invalidate the solution obtained in this limit, because f can always be multiplied by any arbitrary parameter. However, it does invalidate the use of the same canonical variables to cover the whole spacetime.

Therefore we shall now consider in detail the case $w \neq 0$. For this it is better to drop the canonical variables (t, z) and introduce new variables (T, Z) which give a simpler form for σ_1 [286],

$$t = w \cosh(2aZ) \sinh(2aT), \quad z_1 = w \sinh(2aZ) \cosh(2aT), \qquad (4.86)$$

where a is an arbitrary parameter. Note that this coordinate change is of the type that preserves the metric form (1.36). Now $\sigma_1^{\pm} = [\tanh(aT)]^{\pm 2}$ and the metric takes the form

$$\begin{aligned} ds^2 = {} & F(T, Z)(dZ^2 - dT^2) + [\cosh(2aZ) \sinh(2aT)] \\ & \times \{[\cosh(2aZ) \sinh(2aT)]^d [\tanh(aT)]^h dx^2 \\ & + [\cosh(2aZ) \sinh(2aT)]^{-d} [\tanh(aT)]^{-h} dy^2\}, \end{aligned} \qquad (4.87)$$

where

$$\begin{aligned} F(T, Z) \equiv {} & [\cosh^2(2aZ) + \sinh^2(2aT)]^{1-h^2/4} \\ & \times [\cosh(2aZ)]^{(d^2-1)/2} [\sinh(2aT)]^{(d^2+h^2-1)/2} [\tanh(aT)]^{hd}. \end{aligned}$$

Here we have taken $w = 1$, or else w can be absorbed into the variables x, y and the arbitrary constant that can multiply the factor $F(T, Z)$. This solution has been written in a form resembling the Ellis and MacCallum solution (4.62) and it is apparent from comparing these solutions that when $T \ll |Z| \to \infty$, solution (4.87) approaches the Wainwright *et al.* generalization of (4.62). That this is the case is proved by comparing the Riemann tensor components as given in the next paragraph. Note that the coordinates (T, Z) in both solutions defined, respectively, in (4.61) and (4.86) also approach each other in this asymptotic region.

Now it is a simple matter to obtain the curvature tensor for metric (4.87), using the null tetrad (4.11) by changing $(t, z,)$ and f (T, Z) and F, respectively. The corresponding Riemann, or Weyl, scalars (4.12) are

$$\Psi_0 = 2a^2 F^{-1} X^+, \quad \Psi_4 = 2a^2 F^{-1} X^-,$$

$$\left.\begin{array}{l} \Psi_2 = -a^2 F^{-1} \left\{ \dfrac{1}{2}(d^2 - 1)\left[\cosh(2aZ)\right]^{-2} \right. \\[2mm] \left. + \left[hd \cosh(2aT) + \dfrac{1}{2}(d^2 + h^2 - 1)\right][\sinh(2aT)]^{-2} \right\}, \end{array}\right\} \qquad (4.88)$$

where

$$\begin{aligned} X^{\pm} \equiv &-\frac{1}{2}d(d^2 - 1)\left\{1 - \frac{1}{2}[\cosh(2aZ)]^{-2}\right\} \\ &+ h(h^2/4 - 1)\cosh(2aT)[\cosh^2(2aZ) + \sinh^2(2aT)]^{-1} \\ &+ [(d/4)(1 - 3h^2 - d^2) \\ &+ (h/4)(1 - h - 3d^2)\cosh(2aT)][\sinh(2aT)]^{-2} \\ &\pm \sinh(2aZ)[\sinh(2aT)]^{-1} \\ &\times \{h(h^2/4 - 1)\cosh(2aZ)[\cosh^2(2aZ) + \sinh^2(2aT)]^{-1} \\ &+ (1 - d^2)(3h/4 + d/2)\cosh(2aT)[\cosh(2aZ)]^{-1}\}. \qquad (4.89) \end{aligned}$$

For $h = 0$, the solution is, as expected, the homogeneous Kasner solution, in nonstandard coordinates, with a curvature singularity at $T = 0$. Now, for $h \neq 0$ it is easy to see from (4.88) that the metric has in general a curvature singularity at $T = 0$ (cosmological singularity) and space-like curvature singularities at $T \ll |Z| \to \infty$; therefore, its interest as a cosmological model is doubtful. The only exception is the case $h^2 = d^2 + 3$, which has the cosmological singularity only. This is similar to what happens with the Ellis and MacCallum solution (4.62). An important difference, however, is that metric (4.87) is not homogeneous even when the above equality holds, with the exception of when $d = 1, h = 2$ as we see in what follows.

It can be shown that metric (4.87) is inhomogeneous by proving that it admits only a two-dimensional isometry group with the obvious Killing fields ∂_x and ∂_y. The key to the proof [286] is that the Jacobian $\partial(I, J)/\partial(T, Z)$, where I and J are the two curvature scalar invariants (4.13), is different from zero for $d \neq 1$ and $h \neq 2$; so that I and J could be used as coordinates of the metric to furnish an invariant description of the spacetime. Then some results of the 'equivalence problem' [163] ensure that the dimension of the isometry group is 2.

Although the spacetime (4.87) is inhomogeneous, it is related to an interesting family of homogeneous cosmological models. If we take $d = 1$, metric (4.87) is just the vacuum solution corresponding to some LRS Bianchi type III stiff perfect fluid solutions which will be described in section 5.4.2. In fact, corresponding to one of the three inequivalent Abelian two-dimensional subgroups of the LRS Bianchi III symmetry group classified by Jantzen [157], a family of stiff perfect fluid solutions has been given by Kitchingham [174] as

$$
\begin{aligned}
ds^2 \;=\; & [\sinh(2aT)]^2[\tanh(aT)]^h(dZ^2 - dT^2) + [\sinh(2aT)\cosh(2aZ)] \\
& \times\{\sinh(2aT)\cosh(2aZ)[\tanh(aT)]^h dx^2 + [\sinh(2aT)\cosh(2aZ)]^{-1} \\
& \times [\tanh(aT)]^{-h} dy^2\},
\end{aligned} \tag{4.90}
$$

with scalar field $\sigma \equiv (1 - h^2/4)^{1/2} \ln[\tanh(aT)]$ (not to be confused with the pole trajectory in (4.84)). Soliton solutions with massless scalar fields and perfect fluids formed by stiff matter will be considered in section 5.4. Here it suffices to say that metric (4.90) corresponds to a spacetime with a comoving stiff perfect fluid with energy density and pressure $-\sigma_{,\mu}\sigma^{,\mu}$, and four-velocity $u^\alpha = -(-\sigma_{,\mu}\sigma^{,\mu})^{-1/2}\sigma^{,\alpha}$. This is a homogeneous metric with coordinates (T, Z) adapted to spatial homogeneity. The difference between metric (4.90) and metric (4.87) for $d = 1$ lies in the $F(T, Z)$ coefficients; the ratio of those coefficients is

$$
\{1 + \cosh^2(2aZ)[\sinh(2aT)]^{-2}\}^{1-h^2/4}.
$$

This ratio gives a measure of the inhomogeneity of the metric (4.87) with $d = 1$; it shows that this inhomogeneity becomes enhanced at small values of T. Consequently, the inhomogeneous vacuum solution (4.87) becomes homogeneous when a stiff perfect fluid is introduced. Of course, for $h = 2$ the solution (4.90) is a vacuum solution and agrees with metric (4.87) for $d = 1$, $h = 2$. Therefore, the three-dimensional isometry group of metric (4.87) when $d = 1, h = 2$ is the LRS Bianchi III symmetry group and the coordinates (T, Z) are adapted to spatial homogeneity.

More physical insight into the inhomogeneous solution (4.87) can be obtained from the optical scalars (4.18), which give physical information on the behaviour of light propagation in the spacetime. Analysing the expansion, and the shear, of null rays defined by $T = Z + C$ ($-\infty < C < \infty$) one arrives at the conclusion that the inhomogeneities cannot be seen as localized waves propagating on a homogeneous background. The inhomogeneity involves the whole spacetime and is more important near the cosmological singularity. As the spacetime expands it approaches homogeneity.

4.6.2 Cosolitons

These are the solutions related to the imaginary parts of the complex poles as given by (4.81)–(4.82). As for the case of the solutions considered in previous sections, these metrics can be studied by considering solutions with one pole and with two opposite poles. The qualitative features of the general solution (4.81)–(4.82) for an arbitrary number of pole trajectories are given by these two cases.

Cosolitons with one pole. Let us now consider $s = 1$, $h_1 = h$ in (4.81)–(4.82). This solution was first given by Feinstein and Charach [99], who called it a cosoliton. Note that when $c_1 \to 0$, Φ goes to the continuation of the one-pole solution for real poles, equation (4.70), in the region $|z_1| \le t$ and it goes to the Kasner background metric in the region $|z_1| \ge t$.

Let us now analyse the behaviour of this solution in some asymptotic regions. Near the initial singularity ($t \to 0$)

$$\Phi = d \ln t + h\gamma_1, \quad f = 2^{-h(h-1)} t^{(d^2-1)/2} e^{dh\gamma_1} [1 + O(t^2)], \qquad (4.91)$$

and the curvature tensor becomes singular, unless $d^2 = 1$ (Minkowski background). At null infinity ($|z| \simeq t \to \infty$) the solution approaches the Kasner background and the Riemann components become

$$\left.\begin{array}{l} \Psi_0 = -(4c_1^{3/2} f)^{-1} h(1 - h^2/2) t^{-1/2} [1 + O(t^{-1/2})], \\[2mm] \Psi_2 = f^{-1} 0(t^{-3/2}), \quad \Psi_4 = f^{-1} O(t^{-1}), \end{array}\right\} \qquad (4.92)$$

which implies a Petrov type N behaviour for the leading Riemann components.
At time-like infinity,

$$\Phi = d \ln t + h\gamma_1, \quad f = 2^{-h(h-1)} t^{(d^2+h^2-1)/2} e^{dh\gamma_1} [1 + O(t^{-2})],$$
$$\gamma_1 = \pi/2 - (z_1/t)[1 + O(t^{-2})],$$

and the Riemann components behave as $\Psi \sim f^{-1} O(t^{-2})$. The metric does not tend to the Kasner background in this region but to an inhomogenous spacetime. At space-like infinity it does go to the Kasner background and the Riemann components approach the Riemann components of the Kasner metric.

Therefore the cosoliton metrics with one pole represent inhomogeneous cosmological models: they have the cosmological singularity only, unless $d^2 = 1$, and are regular in the whole range of canonical variables. They describe a spacetime that starts almost Kasner-like with a localized inhomogeneity which then grows at the speed of light leaving an inhomogenous spacetime. Thus, as was the case for the 'composite universes' of subsection 4.6.1, the cosolitons do not describe localized inhomogeneities propagating on Kasner backgrounds.

Cosolitons with two opposite poles. Let us now turn to the cosolitons with two opposite poles, i.e. solutions (4.81)–(4.82) with $s = 2$, and $h_1 = -h_2 = h$. Equation (4.81) and the asymptotic properties of γ_k, as given by (4.76)–(4.77), ensure that the poles give localized perturbations only. Here one can describe the collisions of such perturbations (or solitons) as in section 4.6.3, where such a collision is studied in some detail. Let us perform an asymptotic analysis of such solutions.

Near the initial singularity $(t \to 0)$ we have

$$\Phi = d \ln t + h(\gamma_1 - \gamma_2), \quad f = 4^{(1-h)/h} t^{(d^2-1)} \exp[hd(\gamma_1 - \gamma_2)][1 + O(t^2)],$$
$$\gamma_k = \sin^{-1}[c_k(c_k^2 + z_k^2)^{-1/2}] + O(t^2),$$

where $k = 1, 2$. From the Riemann components one sees that the metric becomes singular unless $d^2 = 1$ (Minkowski background).

At null infinity the metric coefficients are determined by

$$\gamma_k = \pi - (c_k)^{1/2} t^{-1/2} [1 + O(t^{-1/2})], \quad k = 1, 2,$$

and approach the Kasner values, but the curvature tensor becomes

$$\left.\begin{aligned}
\Psi_0 &= (4f)^{-1} h(c_1 c_2)^{-3/2} \left[(r_2^{3/2} - r_1^{3/2}) - \frac{h^2}{2}(r_2^{1/2} - r_1^{1/2})^3 \right] \\
&\quad \times t^{-1/2}[1 + O(t^{-1/2})], \\
\Psi_2 &= f^{-1} O(t^{-3/2}), \quad \Psi_4 = f^{-1} O(t^{-1}).
\end{aligned}\right\} \tag{4.93}$$

Thus the leading terms behave as a Petrov type N metric, i.e. pure radiation (cf. (4.92)). At space-like and time-like infinities the metric coefficients and the Riemann components go to the corresponding Kasner values at a rate $O(|z|^{-1})$ or $O(t^{-1})$, respectively. Therefore, these solutions can be interpreted as cosmological models representing an inhomogeneous localized perturbation which evolves towards gravitational radiation on an expanding Kasner background. As in the models described in section 4.6.3, these are examples of the generation of cosmological gravitational radiation from purely classical irregularities in the initial structure of the spacetime.

4.6.3 Soliton collision

From (4.77) we see that all σ_k^- approach unity at future time-like infinity and this means that the soliton solutions approach the background metric there, but since σ_k^- approach zero at space-like infinity the soliton solutions depart from the background metric in general. That was the case, for instance, for the solutions with one pole of section 4.6.1 or the more general case of superpositions of poles with the same sign. However, if we superpose opposite poles, i.e. $h_k = -h_l$, we may get solutions which approach the background metric at space-like infinity also, since $\sigma_k^- \sigma_l^+ \to 1$ there. Note also that all σ_k^- approach unity at null infinity but at a lower rate than in other regions. This allows us to describe localized perturbations on a Kasner background; moreover such perturbations decrease as $O(t^{-1/2})$ at null infinity which is typical of gravitational waves. Another interesting feature of these solutions is the following. Since two poles will in general define two different light cones (if $z_1^0 \neq z_2^0$), we have the possibility of describing the collision of two such perturbations where the light cones intersect, see fig. 4.2.

In this subsection we shall review the properties of solutions with opposite poles. All such solutions have the cosmological singularity only and are interesting as inhomogeneous cosmological models. They represent highly inhomogeneous cosmologies which evolve towards homogeneous Bianchi I models with gravitational radiation. They are an example of the creation of a background of gravitational radiation as a consequence of initial classical inhomogeneities in line with the work of Adams *et al.* [2, 1] or the 'dissipation' of an initial inhomogeneous cosmological singularity in the sense of refs [308, 309]. Throughout we shall use 'solitons' to refer to the (soliton-like) perturbations of these models, but we shall discuss the relation with the solitons of nonlinear physics at the end of this subsection.

Thus we shall now consider solutions (4.79)–(4.80) with $s = 2$ and $h_1 = -h_2 = h$. The pole degeneracy $|h|$ is sometimes called the 'intensity' of the solitons but its specific value does not play an important role here. The case $h = 1$ (i.e. the true soliton solution which may be considered as the limit of a nondiagonal metric) was studied in ref. [150]. The case $h_1 = -h_2 = \cdots = h_{s-1} = -h_s = 1$ has also been studied [151], but the main features are present in the two-pole case, $s = 2$.

It is worth noticing that the solution (4.79)–(4.80) with two opposite poles (with $w \equiv c_1 \equiv c_2$) can be written in terms of the Fourier Bessel integral (4.44) [97] as

$$
\left.
\begin{aligned}
\Phi &= d \ln t + \int_0^\infty A_k \sin\left[k\left(z - \frac{z_1^0 + z_2^0}{2}\right)\right] J_0(kt)dk, \\
A_k &= 2h\frac{e^{-wk}}{k} \sin\left[\frac{k(z_2^0 - z_1^0)}{2}\right],
\end{aligned}
\right\}
\tag{4.94}
$$

which corresponds to the nonchaotic part of the solution, with $B_k = 0$. This leads to the interesting interpretation of (4.94) as the superposition of travelling waves, (4.46), at late times. The superposition amplitude A_k reduces to (4.65) in the real-pole limit $w \to 0$.

First, we should note that whereas in all asymptotic regions the transversal metric coefficients tend to the Kasner values, see (4.79), this is true for the longitudinal coefficient f only at space-like infinity. In fact, let us call $f_0 = t^{(d^2-1)/2}$ (the Kasner value), then at null and time-like infinities f will go to f_0 multiplied by a factor that depends on the soliton parameters c_k. For instance, for solutions for which $c_1^2 \sim c_2^2 \sim w^2 \ll 1$, $f \to f_0 w^{-4}$ at time-like infinity and $f \to f_0 w^{-4}$ at null infinity along the light cones of the inner solitons (those which have collided, see fig. 4.2) but $f \to f_0 w^{-2}$ along the light cones of the outer solitons. It appears that each collision increases the value of the coefficient f relative to the Kasner background value: after each collision

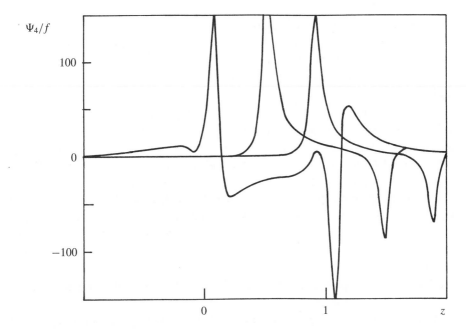

Fig. 4.4. Time evolution of the Riemann tensor component Ψ_4 (divided by f in order to make the solitons which collide apparent, i.e. in the boosted tetrad) giving the gravitational strength of the right moving solitons. The widths and origins of the solitons are $c_1 = c_2 = 0.05$ and $z_1^0 = 0$, $z_2^0 = 1$. The Kasner background has $\Psi_4 = 0$, for $d = 0$. The three curves from left to right according to the positions of their peaks represent the time sequence $t = 0.1$, $t = 0.5$ and $t = 0.9$.

the background suffers an expansion along the direction of propagation of the solitons. In general [151] one finds that along a soliton which undergoes n collisions, $f \to f_0 w^{-2(n+1)}$, and that at time-like infinity, $f \to f_0 w^{-2s}$. The parameters c_k also control the widths of the soliton-like perturbations.

The localized structure of the solitons and the intrinsic properties of the metric can be seen from the Riemann tensor. Thus we choose the tetrad (4.11) and make use of (4.78) as recursion relations to obtain the Riemann components (4.12) in terms of σ_1 and σ_2 only. The results are rather involved and it is best to display a graphical representation of the coefficients. In fig. 4.4 the radiative part of the gravitational field Ψ_4 in the boosted tetrad (4.14) with $A = \sqrt{f}$ (i.e. $A\vec{n}, \vec{m}$ and $\vec{m}^*$ are parallel transported along the null congruence $A^{-1}\vec{l}$) is represented for $d = 0$ (axisymmetric Kasner). This field induces time varying tidal forces to test particles on the plane orthogonal to the propagation of the solitons moving to the right. From this, one can see the localized structure (soliton-like) of the perturbations and the nondispersive nature of the wave. The amplitude of Ψ_4 decreases as $t^{-1/2}$ (typical of gravitational waves) but this is due to the background expansion and does not imply dispersion. Initially the solitons show tails and a complicated structure, but as time increases the tails disappear. The radiative part of the gravitational field Ψ_0 has a similar behaviour for the solitons moving to the left, and the Coulomb component Ψ_2 is stronger near the soliton origins: $z_1 = 0, z_2 = 1$.

The asymptotic values of the Riemann components at null infinity, when calculated analytically, show that, for $z > 0$, $\Psi_0/\Psi_4 \to 0$ and $\Psi_2/\Psi_4 \to 0$ so that the metric in its leading terms behaves as a Petrov type N spacetime: pure gravitational radiation travelling to the right. The metric is, however, Petrov type I. At null infinity for $z < 0$, the dominant term is Ψ_0.

Frame-independent properties can also be seen from the scalar invariants (4.13), I and J. In fig. 4.5 the time evolution of the ratio I/I_0, where I_0 is the Kasner value, is shown for the same model. Before the collision four solitons with small tails are seen, after the collisions the outer solitons are clear but the intensities of the inner solitons are very small due to the w^{-4} factor in the longitudinal expansion. The same factor makes the value of I at time-like infinity very small too.

More physical information on the collision process may be obtained by investigating the focusing effect on null congruences resulting from the collision of solitons. Thus we shall consider the expansion, θ, and shear, σ, defined in (4.18) produced on the null congruence with tangent $\vec{k}$ defined by $\vec{k} = (1/\sqrt{f})\vec{n}$ in section 4.3.1. These quantities give a measure of the energy content of the solitons. The Kasner background produces homogeneous expansion, θ_0, and shear, σ_0, on the above null congruence as a consequence of its overall expansion on the (x, y)-plane. The solitons will produce inhomogeneities on θ and σ and, as a consequence of their energy, they will focus the null rays. The ratio of the expansion θ/θ_0 is shown in fig. 4.6. Initially the focusing produced by the four solitons is clear. By the Raychaudhuri equation [299] this can be interpreted as due to the gravitational 'energy' of the solitons, which may be measured by the θ^2 and σ^2 they produce. After the collisions θ becomes smaller as a consequence of the factor w^{-4} in the f coefficient but is still positive, so that the overall expansion prevents the formation of singularities in the future.

The results for the shear show that it suffers a jump as the congruence crosses a soliton; the left-directed solitons appear more clearly than the right-directed ones (note that the congruence is right-directed). It behaves in a way similar to Ψ_0 [151] and after the passage of the solitons it slowly approaches the Kasner shear. This behaviour is qualitatively very similar to the collisions of plane gravitational waves in an expanding cosmology [52].

Another interesting quantity which one may study is the velocity field (4.23) associated with the Bel–Robinson tensor (4.19). Initially as $t \to 0$, the velocity field satisfies $|v| \sim 0$ almost everywhere. At null infinity $|v| \to 1 - 0(1/t)$ and

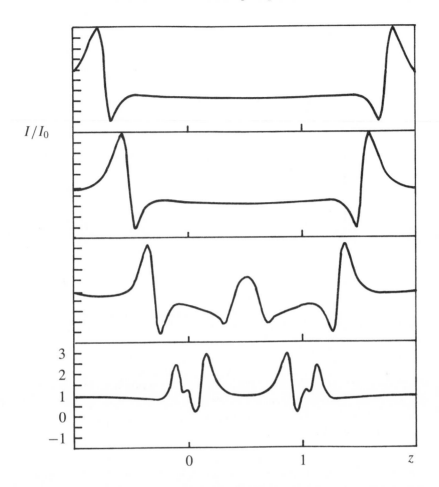

I/I_0

3
2
1
0
−1

0   1 z

Fig. 4.5.   Time evolution of I/I_0 for $d = 0$. The curves are represented against the propagation axis (z axis). The width and origins of the solitons are as in fig. 4.4. The different curves from bottom to top of figure represent the time sequence $t = 0.1$, 0.3 (before collision), 0.5 (collision time, t_c), and 0.7 (after collision).

at space-like and time-like infinities $|v| \sim 0$. This may be interpreted in terms of fluxes of (super)energy. Recall from section 4.3.3 that an observer moving at such a velocity measures no flux. In fig. 4.7 the velocity field of the metric for $d = 0$ is shown. Initially one can see the effect of the four solitons as four localized regions (with some tails) with large velocity fields, two indicating fluxes to the right and two to the left. After the collision the localized regions broaden and we have two main localized regions to the right (with a front tail) both with positive fluxes (to the left we have negative fluxes). The region in between has almost no velocity flux. This means that at large times an observer at rest near the origin measures no energy flux, whereas an observer near the

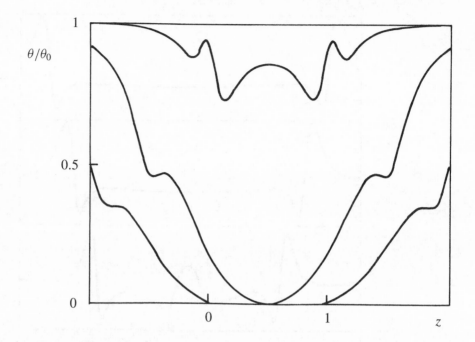

Fig. 4.6.   Time evolution of the expansion ratio θ/θ_0 produced by the solitons on the null geodesics generated by the null vector $\vec{k}$. Same model as in fig. 4.4 but with soliton widths $c_1 = c_2 = 0.1$. The three curves from top to bottom of figure represent the time sequence $t = 0.1$, $t_c = 0.5$ and $t = 0.9$.

solitons needs to go to a speed close to the speed of light to measure no energy flux, except at two points where the observer's speed should be very small. An interpretation of this is that the solitons represent the *resulting net flux of energy*. This interpretation is consistent with solitons in the cylindrical case where a C-energy may be defined and a similar analysis in terms of this energy can be performed [109, 113], see section 6.2.3. In this way we do not need to interpret the solitons as travelling waves, the soliton shapes just reflect the interfering effect of the two competing fluxes. We will now see by a perturbative analysis that, in fact, the maximum of the soliton perturbation moves at a speed greater than light, and the same is true in the cylindrical case, so that to consider the gravitational solitons as travelling waves may not be correct.

Perturbative analysis. One may gain a better understanding of some gravi-soliton features such as shape and motion by performing a perturbative analysis. For this we again consider (4.79)–(4.80) with the restrictions $s = 2$, $h_1 = -h_2 = h$, but now we will take $z_1^0 = z_2^0$, so that instead of the two light cones we had before we have only one and, therefore, we have only one soliton moving

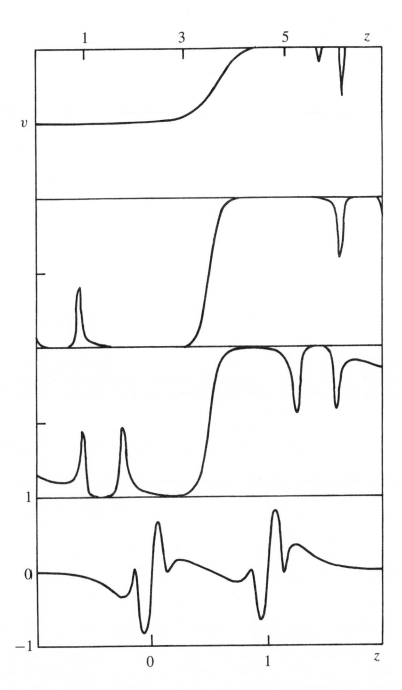

Fig. 4.7. Time evolution of the soliton velocity field v. The parameters and time sequence are the same as in fig. 4.4 but the soliton widths are $c_1 = c_2 = 0.08$. The top curve shows the velocity field at $t = 5.0$.

to the right and another to the left, i.e. there is no soliton collision in this case. Define $\delta w \equiv c_2 - c_1$, and call $w \equiv c_1$, $\sigma \equiv \sigma_2$. The soliton perturbation, or pulse, in (4.79) is proportional to $\ln(\sigma_1/\sigma_2)$ and we may expand this to first order in δw. The ratio σ_1/σ_2 differs from unity only on a small localized region and we have

$$\sigma_1/\sigma_2 \simeq 1 + \delta w \, \partial_w \ln \sigma. \tag{4.95}$$

To find the properties of this function in the (t, z)-plane analytically we first change to the new coordinates (T, Z) defined in (4.86) (taking the parameter $a = 1$), in terms of which σ takes the very simple form $\sigma = \tanh^2 T$. Let us now derive the trajectory of the maximum of $\partial_w \ln \sigma$ along the z-direction: $\partial_z \partial_w \ln \sigma = 0$. This is easily done by computing $\partial_z T$ and $\partial_z Z$ from the coordinate change. In terms of the (T, Z) coordinates this trajectory is

$$\cosh^4(2T) - 3\cosh^2(2Z)\cosh^2(2T) + \cosh^2(2Z) - 1 = 0. \tag{4.96}$$

It is not difficult to find the trajectory in the (t, z)-plane using $\cosh^2(2T)$ and $\cosh^2(2Z)$ as intermediate coordinates. The result is

$$(z^2 - t^2 + w^2)\left[(z^2 - t^2 + w^2) - \sqrt{(z^2 - t^2 + w^2)^2 + 4w^2t^2}\right] = 4w^2(z^2 + w^2). \tag{4.97}$$

This trajectory has $z = 0$ at $t = \sqrt{2}w$ and approaches, for z positive, the asymptote $z = t - w/\sqrt{3}$ as $t \to \infty$ (we have a symmetrical situation for z negative). Therefore the maximum of the soliton perturbation is space-like and approaches the speed of light when $t \to \infty$. Note that this would be the trajectory of the maximum according to cosmological comoving observers. Thus we cannot interpret the soliton perturbation (or more precisely its maximum) as a travelling pulse.

On the other hand, one may also compute the trajectory of the maximum of the perturbation along the t-direction, i.e. $\partial_t \partial_w \ln \sigma = 0$. In this case one needs $\partial_t T$ and $\partial_t Z$ and finds

$$3\sinh^2(2Z) - \cosh^2(2T) = 0, \tag{4.98}$$

which in (t, z) coordinates are

$$t^2 = z^2 + \frac{2w}{\sqrt{3}}z - w^2. \tag{4.99}$$

This trajectory starts at $t = 0$ at the point $z = w/\sqrt{3}$ and approaches the same asymptote $z = t - w/\sqrt{3}$ for z positive as $t \to \infty$; it is thus a time-like trajectory. The gravisoliton perturbation is localized around the two trajectories (4.97) and (4.99); in particular at large t it approaches the speed of light. This is in qualitative agreement with the previous numerical analysis. In the cylindrically symmetric context, where an unambiguous definition of energy can be given (the C-energy), we saw in section 6.2.3 that the flux of energy just vanishes along these two previous trajectories which have a different character in such a context: namely the first one is time-like and the second one is space-like. The maximum of the energy flux is found between these trajectories. This is also compatible with our results on the velocity field introduced above based on the superenergy flux and reinforces our interpretation that the soliton perturbations can be understood as a superposition effect produced by the interference of energy fluxes propagating in opposite directions.

To summarize, the above results suggest interpreting the gravitational solitons (gravisolitons) as localized perturbations of the gravitational field on an expanding Kasner background. They start with an important Coulomb component and may be interpreted as the superposition effect produced by the interference of (super)energy fluxes propagating in opposite directions. The Coulomb field they create gives them some extended particle-like character. As they evolve the Coulomb component of their gravitational fields becomes small and the radiative components dominate, indicating the presence of gravitational radiation. The speed of their maximum is greater than the speed of light indicating that they should not be interpreted as travelling waves. Although we have considered only vacuum solutions a similar conclusion can be reached for soliton solutions for a stiff fluid [67].

Comparing with solitons in nonlinear physics. Some further comments relating the gravitational solitons generated by the ISM and the classical solitons which are often found in nonlinear physics seem in order. Note that classical solitons appear only in a very special class of nonlinear equations, for instance, soliton waves propagating in a nondissipative medium are the result of a balance between nonlinear effects and wave dispersion [185]. Generally classical solitons are characterized by their localizability and shape invariance, by a peculiar behaviour under collisions and by carrying energy with some associated velocity of propagation [259, 245]. As we have seen, gravitational solitons share some properties with classical solitons, although in a strict sense they cannot be considered true solitons. This is because, whereas all soliton solutions of nonlinear physics can be obtained by the ISM (see, for instance, Scott *et al.* [259] for a concise review), the ISM described in chapter 1 is not a standard ISM. As remarked in chapter 1, the poles are not fixed numbers, but pole trajectories.

On more physical grounds, we have difficulties due to the curved nature of the spacetime in identifying or defining some of the features of the gravitational solitons such as amplitude, velocity of propagation, energy and even shape. For instance, it is clear that the amplitude of the gravitational solitons decreases in time as $t^{-1/2}$, but this is a consequence of the background expansion and may not imply dispersion. Also, it has not been possible to define a velocity or an energy for the gravitational solitons that is clearly distinguished from the background, but it is clear from our previous discussion that they carry energy in some sense, although not necessarily in a particle-like way. It is also clear that when two gravitational solitons collide the shape of the solitons does not seem to change; this is an important feature of classical solitons. An attempt to characterize the soliton energy has been made by Feinstein [97], who using the Fourier Bessel decomposition (4.94) and its asymptotic behaviour at large t, (4.46), has defined an effective energy for the gravitational solitons in terms of the superposition amplitudes A_k.

A point that deserves some attention is connected with the time shift in the trajectories of colliding solitons. Some classical solitons, for instance the soliton solutions of the Korteweg–de Vries equation, suffer such time shifts under collisions. The shift is related to the soliton widths and it is interpreted as due to the nonlinear interaction of the solitons during the collision. Boyd *et al.* [40] have emphasized that the character of the gravitational solitons discussed in this subsection is not classically solitonic on the basis of the linear equation (4.41) satisfied by the field Φ, which is the main ingredient for the construction of the metric coefficients. This is obviously true although nonlinear aspects of gravity appear as a consequence of the nonlinear relation between Φ and f given by (4.42). In fact, we have seen that as a result of each collision f is 'shifted' by factors of w^{-2} and Dagotto *et al.* [71, 73, 121] have emphasized that the gravitational solitons and other solutions related to the linear equation (4.41) do suffer a time shift when the soliton peaks are compared with test null rays propagating near the solitons. This effect is clearly nonlinear and a consequence of the fact that the spacetime is described by both Φ and f.

However, the closest analogue of the time shift in the collision of Korteweg–de Vries, or sine-Gordon, solitons must be sought in nondiagonal metrics. In the nondiagonal gravitational field there is a nonlinear coupling of the two modes of polarization. This may imply conversion of one mode into the other and vice versa. This is analysed in section 5.3 in the collision of a pulse wave and a gravisoliton, where a small time shift is detected, and in section 6.3 where this nonlinear interaction is interpreted as the analogue of the electromagnetic Faraday rotation. This does not invalidate the use of diagonal metrics to see some of the gravisoliton features. In fact, the main ingredients of Φ are the pole trajectories $\sigma_k(t, z)$, which are also the main ingredients for the coefficients on a nondiagonal soliton solution. Since the pole trajectories are the main ingredients to describe the shape and the propagation of the solitons, the diagonal metrics are very useful for studying this. Disentangling the soliton interaction from the background geometry and looking for physically meaningful quantities is not, however, a simple matter. We believe that the approach developed in section 2.3, see also ref. [14], also gives some insight into this problem.

5

Cosmology: nondiagonal metrics and perturbed FLRW

In this chapter we continue describing soliton solutions in cosmological models but now we concentrate on nondiagonal metrics and on backgrounds other than Kasner.

In section 5.1 soliton solutions with two polarizations are discussed. Although in this case explicit expressions for the metric coefficients cannot be displayed in general, a fairly complete understanding of these metrics and their relevance as cosmological models is possible. As in chapter 4, the metrics are also classified in terms of real and complex poles. In section 5.2 soliton solutions obtained from anisotropic Bianchi type II metrics are considered. In section 5.3 a solution describing the nonlinear interaction between a gravitational pulse wave and soliton-like waves is described. A polarization angle and wave amplitude are defined and used to characterize the interaction. As a consequence of the nonlinear interaction of the waves a time shift in the pulse-wave trajectory is observed. Finally, in section 5.4 we discuss soliton solutions which describe finite cylindrical perturbations on FLRW isotropic cosmological models. Models representing perturbations on the late time behaviour of low density open FLRW are derived and studied. Soliton solutions when a massless scalar field is coupled to the gravitational field, and their interpretation either as perfect fluids of stiff matter or as anisotropic fluids are described, together with some solutions representing perturbations on an FLRW model with stiff matter. Perturbations on more realistic radiative FLRW are also discussed, as well as related solutions of the Brans–Dicke theory. Some of these solutions are obtained by using soliton solutions in five-dimensional gravity.

5.1 Nondiagonal metrics

We turn now to the nonlinear case, i.e. to the solutions obtained by the ISM when (1.39), or (4.25), is truly nonlinear. Such solutions are simply obtained from a diagonal background when the arbitrary parameters $m_{0b}^{(k)}$ of (1.80), or (4.34), are different from zero. The soliton metric now has two polarizations and following Adams *et al.* [2] one may identify from its coefficients a wave amplitude and the two independent wave polarizations. Unfortunately, one can no longer give explicit expressions for the metric coefficients for an arbitrary number of pole trajectories. This is largely due to the problem of inverting the $n \times n$ matrix Γ_{kl}, defined in (1.83), which is required in finding the components (1.87), although some convenient expressions can be obtained for diagonal backgrounds [191]. However, information can still be obtained from asymptotic expressions and some limits in which the solutions lose one polarization, thus becoming diagonal metrics. This will also provide a connection with the diagonal metrics discussed in the previous chapter. The intrinsic properties of these metrics are more difficult to study than for diagonal metrics because the Riemann components are not so easily computed. Moreover, the nonlinear coupling between the modes of polarization is difficult to characterize in a precise way. We shall first consider the case of real-pole trajectories and then solutions with complex poles.

5.1.1 Solutions with real poles

From the point of view of exact cosmological models these solutions are not always of interest because they are only defined in a limited region of the canonical coordinate patch and the solution may be singular on the light cones $z_k^2 = t^2$. If they are not singular there one may try to match them to the background metrics as we did for diagonal metrics. Sometimes it is useful to think of those with fused double poles as the limiting case of complex-pole solutions which are defined on the whole canonical coordinate patch (but see the first paragraph of section 4.6.1). Note also that some of the most interesting solutions with real poles in the diagonal case were generalized soliton solutions. Such generalizations that were possible due to the essentially linear nature of the solutions, see (4.41)–(4.42), are not possible here. This means that the generalized soliton solutions generally have no counterpart in the nondiagonal case. For these reasons nondiagonal real-pole solutions have not been much studied in the cosmological context.

One-soliton solution. With real poles the nondiagonal solution on the Kasner background for one pole, $n = 1$, was obtained in ref. [23]:

$$ds^2 = Bt^{d^2/2} \cosh(dr/2 + C)(z^2 - t^2)^{-1/2}(dz^2 - dt^2) + [\cosh(dr/2 + C)]^{-1}$$
$$\times \{t^{1+d} \cosh[(1+d)r/2 + C]dx^2 + t^{1-d} \cosh[(1-d)r/2 + C]dy^2$$
$$- 2t \sinh(r/2)dxdy\}, \tag{5.1}$$

where $r = 2\ln(\mu_1/t)$ and B and C are arbitrary parameters. It is defined for $z_1^2 \geq t^2$. This solution follows from (2.18)–(2.19) when ρ_1 is given by (2.27) with $C_1 = C$; one also needs the identity (4.55).

This solution has the cosmological singularity at $t \to 0$ and is singular at $|z| \to \infty$ too. If one considers the metric component g_{11} and determines the position of its extremum, with respect to the space-like variable z for various fixed instants of time t, it can be seen that the world line of the extremum has the equation $z = t \cosh(r_0/2)$, $z \geq t$, where r_0 is a constant. So this maximum cannot be considered to represent the propagation of a physical effect; this solution simply describes the evolution of some initial condition. Taking the limit $C \to \infty$ (assuming $B \sim \exp(-C)$), the metric (5.1) becomes the diagonal one-soliton solution (4.58)–(4.59) with $h = 1$ which is singular at $|z| \to \infty$ but not on the light cone $z_1^2 = t^2$.

When $d = 1$, i.e. the Minkowski background, this solution has been studied in detail by Gleiser [118] for both $B > 0$ and $B < 0$. He showed that the solution can be extended to the region $z_1^2 \leq t^2$ by using appropriate new coordinates (T, Z) related to the canonical ones by $z_1 = (T^2 + Z^2)/2$, $t = ZT$. He also considered the matching of (5.1) with flat spacetime (the background metric) through the null hypersurfaces $z_1^2 = t^2$ and showed that there are three possible ways in which this matching is possible. These correspond to the three different ways in which the Minkowski spacetime may be written in the form (4.2). Two such matched solutions are seen to contain a null fluid on the light cones, i.e. they have distributional valued curvature tensors. One of these solutions is interpreted as a cylindrically symmetric spacetime. The third solution represents a pure gravitational shock front (no fluid) and can be thought of as describing a singular straight line parallel in Minkowski space to the Z axis moving at the speed of light along such axis. The shock front has the shape of an infinite cylinder one of whose generatrices is the singular line. Therefore these solutions with $d = 1$ in general have no cosmological interpretation as one might expect from the fact that the background is not cosmological (Minkowski background).

Explicit expressions for the metric coefficients for two-pole trajectories ($n = 2$) are given by Economou and Tsoubelis [88], but are not discussed in the cosmological context.

5.1.2 Solutions with complex poles

Two-soliton solution. For complex-pole trajectories the soliton solutions are defined on the whole canonical coordinate patch. We have seen in section 1.4 that complex poles always go in pairs in the nondiagonal case, because the metric must be real. Therefore the simplest nondiagonal metric with complex poles from the Kasner background is obtained from (1.87), (1.100), (1.110) and (1.111) with $n = 2$ and $\mu_2 = \bar{\mu}_1$. This leads to the formulas (2.20)–(2.25), adapted to the Kasner background: $\alpha = t$, $u_0 = d \ln t$ and ρ_1 and ρ_2 defined in (2.27), with $\rho_2 = \bar{\rho}_1$. The final exact form of the metric can be represented, in slightly different notation [15, 51], as follows (note that now due to the constraint $w_2 = \bar{w}_1$ there is only one-soliton origin, see (4.73), $z_1^0 = z_2^0$ and

$c_1 = -c_2 = w$):

$$g_{11} = \frac{t^{1+d}}{E}[(\sigma + \sigma^{-1} - 2)\sin^2(\gamma + \psi) \\ + (L_0^2\sigma^{-(1+d)} + L_0^{-2}\sigma^{(1+d)} + 2)\sin^2\gamma],$$

$$g_{22} = \frac{t^{1-d}}{E}[(\sigma + \sigma^{-1} - 2)\sin^2(\gamma - \psi) \\ + (L_0^2\sigma^{(1-d)} + L_0^{-2}\sigma^{-(1-d)} + 2)\sin^2\gamma], \qquad (5.2)$$

$$g_{12} = \frac{2w}{E}\{L_0\sigma^{-(1+d)}[\sin(\gamma - \psi) + \sigma\sin(\gamma + \psi)] + L_0^{-1}\sigma^{-(1+d)/2} \\ \times [\sin(\gamma + \psi) + \sigma\sin(\gamma - \psi)]\},$$

$$f = Ct^{(d^2-5)/2}\sigma^2 E H^{-1}(1 - \sigma)^{-2}(\sin\gamma)^{-2},$$

where we use the usual notation of section 4.6 (although we drop the index 1 in $\sigma_1(t, z)$ and $\gamma_1(t, z)$), and

$$E \equiv (\sigma + \sigma^{-1} - 2)\sin^2\psi + (L_0^2\sigma^{-d} + L_0^2\sigma^d + 2)\sin^2\gamma, \quad \psi \equiv d\gamma + \psi_0, \quad (5.3)$$

C, L_0 and ψ_0 are arbitrary real constants, the last two are related to C_1 of (2.27) by $\mathrm{Re}\, C_1 = -\ln L_0$ and $\mathrm{Im}\, C_1 = \psi_0$.

We can now determine the connection with the diagonal (one-polarization) two-soliton solution. By taking the limits $L_0 \to 0$, $CL_0^2 \to C$ (finite), the metric (5.2) becomes $g = g_d[1 + O(L_0^2)]$ and $f = f_d[1 + O(L_0^2)]$, where (f_d, g_d) is given by (4.84) with $h = 2$ and with the corresponding f coefficient of (4.80). Thus the solution (5.2) can be regarded as a generalization of the diagonal metric (4.84) and the parameter L_0 can be interpreted as a 'polarization' parameter.

All the diagonal metrics, whatever the Kasner background, have the common feature that they evolve towards the background at the asymptotic time-like and null infinity regions. The same holds for the nondiagonal two-soliton solution. Taking the limits (4.77), we have

$$g = g_0[1 + O(t^{-1})], \quad f = C\frac{(L_0 + L_0^{-1})^2}{16w^2}f_0[1 + O(t^{-1})], \qquad (5.4)$$

at time-like infinity, where (g_0, f_0) stands for the Kasner background, and

$$g = g_0[1 + O(t^{-1/2})], \quad f = C\frac{[4\sin^2\psi_0 + (L_0 + L_0^{-1})^2]}{32w^2}f_0[1 + O(t^{-1/2})], \qquad (5.5)$$

at null infinity. One should not deduce from (5.5) that the Riemann tensor behaves in the same way at null infinity as in the background solution, because the z dependence in (5.5) has been hidden in the assumption $|z| \sim t$.

We can now study the asymptotic behaviour at space-like infinity. In that limit the behaviour depends crucially on the background metric. For $1 > d \geq 0$ the metric becomes the background metric

$$1 > d \geq 0, \quad g \rightarrow g_0, \qquad \text{space-like infinity.} \tag{5.6}$$

Thus, when the background is contracting in the z-direction the two-soliton solution can be interpreted as two localized perturbations (gravisolitons) along the z axis. Since the solitons have their maximum amplitude on the light cone for large t, they 'move' in opposite directions with a speed asymptotically approaching the speed of light. Again as in subsection 4.6.3 we do not need to interpret them as travelling waves but rather as a resulting net flux of energy. The main difference with the gravisolitons in diagonal metrics is that these waves have two polarizations.

For $d > 3$ all metrics become the diagonal solution (4.84),

$$d > 3, \quad g \rightarrow g_d, \qquad \text{space-like infinity.} \tag{5.7}$$

This means in agreement with subsection 4.6.1 that these metrics have curvature singularities at space-like infinity and therefore their interest as cosmological models is doubtful.

For $3 \geq d \geq 1$ the asymptotic behaviour at space-like infinity is more complicated. See ref. [51] for details.

In fig. 5.1 the evolution of the transversal component g_{yy} of metric (5.2) is shown when the background metric is axisymmetric Kasner, $d = 0$. We take the width of the soliton to be relatively small ($w = 0.01$) and the parameters in (5.2)–(5.3) to be $L_0 = 1$ and $\cos \psi_0 = (1.01)^{-1/2}$. In the representation in fig. 5.1, the x and y axes have been rotated through $\pi/4$. It is clear that the two-soliton solution tends to Kasner in the causal and far regions if the propagation axis is contracting. However, we should note that, if we take $C \equiv w^2 / \sin^2 \psi_0$, the f coefficient in the far region becomes the corresponding f for the Kasner background (4.3) whereas, in the causal and light-cone regions, it will include a different constant (see (5.4)–(5.5)). Thus, in those asymptotic regions, the existence of solitons modifies the 'longitudinal expansion' with respect to the Kasner background. This behaviour was also seen in the diagonal metric. The solution just described has only the cosmological singularity and it may have cosmological interest as an inhomogeneous model.

n-soliton solution. Although in general explicit expressions cannot be given in this case, we shall now discuss the general n-soliton solutions with complex poles (n even) in the asymptotic regions. We will find that, for some backgrounds they share many of the asymptotic properties of the two-soliton and diagonal n-soliton solutions. In particular, as in all the solutions considered so far (except the cosoliton solutions), the n-soliton solutions evolve towards the

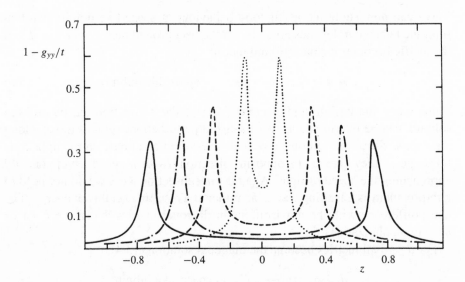

Fig. 5.1.   This shows the time evolution of $g_{yy}(t, z)$ for the two-soliton solution (5.2) generated from the axisymmetric Kasner background ($d = 0$). The Kasner background has been subtracted and 'normalized' by the factor t; we have also made a $\pi/4$ rotation of the x and y axes. The soliton origin is $z = 0$ and its width is $w = 0.01$. The dotted line is $t = 0.1$, the dashed line $t = 0.3$, the dot-dashed line $t = 0.45$ and the continuous line $t = 0.7$.

Kasner background, with the perturbation decreasing as t^{-1} at time-like infinity. This is a consequence of the value of σ_k in that region as given by (4.77) and is also true for any other background, i.e. the n-soliton solutions approach the background metric at time-like infinity.

This can be seen explicitly from the asymptotic values of matrix of (1.83), Γ_{kl}; see ref. [51] for details. One gets

$$g = g_0[1 + O(t^{-1})] \quad \text{time-like infinity,} \tag{5.8}$$

$$g = g_0[1 + O(t^{-1/2})] \quad \text{null infinity,} \tag{5.9}$$

where g_0 stands for the background metric.

Again the situation is different at space-like infinity. However, some features can be deduced from the diagonal and nondiagonal two-soliton metrics. These depend essentially on the background metric.

For $1 > d \geq 0$ (z axis contracting), the n-soliton solution will always tend to the background metric at space-like infinity. This is because one can find the n-soliton solution step by step, using the $(n - 2)$-soliton solution as background, etc. It is clear from (5.6) that at each step we will recover the background metric in this limit. Consequently, the general n-soliton solution can

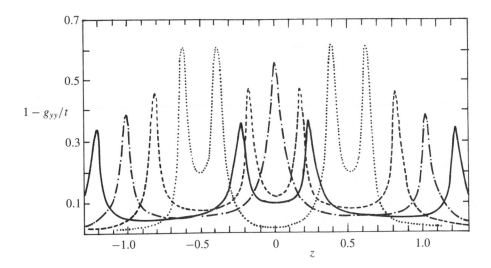

Fig. 5.2. This shows the time evolution of the $g_{yy}(t, z)$ component for the four-soliton solution generated from the axisymmetric Kasner background ($d = 0$). The same conventions as in fig. 5.1 apply. The soliton widths are $c_1 = c_2 = 0.01$ and the separation of their origins is $|z_1^0 - z_2^0| = 1$. The dotted line corresponds to the time $t = 0.1$ and the dashed line to the time $t = 0.3$, both before collision. The dot-dashed line corresponds to the time $t = 0.45$ during the collision of the inner solitons. After the collision these inner solitons move unperturbed, as seen by the continuous line which corresponds to the time $t = 0.7$.

be considered as n gravisolitons on a Kasner background which is contracting along the propagation axis. The speed of the solitons asymptotically approaches the speed of light. Since (as in the diagonal case) the radiative part of their gravitational field dominates at large times they may represent gravitational radiation with two polarizations at null infinity. If we take $|z_k^0 - z_{k-1}^0| = s$ for all k, so that the solitons are equally spaced, the wave 'period' will be s.

As an example we show in fig. 5.2 the g_{yy} coefficient of the four-soliton solution for the axisymmetric Kasner background ($d = 0$). The structure of four solitons propagating on a background is clear. One can also observe the collisions of the two inner solitons. The amplitude of the colliding pair is much greater, implying larger curvature, than that of the other pair, but the two solitons leave the collisions unmodified (as is typical of classical solitons). This is similar to what we have seen in subsection 4.6.3. The difference here is that these are solutions of a nonlinear system and one may expect some nonlinear interaction effect such as a time shift [40] in the colliding perturbations. This question, however, has not been investigated in this solution. Notice that there is a 'hierarchy' effect, in that at large t the gross features of the four soliton solution will resemble those of the two-soliton one.

For background metrics with $d > 3$, the general n-soliton solution still tends asymptotically to the diagonal n-soliton solution (4.79). We have already seen that this is the case in the two-soliton solution, and one can prove the general result by induction [51]. This can be interpreted as the loss of one of the polarizations in the asymptotic regions. Thus the asymptotic results of section 4.6 for the diagonal metrics apply; some of those metrics, depending on d and n, will become singular at $|z| \to \infty$ and will have no clear cosmological interpretation. Of course in the 'near regions' (i.e. small values of t) these metrics are very different from the diagonal ones. They have a much richer fine structure, with two polarizations and many extra parameters.

The solutions with $1 \leq d \leq 3$, which include the Minkowski background, have not been studied in detail. They are more complicated because they do not tend towards diagonality or the background metric as $|z| \to \infty$.

5.2 Bianchi II backgrounds

As we have remarked in chapter 1, the integration of (1.51) to find the generating matrix $\psi_0(\lambda, t, z)$ once a background metric $g_0\,(t, z)$ is given, may not be an easy task when $g_0(t, z)$ is not a diagonal metric. Belinski and Francaviglia [16] have given the general formalism for calculating generating matrices for all Bianchi types from I to VII, which include several nondiagonal metrics.

The first example of nondiagonal metrics to which the ISM was applied are the Bianchi type II metrics. In ref. [16] the matrix $\psi_0(\lambda, t, z)$ was found for a Bianchi II metric, and in ref. [17] the corresponding one-soliton solution was given; see also Letelier [194]. However, as has been shown by Kitchingham [172] such a solution is not truly nondiagonal from the viewpoint of the ISM because the generating matrix can be found from a diagonal background, Bianchi type I, followed by an Ehlers transformation. The Ehlers transformation exploits the symmetries of the Ernst formulation of Einstein equations for spaces with two commuting Killing vectors and transforms a given Ernst potential into another Ernst potential, thus providing a different solution to Einstein equations [179]. In particular when applied to the Kasner background, whose Ernst potential is real, it leads to a new Ernst potential with an imaginary part (twist potential). As shown by Kitchingham the generating matrix may also be transformed by the Ehlers transformation and the generating matrix of the transformed background is the same as the transformed generating matrix of the background. When this is applied to the present problem the generating matrix constructed in ref. [16] can be obtained by simply transforming the corresponding generating matrix $\psi_0(\lambda, t, z)$ given in (4.35) for the Kasner metric. Here we shall discuss the one-soliton solution found in ref. [17].

The homogeneous Bianchi type II background metric is given after an Ehlers transformation of the Kasner metric as follows. The coefficient f_0 is

$$f_0(t) = \frac{C_0}{4a_0^2} t^{(d^2-1)/2}(1+p), \quad p = 4\chi^2 a_0^2(d+1)^{-2}t^{2(d+1)}, \qquad (5.10)$$

where C_0, a_0, d and χ are arbitrary real parameters. As always d is the Kasner parameter and χ is related to the Ehlers transformation. When $\chi = 0$ we recover the Kasner metric. The matrix g_0 is $g_0 = l^T \gamma_0 l$, where γ and l are 2×2 matrices (l^T is the transpose matrix) defined by

$$\gamma_0 = \begin{pmatrix} a^2(t) & 0 \\ 0 & b^2(t) \end{pmatrix}, \qquad l = \begin{pmatrix} 1 & \chi z \\ 0 & 1 \end{pmatrix}, \qquad (5.11)$$

where

$$a^2(t) = 4a_0^2(1+p)^{-1}t^{d+1}, \quad b^2(t) = t^2 a^{-2}(t).$$

Note that here $\det g_0 = t^2$ as in (1.38), so that we are dealing with canonical coordinates.

To find the one-soliton solution we must choose a real-pole trajectory, μ_1, as in (4.49). Here, we have again the problem of discontinuities across the light cone $(z_1^0 - z)^2 = t^2$. Only in the region $(z_1^0 - z)^2 \geq t^2$ does the soliton solution have meaning. Inside the light cone one may try to match the solution with the homogeneous background metric (5.10)–(5.11).

The metric coefficient f for the one-soliton solution is

$$f = C_1 f_0 \frac{\mu^2 Q}{(t^2 - \mu^2)\sqrt{t}}, \qquad (5.12)$$

where C_1 is an arbitrary parameter and

$$Q = a^{-2}\Lambda_1^2 + b^{-2}\Lambda_2^2, \quad \Lambda_1 = a^2 \left(-\chi q_1 \mu^{(d+1)/2} + q_2 \mu^{-(d+1)/2} \right),$$
$$\Lambda_2 = t^2 \left(\chi q_1 \mu^{(d-1)/2} p^{-1} + q_2 \mu^{-(d+3)/2} p \right),$$

with q_1, q_2 arbitrary parameters, and $\mu \equiv \mu_1$.

The matrix g of the one-soliton solution is

$$g = l^T \gamma l, \quad \gamma_{ab} = \frac{|\mu|}{t}(\gamma_0)_{ab} + \frac{t^2 - \mu^2}{t|\mu|}\frac{\Lambda_a \Lambda_b}{Q}, \quad a, b = 1, 2. \qquad (5.13)$$

When we take $\chi = 0$ the soliton solution (5.12)–(5.13) reduces to the nondiagonal one-soliton solution on a Kasner background with real poles (5.1). This can be considered a generalization of (5.1).

This solution has the cosmological singularity at $t = 0$ and also an apparent space singularity when $|z| \to \infty$, which turns out to be a coordinate artefact

[41]. The matching of the solitonic and the background regions has been studied in refs [41, 25] and it is found that, as in the case of Kasner backgrounds, a null fluid with negative energy density is needed along the matching hypersurfaces. Its cosmological interest is not clear. Solutions with an increasing number of real poles have been considered [68, 69, 42] by using algebraic computing and numerical analysis. To obtain a solution that goes to the background at space infinity one may take two opposite poles, say μ_1^+ and μ_2^-, as in the Kasner case. Such a solution could have a cosmological interpretation as pulse waves on a Bianchi II background, as in subsection 4.5.2, provided the appropriate matchings are performed. As in the case of the Kasner background, this spacetime swept by two pulse waves travelling in opposite directions which leave the region between the waves in the homogeneous background is also called a 'cosmic broom'. To avoid matching discontinuities one may take complex-pole trajectories with two opposite poles. In this case the cosmological interpretation as localized perturbations with two polarizations on a Bianchi II background (5.10)–(5.11) is clear. These, however, generally cannot be seen as travelling waves since they could travel at speeds faster than light, but they can be understood as the result of the interference of energy fluxes propagating in opposite directions.

5.3 Collision of pulse waves and soliton waves

Up to this point all soliton solutions of possible cosmological interest have been deduced on homogeneous backgrounds, mainly Bianchi I but also Bianchi II, V and VI$_0$. Moreover, all background metrics have been diagonal or easily related to diagonal metrics, although the resulting soliton solutions can be diagonal or nondiagonal. In this section a nonhomogeneous and truly nondiagonal background metric is considered.

This comes about when one tries to describe the collision between a soliton-like wave and gravitational waves on cosmological backgrounds [54]. We have seen in subsection 4.6.3 how we could generate soliton-like waves on cosmological backgrounds and how we could describe the collision of two such waves. It is clear that in order to describe the collision of a soliton wave and a gravitational wave on a cosmological background we must take a background metric already containing gravitational waves.

Some solutions with gravitational waves are found in spatially homogeneous models: the Lukash Bianchi type VII$_h$ [205] or Siklos plane waves of types IV, VI$_h$ and VII$_h$ [260, 261]. The Lukash solution is rather complicated and Siklos plane waves can be integrated to find the generating functions, but they are given in terms of a rather complicated combination of hypergeometric functions [173]. Moreover, plane-wave solutions are a class of their own, as we shall see in section 7.2. As a consequence, it is best to look directly for inhomogeneous solutions. Wainwright and Marshman [298] found a family of inhomogeneous nondiagonal solutions that depend on an arbitrary function of one null coordinate. Those solutions can be interpreted as cosmological models with gravitational waves. Furthermore, Kitchingham [172] has found the generating functions for such solutions. One solution of the Wainwright and Marshman family has been interpreted as a gravitational wave pulse [298] propagating on a Kasner background. It turns out that this is the simplest solution with a gravitational wave that can be constructed with the Kasner background; this has been proved by Stachel [262] in the cylindrical wave context.

The background solution. This metric can be written [293] as

$$ds^2 = t^{-3/8}e^n(dz^2 - dt^2) + t^{1/2}[dx^2 + (t + w^2)dy^2 + 2w\,dx\,dy], \quad (5.14)$$

where $w(t + z)$ is an arbitrary function and the function $n(t + z)$ satisfies the differential equation $n' = (w')^2$. The coordinate range is $0 \leq t \leq \infty$, $-\infty \leq x, y, z, \leq \infty$. When $w = 0$ (or constant) this metric reduces to a member of the Kasner family. To define a pulse wave we choose w localized in a small region of the spacetime in the following way:

$$w(u) = -A\{1 - \cos[2\pi(u - u_F)/(u_B - u_F)]\}, \quad u_F \leq u \leq u_B, \quad (5.15)$$

and $w(u) = 0$ otherwise, i.e. when $u \leq u_F$ or $u_B \leq u$, where $u \equiv t + z$ and A, u_F and u_B are arbitrary constants; A may be interpreted as the amplitude of the pulse wave and $u_B - u_F$ as its width. This solution represents a gravitational wave pulse propagating at the speed of light on a Kasner background (4.3) with Kasner parameter $d = 1/2$.

The background metric and the new metric we generate can be seen as a generalization of a Bianchi type I metric in which we break the homogeneity in the z-direction. Following Adams *et al.* [2], they can be written generically as

$$ds^2 = f(dz^2 - dt^2) + e^{2b}\left\{\left[\cosh(2\phi) + \frac{\psi}{\phi}\sinh(2\phi)\right]dx^2\right.$$

$$\left. + \left[\cosh(2\phi) - \frac{\psi}{\phi}\sinh(2\phi)\right]dy^2 + \frac{\gamma}{\phi}\sinh(2\phi)dxdy\right\}, \quad (5.16)$$

$$\phi \equiv (\psi^2 + \gamma^2)^{1/2}, \quad (5.17)$$

and all functions f, b, ψ, and γ depend on t and z. The two-dimensional metric with dx and dy (∂_x and ∂_y are the two Killing vectors) has only two independent components ψ and γ. These will be identified as the two independent polarizations of the gravitational waves: ψ corresponding to the + mode and γ corresponding to the × mode, relative to the invariant basis ∂/∂_x and ∂/∂_y.

We can now define a phase angle θ,

$$\tan(2\theta) = \gamma/\psi, \tag{5.18}$$

and then we may use the functions ϕ, θ instead of ψ, γ, since $\gamma = \phi \sin(2\theta)$ and $\psi = \phi \cos(2\theta)$. It is possible to give a physical meaning to the ϕ and θ functions defined in (5.17) and (5.18). In fact, performing a rotation of the invariant basis at any spacetime point with angle θ,

$$\omega^1 = \cos\theta \, dx + \sin\theta \, dy, \quad \omega^2 = -\sin\theta \, dx + \cos\theta \, dy, \tag{5.19}$$

the two-dimensional metric becomes $e^{2\phi}(\omega^1)^2 + e^{-2\phi}(\omega^2)^2$, which has the form of a pure $+$ wave of amplitude ϕ. It is therefore clear that ϕ in (5.18) represents the total amplitude of the gravitational wave, while θ in (5.18) is the physical angle between dx and the direction of polarization of the wave.

For the pulse-wave solution (5.14) the polarization angle θ and the wave amplitude are given by

$$\tan(2\theta) = 2w/(1 - w^2 - t), \quad \phi = \cosh^{-1}\left[(1 + w^2 + t)/2t^{1/2}\right]. \tag{5.20}$$

For the value of w taken in (5.15) the polarization angle is null except along the null rays $u \in (u_B, u_F)$. It is interesting to see how θ changes along the null ray $u = 0$, say, from $t = 0$, where it takes a finite value, to $t \to \infty$, where it goes like $\tan(2\theta) \to -2w/t$. This indicates that the metric approaches the Kasner background when $t \to \infty$; i.e. it becomes diagonal. The wave amplitude ϕ decreases like $t^{-1/2}$ as is typical of gravitational waves in homogeneous backgrounds.

Four-soliton solution. The generating matrix $\psi(\lambda)$ of (1.51) for metric (5.14) has been given by Kitchingham [172]; but see ref. [54] for corrected misprints. Now since coordinates (t, z) in (5.14) are canonical coordinates we need to use complex-pole trajectories to avoid discontinuous first derivatives on the light cones. Moreover, to get localized solutions we need to use opposite pole trajectories (i.e. with modulus less and greater than t). This means that, since complex poles come in pairs, our solution will have four poles and thus the matrix Γ_{kl} in (1.83) is a 4×4 complex matrix. Therefore it is not practical to give the explicit form of the four-soliton solution from the background (5.14). We take $\mu_3 = \bar{\mu}_1$, $\mu_4 = \bar{\mu}_2$ and the metric takes the form (4.2), with

$$f = f_0 \left(\frac{\mu_1\mu_2}{w_1 w_2}\right)^{3/2} \frac{(\mu_2 - \mu_1)^{-2}}{(\mu_1^2 - t^2)(\mu_2^2 - t^2)}$$

$$\times \left\{\frac{t(\mu_1 - \mu_2)^2}{(\mu_1\mu_2 - t^2)^2}[tc_1c_2 + (\mu_1\mu_2)^{1/2}s_1s_2]^2 + [\mu_2^{1/2}c_1s_2 - \mu_1^{1/2}s_1c_2]^2\right\},$$

$$\tag{5.21}$$

$$g_{ab} = (|\mu_1||\mu_2|)^2 t^{-4} \left[(g_0)_{ab} - \sum_{k,l=1}^{4} (\Gamma^{-1})_{kl} \phi_a^{(k)} \phi_b^{(l)} (\mu_k \mu_l)^{-1} \right], \tag{5.22}$$

where f_0 is the corresponding coefficient of the background metric and

$$\phi_1^{(k)}(t, z) \equiv t^{1/2} (2w_k \mu_k)^{-3/4} \mu_k^{1/2} s_k,$$

$$\phi_2^{(k)}(t, z) \equiv t^{1/2} (2w_k \mu_k)^{-3/4} (w \mu_k^{1/2} s_k + t d_k),$$

$$s_k(t, z) \equiv \sin[Y(\mu_k, t, z) + \phi_k], \quad d_k(t, z) \equiv \cos[Y(\mu_k, t, z) + \phi_k],$$

$$Y(\mu_k, t, z) = (1/2w_k)^{1/2} \int w'(t - u/w_k)^{-1/2} du.$$

The complex parameters ϕ_k have been introduced instead of $m_{0c}^{(k)}$ of (1.80): $(2w_k)^{1/2} m_{01}^{(k)} \equiv \epsilon_k \sin \phi_k$, $m_{02}^{(k)} \equiv \epsilon_k \cos \phi_k$, and w_k are the arbitrary constants in the pole trajectories (1.67). The parameters ϵ_k are absorbed into the arbitrary constant that multiplies the coefficient (5.21). The metric (5.21)–(5.22) represents two perturbations that are soliton-like propagating on the background of (5.14), i.e. a Kasner background with a pulse gravitational wave. One of the solitons collides with the pulse wave. This can be seen in fig. 5.3, where the different transversal metric components (5.22) have been represented. The two-soliton waves start at the same origin. In fig. 5.3(c) both the soliton waves and the pulse wave are seen with the same amplitudes, whereas in fig. 5.3(a) the soliton wave dominates.

An analytic study of this solution may be performed in the asymptotic regions and it is useful for this purpose to use the wave amplitude ϕ and polarization angle θ defined in (5.17) and (5.18). At space-like and time-like infinities the amplitude and polarization angle of the background metric are recovered as expected. However, at null infinity one finds,

$$\tan 2\theta \rightarrow \frac{4\sqrt{2} C(\chi_{12})}{1 - 2\sqrt{2} C(\chi_{12})} t^{-1/2} [1 + O(t^{-1/2})], \tag{5.23}$$

where $C(\chi_{12})$ takes different values if the light cone is $z = -t + b$ or $z = t + a$ (a, b are real parameters) which represent the asymptotic trajectories of the soliton waves [54]. The first one just propagates on the homogeneous background, but the second one collides with the pulse wave. The change of polarization angle can be understood in terms of the nonlinear interaction of the two polarization modes which is typical of the nondiagonal metrics, and may be compared with a similar phenomenon in the cylindrical waves described in section 6.3, where an analogy with the electromagnetic Faraday rotation is noted [243].

Comparison with the background value indicates that along the pulse wave the polarization angle is greater in the solution (5.22). Calculating the polarization

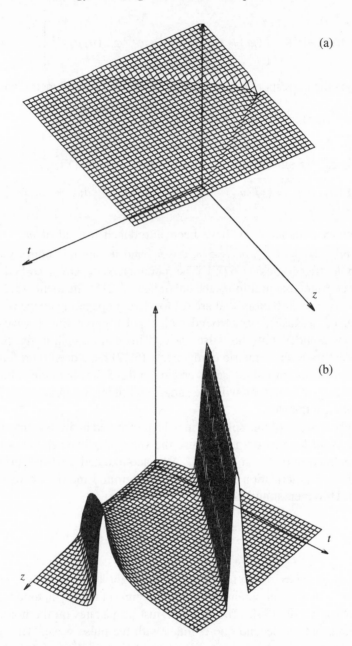

Fig. 5.3.   Metric coefficients g_{ab} of (5.22) in the (t, z)-plane with the parameters $A = 8 \times 10^{-5}$, $u_F = 1.8$, $u_B = 2.4$, $\mathrm{Re}(w_1) = \mathrm{Re}(w_2) = -0.8$, $\mathrm{Im}(w_1) = 0.043$, $\mathrm{Im}(w_2) = 0.045$, $\phi_1 = \phi_2 = 0$: (a) $g_{11}/t^{1/2}$, (b) g_{12}/t and (c) $g_{22}/t^{3/2}$. The pulse wave travels along $z = -t + \text{constant}$, at $t = 0$ it is located between $z = 1.8$ and $z = 2.4$. The two solitons start at $z = -0.8$. Note that in (a) and (c) the z axis indicates only the z-direction and the origin of the t axis is not $z = 0$, whereas in (b) a change of z to $-z$ is assumed.

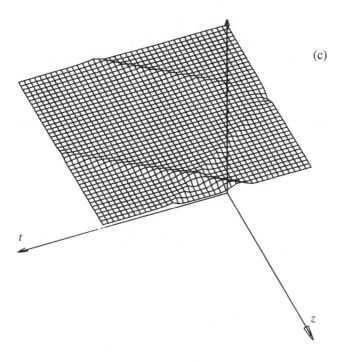

(c)

Fig. 5.3.   Continued.

angle along the soliton waves indicates that they give an angle of polarization comparable to that of the pulse wave. Recall that in the background solution this angle is zero along the light cones $|z|^2 \sim t^2$. The wave amplitude of the metric (5.22) goes like

$$\phi \to \cosh^{-1}\left[\frac{1 - 2\sqrt{2}C(\chi_{12})}{2}t^{1/2}\right][1 + O(t^{-1/2})], \qquad (5.24)$$

which grows like the background metric but with different parameters along each null line, similar to the behaviour of the diagonal coefficients.

The effect of the collision of the soliton wave with the pulse wave is seen in the different values that $C(\chi_{12})$ takes along the different light cones; see ref. [54] for details. This is clearly seen in fig. 5.3(b) where the two-soliton waves have different amplitudes, reflecting the fact that one of them interacts strongly with the pulse wave. Another effect of such (nonlinear) interactions is seen in the small time shift suffered by the peak of the pulse wave after colliding with the soliton wave. This may be relevant in connection with the discussion at the end of subsection 4.6.3.

To summarize, the soliton solution (5.21)–(5.22) can be interpreted as giving the propagation and interaction of a gravitational wave pulse and two-soliton-like waves on the same Kasner background. This metric, which is of Petrov type I, has the cosmological singularity only ($t = 0$) like the corresponding Kasner metric. Thus it may be used as a cosmological model which starts highly inhomogeneous and evolves to a Kasner background with small localized waves of decreasing amplitudes propagating on it.

5.4 Solitons on FLRW backgrounds

So far we have seen that the ISM can be used to derive solutions that represent finite perturbations on some homogeneous but anisotropic cosmological backgrounds, and that such perturbations in some cases become cosmological waves as the background expands. However, the physically most relevant cosmological models are the homogeneous and isotropic FLRW models. It is of obvious interest to be able to describe finite soliton-like perturbations on such backgrounds.

The study of small perturbations on FLRW, which was initiated by the pioneering work of Lifshitz [204], has played an important role in cosmology in problems of galaxy formation, cosmological stability and the propagation of gravitational waves [301, 238]. Those studies have also triggered investigations of more general, anisotropic, and inhomogeneous cosmological models. These investigations have been carried out mainly by the search for and interpretation of exact solutions; see the reviews by Carmeli *et al.* [48], MacCallum [210], and Krasiński [180]. Of course, the snag with exact solutions is that only a few of them can be viewed as realistic cosmological models since they are obtained after imposing strong restrictions on the symmetry of the spacetime. On the other hand, exact solutions can partially illustrate important physical features of the real Universe. For instance, some exact solutions illustrate the generation of gravitational waves of cosmological origin [2, 1] or the models we have considered in the previous sections. Other models illustrate the evolution of primordial density fluctuations on different backgrounds [202, 47]. In this context it is of great interest to find exact cosmological solutions, anisotropic and inhomogeneous, which evolve towards FLRW models. Although these models admit two commuting Killing vectors the ISM cannot be applied directly to the most relevant ones, namely the radiative and matter dominated models.

In this section we examine three classes of solutions that in one way or another describe finite perturbations on FLRW backgrounds and which have been obtained by the ISM. In the first class the background is a vacuum FLRW, in the second class the backgrounds are FLRW spacetimes with stiff fluid, and in the third class the backgrounds are FLRW spacetimes with radiation, or a mixture of radiation and stiff matter. Some of these spacetimes also describe solutions of Brans–Dicke theory. The necessary modifications to deal with fluid or scalar fields will be given here.

5.4.1 Solitons on vacuum FLRW backgrounds

In this subsection we describe finite soliton-like perturbations on a Milne universe background which evolve towards gravitational radiation. The Milne model, which is the region of flat space defined by the forward light cone from the origin, can be interpreted as a vacuum open FLRW universe, since all open FLRW models evolve towards it when the influence of matter can be neglected. It is thus used to approximate low density open cosmological models at late times. The Milne model has a space-like hypersurface of constant negative curvature and is of Bianchi type V or VII$_h$ [261]. We may note that soliton-like perturbations on the Milne universe can again be directly related to perturbations of the Kasner solutions since the Milne metric itself can be considered as the diagonal one-soliton solution from a locally rotationally symmetric Kasner background [174].

Milne's model is described by the metric [219]

$$ds^2 = -d\tau^2 + \tau^2 dl^2, \quad dl^2 = d\chi^2 + \sinh^2 \chi \, (\sin^2 \theta \, d\varphi^2 + d\theta^2), \quad (5.25)$$

where $0 \leq \tau \leq \infty, 0 \leq \chi \leq \infty, 0 \leq \theta \leq \pi$ and $0 \leq \varphi \leq 2\pi$. This is obtained from flat space in spherical coordinates T, R, θ, φ by the coordinate transformation $T = \tau \cosh \chi$, $R = \tau \sinh \chi$. The meaning of these coordinates is the following. Let us assume a set of particles propagating isotropically and with arbitrary constant velocities from the origin $T = R = 0$. The lines $\chi = $ constant, emanating from the origin, represent the world lines of these particles and τ measures their proper times. These particles define a set of inertial observers who see a 'universe' expanding from the origin at $\tau = 0$, where χ are the comoving coordinates. The hypersurfaces $\tau = $ constant are space-like homogeneous hypersurfaces of constant negative curvature as the line element dl^2 in (5.25) shows; thus τ corresponds to the 'cosmological' time. These hypersurfaces are hyperboloids in the (T, R)-plane asymptotic to the forward light cone from the origin $T^2 = R^2$ $(T \geq 0)$.

The spherical coordinates in which the Milne model is written are not suitable for the application of the ISM and we have to adapt the metric to the two commuting Killing vectors. This can be done by the coordinate change [20]

$$\sinh \rho = \sinh \chi \sin \theta, \quad \cosh \rho \sinh z = \sinh \chi \cos \theta, \quad (5.26)$$

where $0 \leq \rho \leq \infty$ and $-\infty \leq z \leq \infty$. Then the line element dl^2 is written as

$$dl^2 = d\rho^2 + \sinh^2 \rho \, d\varphi^2 + \cosh^2 \rho \, dz^2. \quad (5.27)$$

Instead of Milne's time τ we shall often use t defined as $t = \ln \tau, -\infty \leq t \leq \infty$. Then the Milne metric (5.25) reads:

$$ds^2 = e^{2t}(d\rho^2 - dt^2) + e^{2t}(\sinh^2 \rho \, d\varphi^2 + \cosh^2 \rho \, dz^2). \quad (5.28)$$

Although it is now written in the form (1.36), (t, ρ) are not canonical coordinates since

$$\det g_{ab} = e^{4t} \sinh^2 \rho \cosh^2 \rho \qquad (5.29)$$

is not equal to t^2. Canonical coordinates (t', z') can be defined by a coordinate transformation $t' = f_1(\rho + t) - f_2(\rho - t)$, $z' = f_1(\rho + t) + f_2(\rho - t)$. When one evaluates the real-pole trajectories in terms of (t, ρ) one sees that the poles are well defined in the whole range of (t, ρ), which corresponds to the region $z'^2 \geq t'^2$ in canonical coordinates. In order to describe localized waves propagating on the Milne universe we take two opposite (real) poles. The two-soliton metric is given by ref. [154]:

$$ds^2 = f(\tau^2 d\rho^2 - d\tau^2) + \tau^2 dl^2, \qquad (5.30)$$
$$f(\tau, \rho) = C s_1^3 s_2^3 \tau^{-8} (\sinh \rho)^{-2} (\cosh \rho)^{-6} (s_1 + \tanh \rho)^{-2} (s_2 + \tanh \rho)^{-2}$$
$$\times (s_1^2 - 1)^{-1} (1 - s_2^2)^{-1} (s_1 s_2 - 1)^{-2},$$
$$dl^2 = [\sinh^2 \rho / (s_1 s_2)] d\varphi^2 + s_1 s_2 \cosh^2 \rho \, dz^2 \, ;$$
$$s_1 = \beta_1/\alpha - [(\beta_1/\alpha)^2 - 1]^{1/2}, \quad s_2 = \beta_2/\alpha + [(\beta_2/\alpha)^2 - 1]^{1/2},$$
$$\beta_k = w_k - \frac{1}{2}\tau^2 \cosh 2\rho, \quad k = 1, 2, \quad w_1 \neq w_2; \quad \alpha = \frac{1}{2}\tau^2 \sinh 2\rho.$$

To avoid light-cone discontinuities we must take $w_k < 0$ ($w_k = -\tau_k^2$). In (5.30) we use Milne's time τ; C and τ_k are arbitrary real constants, but we take

$$C = 16\tau_2^4 (\tau_1^2 - \tau_2^2)^2, \qquad (5.31)$$

because now metric (5.30) is regular on the symmetry axis: $\rho = 0$. That is, it satisfies the regularity condition $\lim_{\rho \to 0} X_{,\mu} X^{,\mu}/(4X) = 1$, where X is the modulus of the Killing vector ∂_φ [179]. This means that (ρ, φ) are cylindrical coordinates.

Metric (5.30) becomes the Milne metric in the asymptotic regions. It turns out that the maximum deviations from the Milne background are concentrated around the future light cone $\rho = \ln \tau$, for large τ, and the past light cone $\rho = -\ln \tau$, for small τ. The asymptotic analysis can be made in terms of the Riemann components (4.12) with the tetrad $\vec{n} = (2f)^{-1/2}\tau^{-1}\partial_u$, $\vec{l} = (2f)^{-1/2}\tau^{-1}\partial_v$, $\vec{m} = (2)^{-1/2}((g_{\varphi\varphi})^{-1/2}\partial_\varphi + i(g_{zz})^{-1/2}\partial_z)$ and $\vec{m}^*$, where $u = \ln \tau + \rho$, $v = \ln \tau - \rho$ are null coordinates. It is found that at future time-like infinity (ρ finite, $\tau \to \infty$), $\Psi_0 \sim \Psi_2 \sim \Psi_4 \sim 0(\tau^{-2})$, and so the metric approaches Milne (flat space). Similarly at space-like infinity (τ finite, $\rho \to \infty$), $\Psi_0 \sim \Psi_2 \sim \Psi_4 \sim 0(e^{-\rho})$.

On the other hand, at past time-like infinity (ρ finite, $\tau \to 0$) all Riemann components become constant, for instance the Coulomb field is $\Psi_2 = [2\tau_2^4(\tau_1^2 - \tau_2^2)]^{-1}[1 + 0(\tau^2)]$ so that the metric does not approach flat space in this region, although it has no 'cosmological' curvature singularity. At future null infinity ($v = a = $ constant, $u \to \infty$) we find $\Psi_4 \sim 0(e^{-2u})$, $\Psi_0 \sim \Psi_2 \sim 0(e^{-3u})$

so that the radiative component Ψ_4 dominates. To draw conclusions we must approach this limit in a frame parallel propagated along a null geodesic. The vector $\vec{n}$ is tangent to the geodesic congruence defined by $v = a$ but it does not give an affine parametrization. The boost (4.14) with $A = f^{-1/2}\tau^{-1}$ gives an affine parametrization to the geodesic defined by $d/d\lambda = A\vec{n}$ ($\lambda=$ affine parameter) and the new tetrad is parallel propagated. The new Riemann components (4.15) now give $(0(e^{-4u}), 0(e^{-3u}), 0(e^{-u}))$ which means that the spacetime becomes flat at future null infinity ($\lambda \to \infty$ in this region)

At the initial null hypersurface ($u = a$, $v \to -\infty$) if we approach the null congruence $u = 0$ with the parallel propagated boosted tetrad one finds that the new Riemann components (4.15) are: $(0(1), 0(1), 0(1))$, i.e. they are bounded. The affine parameter $\mu : d/d\mu = A\vec{l}$ vanishes on such a hypersurface. This indicates that it may be possible to extend the spacetime through such a regular null hypersurface. Note that there is a similar situation with the Milne model itself (a region of flat space) which may naturally be extended through the same hypersurface. However, the cosmological region is that described above.

The conformal diagram of metric (5.30), together with that of the Milne background diagram, is given in fig. 5.4. The full curve represents the trajectory of the soliton perturbation. It starts as gravitational radiation and propagates as a cylindrical wave towards the axis $\rho = 0$, where it behaves in a particle-like fashion propagating at a velocity less than the velocity of light. Finally the perturbation propagates away from the axis and becomes gravitational radiation of decreasing amplitude on an expanding Milne background. At this stage the universe is an open FLRW with cylindrical gravitational waves.

The extension of metric (5.30) through the initial null hypersurface can be done by introducing new null coordinates $U = e^u$, $V = e^v$ and allowing them to take the values $-\infty < U, V < \infty$. Then the affine parameter μ of the null geodesic has the range $(-\infty, \infty)$. The Riemann tensor of the extended metric goes to zero at past null and future null infinities. At past null infinity the Ψ_4 component dominates over Ψ_0 and Ψ_2 and this means that we have incoming gravitational radiation. These waves collide at finite values of U and V and then radiate at future null infinity in the way described above for the perturbated Milne universe. Note that such spacetime has no curvature singularities.

5.4.2 Solitons with a stiff perfect fluid on FLRW

Under the appropriate symmetry restrictions, the ISM can also be applied to solve Einstein equations coupled to a massless scalar field σ (a massless Klein–Gordon field). These equations read

$$R_{\mu\nu} = \sigma_{,\mu}\sigma_{,\nu},\tag{5.32}$$

$$\sigma_{,\mu}{}^{;\mu} = 0.\tag{5.33}$$

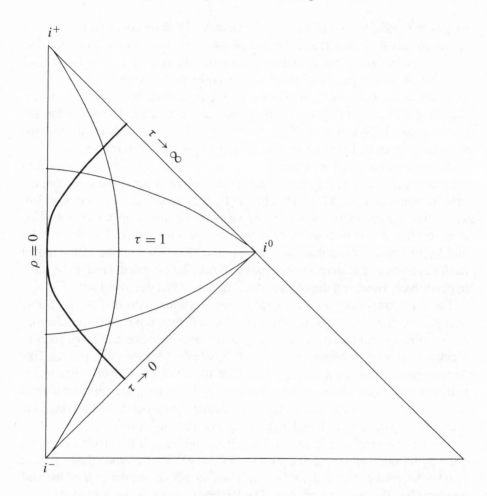

Fig. 5.4.   Structure of the soliton solution on the conformal diagram for the Milne universe. It is drawn for the plane $z = 0$, $\varphi = $ constant.

It is well known that a solution of this system may have a fluid interpretation [267, 297, 188, 12]. Given σ this is done in the following way. If $\sigma_{,\mu}$ is a time-like vector $\sigma_{,\mu}\sigma^{,\mu} < 0$, σ may be considered as the potential of a perfect fluid with a stiff equation of state $p = \epsilon$ ($p =$ pressure, $\epsilon =$ energy density). This is achieved defining the density, pressure and four-velocity of the fluid as

$$\epsilon = p = -\frac{1}{2}\sigma_{,\mu}\sigma^{,\mu}, \quad u_\mu = \sigma_{,\mu}/\sqrt{-\sigma_{,\mu}\sigma^{,\mu}}. \tag{5.34}$$

The stress-energy tensor of the fluid is identified from the right hand side of (5.32) as $T_{\mu\nu} - \frac{1}{2}g_{\mu\nu}T$, where

$$T_{\mu\nu} = \epsilon(2u_\mu u_\nu + g_{\mu\nu}), \tag{5.35}$$

which is a perfect fluid, i.e. $T_{\mu\nu} = (\epsilon + p)u_\mu u_\nu + pg_{\mu\nu}$, with a stiff equation of state ($\epsilon = p$). Such an equation of state implies that the velocity of sound equals the velocity of light and was proposed by Zeldovich [313] to describe the matter content of the Universe in its earlier stages [9].

If $\sigma_{,\mu}$ is a space-like vector, $\sigma_{,\mu}\sigma^{,\mu} > 0$, the above identification is still formally valid but now u_μ is a space-like vector and the perfect fluid interpretation does not hold. Following Tabensky and Taub [267] we can see that the right hand side of (5.32) can be identified with an anisotropic fluid. For this we define an orthonormal tetrad $(\hat{\tau}_\mu, \hat{\sigma}_\mu, \hat{x}_\mu, \hat{y}_\mu)$, where $\hat{\tau}_\mu$ is a time-like vector, $\hat{\sigma}_{,\mu} \equiv u_{\mu,}$, and $\hat{x}_\mu, \hat{y}_\mu$ are space-like vectors. Now $g_{\mu\nu} = -\hat{\tau}_\mu\hat{\tau}_\nu + \hat{\sigma}_{,\mu}\hat{\sigma}_{,\nu} + \hat{x}_\mu\hat{x}_\nu + \hat{y}_\mu\hat{y}_\nu$ and the right hand side of (5.32) can be written as $T_{\mu\nu} - \frac{1}{2}g_{\mu\nu}T$, with

$$T_{\mu\nu} = \frac{1}{2}\sigma_{,\lambda}\sigma^{,\lambda}(\tau_\mu\tau_\nu + \sigma_{,\mu}\sigma_{,\nu} - x_\mu x_\nu - y_\mu y_\nu), \qquad (5.36)$$

which corresponds to the stress-energy tensor of an anisotropic fluid with energy density $\epsilon = \frac{1}{2}\sigma_{,\lambda}\sigma^{,\lambda}$ and vanishing heat-flow vector. The weak and strong energy conditions [143] are satisfied and the fluid interpretation is reasonable.

We assume as usual that the spacetime admits an orthogonally transitive two-parameter group of isometries. We have seen in section 1.2 that the ISM is aimed, basically, at the solution of the field equations $R_{ab} = 0$, i.e. (1.39). This is the case for vacuum, for instance. When we assume the presence of matter, i.e. a given stress-energy tensor $T_{\mu\nu}$, the above equations will not be true in general. However, when a massless scalar field is coupled to the gravitational field, as in (5.32), the Ricci components R_{0a} and R_{3a} vanish ($R_{0a} = R_{3a} = 0$), due to the block diagonal form of the metric (1.36) and, from (5.32) it follows that $\sigma_{,a} = 0$ and, consequently, $R_{ab} = 0$. Therefore the ISM for (1.39) still applies.

The remaining equation (1.40)–(1.41) will be modified now but only the metric coefficient f of (1.36) will be different from the vacuum case [12]. If we write f as the product $f = f_v F$, (5.32)–(5.33) can be divided into four groups. The first and second of these exactly repeat the Einstein equations in vacuum, (1.39) and (1.40)–(1.41), where f is changed by f_v. The third group is just a wave equation for the scalar field σ, which in canonical coordinates ($\alpha = t$) is

$$\sigma_{,tt} + \frac{1}{t}\sigma_{,t} - \sigma_{,zz} = 0, \qquad (5.37)$$

and the fourth group determines the factor F:

$$(\ln F)_{,t} = t(\sigma_{,z}^2 + \sigma_{,t}^2), \quad (\ln F)_{,z} = 2t\sigma_{,t}\sigma_{,z}. \qquad (5.38)$$

Hence, to solve the problem we must first construct a vacuum soliton solution (f_v, g) as we explained in chapter 1. After this we must determine the scalar field σ from (5.37) and then use (5.38) to determine the function F. The final result will be a metric (f, g) with $f = f_v F$. If we want a fluid interpretation we must identify the fluid properties according to (5.35) or (5.36) from the scalar field. Equations (5.37)–(5.38) are identical to (4.41)–(4.42) for a vacuum diagonal metric: the field $\sigma\sqrt{2}$ can be thought of as describing the transversal metric components and $t^{-1/2}F$ as the longitudinal component. Therefore the ISM and its generalizations can be applied to the field σ. One can take different numbers of pole trajectories for the pair (f_v, g) say n, and the pair $(t^{-1/2}F, \sigma\sqrt{2})$ say m. Such solutions are called (n, m)-soliton solutions.

Letelier [190] generalized (5.37)–(5.38) to the case of an anisotropic fluid described by two or more perfect fluids. The pressure on the direction of propagation of the wave equals the energy density. He found the $(0, 1)$-, $(1, 0)$- and $(1, 1)$-soliton solutions with real poles on a Kasner background. In the last case both the gravitational field and the matter have a soliton-like behaviour. A soliton solution that has an FLRW with stiff fluid limit was obtained in ref. [198] with a source formed by three scalar fields. In the limit of one scalar field it produces some known nonsolitonic stiff fluid inhomogeneous metrics, which generalize the flat and open FLRW models [197, 187, 188, 297]. We should add here that Einstein equations generalizing (5.32)–(5.33) and representing the coupling of gravity with selfdual SU(2) gauge fields [305] have also been considered and adapted, under the appropriate symmetry restrictions, to the ISM. The corresponding stress-energy tensors in some cases admit fluid interpretations, generally as anisotropic fluids with different perfect fluid components [190, 195, 196]. Soliton solutions with cylindrical symmetry [199], as well as axial symmetry [194, 200], have been obtained and interpreted in different ways.

Note that, for solitonic behaviour, the existence of a unique velocity defined on the system in the direction of propagation of the wave seems essential. In vacuum we have only the gravitational field (speed of light). With stiff matter, the matter and the gravitational field have a coherent coupling: speed of sound = speed of light. One might expect that for a perfect fluid with a different speed of sound the solitonic behaviour will not persist and the soliton perturbation will be dispersed. We shall see this in the next subsection.

Soliton solutions with a stiff fluid. In ref. [12] the $(1, 0)$-soliton solutions were found (the scalar field remains unperturbed) with nondiagonal metrics for the flat, open and closed FLRW models. Although there is only one pole and therefore it must be real, there is no problem of discontinuities because the physically meaningful range of variables is reduced to the region $(z_1^0 - z)^2 \geq t^2$; see the details in ref. [12]. As an explicit example we consider such a $(1, 0)$-soliton solution in the flat FLRW background with stiff matter. Instead of canonical coordinates it is more convenient to use for this solution coordinates

τ, ρ, θ, z in which the background metric has the form

$$ds^2 = \tau(d\rho^2 - d\tau^2) + \tau(\rho^2 d\theta^2 + dz^2), \quad \sigma = \sqrt{3/2}\ln\tau, \quad \epsilon = p = (3/4)\tau^{-3},$$
(5.39)

with $0 \le \rho \le \infty, 0 \le \theta \le 2\pi, -\infty < z < \infty$ and $\tau \ge 0$. In these coordinates α and β defined in (1.45) and (1.46) are $\alpha = \tau\rho$ and $\beta = (\tau^2 + \rho^2)/2$. Choosing the arbitrary constant w_1 in (1.67) to be negative, i.e. $w_1 = -l^2/2$, where l is an arbitrary real parameter, we get for the pole trajectory μ (which we take for definiteness to be μ_1^{in}):

$$\mu = -\frac{1}{2}(l^2 + \tau^2 + \rho^2) + \frac{1}{2}[(l^2 + \tau^2 + \rho^2)^2 - 4\tau^2\rho^2]^{1/2},$$
(5.40)

which is well defined in the whole (τ, ρ) region. The $(1, 0)$-soliton solution is then

$$
\left.
\begin{aligned}
f &= \frac{l^2\tau[s^2\tau^2 + (\tau^2 + \mu)^2]}{s^2[l^2\tau^2 + (\tau^2 + \mu)^2]}, \\[2mm]
g &= \tau[s^2\tau^2 + (\tau^2 + \mu)^2]^{-1} \\
&\times \begin{pmatrix} s^2\tau^2\rho^2 + \rho^2(\tau^2 + \mu)^2 + q\rho^2(\tau^2 + \mu) - q^2\mu & qs\mu \\ qs\mu & s^2\tau^2 + (\tau^2 + \mu)^2 - q(\tau^2 + \mu) \end{pmatrix},
\end{aligned}
\right\}
$$
(5.41)

with s an arbitrary parameter and $q \equiv s^2 - l^2$. The matter content is given by

$$\sigma = \sqrt{3/2}\ln\tau, \quad \epsilon = p = \frac{3s^2[l^2\tau^2 + (\tau^2 + \mu)^2]}{4l^2\tau^3[s^2\tau^2 + (\tau^2 + \mu)^2]}.$$
(5.42)

This model has the cosmological singularity only. For small τ a soliton-like perturbation is concentrated near the symmetry axis $\rho = 0$. After a critical time $\tau \sim s$ (for simplicity we consider the parameters l and s to be of the same order of magnitude) this cylindrical perturbation leaves the horizon and propagates outward, with decreasing amplitude and, at large τ, it propagates at the speed of light. This is another example of the production of gravitational waves. The qualitative behaviour of this solution is very similar to that of the late stages of the soliton solutions on a Milne universe discussed in subsection 5.4.1. This solution was further discussed by Gleiser *et al.* [119], who also showed that it could be analytically continued to values of l^2 negative. The $(1, 0)$-soliton solutions on the open FLRW models are qualitatively similar to (5.41)–(5.42) and in the closed model the soliton perturbations start decreasing but then increase again in the final stages [12].

Generalized soliton solutions with stiff fluid. The generalized soliton solutions discussed in section 4.4.1 can easily be used to find diagonal solutions with a stiff fluid. As we have emphasized earlier, these diagonal metrics are not the limit of truly nondiagonal soliton solutions in general. The construction of these solutions is obvious if one compares (5.37)–(5.38) with (4.41)–(4.42). One example of a generalized soliton solution which has the FLRW limit is a cosmological solution for a stiff fluid by Wainwright *et al.* [297]. It is the stiff fluid version of the vacuum solution (4.58)–(4.59). The potential Φ and f_v are given by (4.58)–(4.59), and from this we can write the generalized soliton solution for σ and F, recalling the correspondence $\Phi \to \sigma\sqrt{2}$ and $f_v \to t^{-1/2}F$. Then the new metric coefficient f is $f = f_v F$. Introducing two new parameters α and β we can finally write,

$$\sigma = \alpha \ln t + \beta \cosh^{-1}\left(\frac{z_1}{t}\right), \quad |z_1| \geq t, \tag{5.43}$$

$$\Phi = d \ln t + h \cosh^{-1}\left(\frac{z_1}{t}\right), \quad |z_1| \geq t, \tag{5.44}$$

$$f = t^{[(d-h)^2+2(\alpha-\beta)^2-1]/2}\mu_1^{hd+2\alpha\beta}(z_1^2 - t^2)^{-(h^2+2\beta^2)/2}. \tag{5.45}$$

Defining new coordinates (T, Z) by (4.61) we have that $\mu_1/t = [\tanh(aT)]^{-1}$ and the metric can be written as

$$\begin{aligned}
ds^2 &= [\sinh(2aT)]^{(d^2+h^2+2\alpha^2+2\beta^2-1)/2}[\tanh(aT)]^{-(dh+2\alpha\beta)} \\
&\quad \times e^{(h^2-d^2+2\beta^2-2\alpha^2-3)aZ}(dZ^2 - dT^2) \\
&\quad + [\sinh(2aT)]^{1+d}[\tanh(aT)]^{-h}e^{-2a(1+d)Z}dx^2 \\
&\quad + [\sinh(2aT)]^{1-d}[\tanh(aT)]^{h}e^{-2a(1-d)Z}dy^2,
\end{aligned} \tag{5.46}$$

which is in the form given in ref. [297]. This solution is the open FLRW model when $d = h = \alpha = 0$ and $\beta^2 = 3/2$, and it is the flat FLRW when $\alpha = -\beta^2 = \sqrt{3/8}$ and $d = h = \pm 1/2$, see ref. [180]. When $\alpha = \beta = 0$ it reduces to a vacuum solution and in this case it becomes the Ellis and MacCallum homogeneous and anisotropic solution (4.62) when $h^2 = d^2 + 3$. When $h^2 = 0$, $h^2 = 1$ and $3/2 \leq h^2 < 2$ the spacetime can be extended beyond a coordinate boundary that exists when $h^2 < 2$; see ref. [76]. These metrics can be generalized using the solutions (4.48), where $g_i(\lambda)$ are arbitrary bounded functions, With such solutions new metrics representing gravitational and density pulses propagating on the spacetime (5.46) can be constructed [48].

Generalized cosoliton solutions with stiff fluid. Analogously one may construct cosoliton solutions as in section 4.5.3. As an example we consider a stiff fluid solution which has an FLRW limit. Following steps similar to those used in the previous case we may construct from (4.70)–(4.71) the following solution [234]:

$$\sigma = \alpha \ln t + \beta \cos^{-1}\left(\frac{z_1}{t}\right), \quad |z_1| \leq t, \tag{5.47}$$

$$\Phi = d \ln t + h \cos^{-1}\left(\frac{z_1}{t}\right), \quad |z_1| \leq t, \tag{5.48}$$

$$f = t^{(d^2-h^2+2\alpha^2-2\beta^2-1)/2}(t^2 - z_1^2)^{(h^2+2\beta^2)/2} \exp\left[(dh + 2\alpha\beta)\cos^{-1}\left(\frac{z_1}{t}\right)\right].$$

(5.49)

This is a solution of Einstein's equations coupled to a massless scalar field, (5.32)–(5.33). It reduces to the Tabensky and Taub plane-wave symmetric metric when $d = h = 0$ [267]. For large t the metric approaches a spatially homogeneous metric.

The spacetime regions where $\sigma_{,\mu}$ is, respectively, time-like and space-like are divided by the straight line $t = -(\alpha^2 + \beta^2)(\alpha^2 - \beta^2)^{-1}z_1$. According to the previous discussion, see (5.35) and (5.36), we have a perfect stiff fluid in the spacetime region between that straight line and $t = z_1 > 0$ and an anisotropic fluid in the complementary region. The presence of a fluid makes this metric easy to study and interpret because one may adapt the coordinate system to the fluid and introduce comoving coordinates [266]. In the region where $\sigma_{,\mu}$ is time-like we may use $\sigma(t, z)$ as the time coordinate and define a space coordinate $Z(t, z)$ by $dZ = \alpha^{-1}t(\sigma_{,z}dt + \sigma_{,t}dz)$; this ensures that $Z_{,\mu}\sigma^{,\mu} = 0$ and that $Z_{,\mu}$ is space-like. When this equation is integrated we have the new space coordinate

$$Z = z_1 - \alpha^{-1}\beta(t^2 - z_1^2)^{1/2}.$$

(5.50)

The fluid lines are the hyperbolas defined by $Z = $ constant, which approach straight lines when $t \to \infty$. Then the time coordinate may be defined as

$$T = \exp\left[\alpha^{-1}\sigma(t, z) - \alpha^{-1}\beta\cos^{-1}\left(b/\sqrt{\alpha^2 + \beta^2}\right)\right],$$

(5.51)

where the parameters have been introduced for convenience. In the region where $\sigma_{,\mu}$ is space-like, T and Z are space and time coordinates, respectively, and the fluid lines are defined by $\sigma = $ constant. The coordinate change (5.50)–(5.51) is not explicitly invertible. However, for large t it is

$$t \simeq T + Z\beta/\sqrt{\alpha^2 + \beta^2}, \quad z \simeq Z + T\beta/\sqrt{\alpha^2 + \beta^2}.$$

(5.52)

In this case the metric (5.48)–(5.49) can be approximately written in comoving coordinates as

$$
\begin{aligned}
ds^2 \simeq{} & T^{d^2+h^2+2\alpha^2+2\beta^2-1} \\
&\times\left[1 + \frac{Z}{T\sqrt{\alpha^2 + \beta^2}}\left[\frac{\beta}{2}(d^2 - h^2 - 2\alpha^2 - 2\beta^2 - 1) - \alpha hd\right]\right] \\
&\times\left[\left(1 - 2\frac{\beta Z}{T\sqrt{\alpha^2 + \beta^2}}\right)dZ^2 - dT^2\right] \\
&+ T\left[1 + \frac{\beta Z}{T\sqrt{\alpha^2 + \beta^2}}(T^d A dx^2 + T^{-d}A^{-1}dy^2)\right],
\end{aligned}
$$

(5.53)

where $A = 1 + (Z/T\sqrt{\alpha^2 + \beta^2})(\beta d - \alpha h)\exp[h\cos^{-1}(\beta/\sqrt{\alpha^2 + \beta^2})]$. For the set of values $[d = h = 0, 2(\alpha^2 + \beta^2) = 3]$ and for the set $[d = 0, 2(\alpha^2 + \beta^2) = 3 - h^2]$ this metric approaches the flat FLRW metric with a stiff perfect fluid as $T \to \infty$:

$$ds^2 = T(dx^2 + dy^2 + dZ^2 - dT^2).$$

For all other values of the parameters the metric approaches a spatially homogeneous but anisotropic model. To finite values of time the metric is spatially inhomogeneous and may be interpreted as representing inhomogeneous finite perturbations on homogeneous backgrounds. Therefore this solution is another example of inhomogeneous cosmologies that become spatially homogeneous, and even isotropic (FLRW) for some values of the parameters, as a result of cosmological evolution. Since metric (5.48)–(5.49) can be seen as the analytical continuation of (5.44)–(5.45) it has also been interpreted [180] as being determined on another region of the manifold that underlies the solution (5.46).

5.4.3 The Kaluza–Klein ansatz and theories with scalar fields

It would be of interest to extend the ISM to FLRW backgrounds with more realistic equations of state. To some extent this is possible for radiation perfect fluids if we accept anisotropic perturbations on the fluid. As we have remarked in the previous subsection soliton-like perturbations are maintained when there is only one characteristic speed in the problem. For vacuum and for a stiff fluid the speed of propagation of the gravitational field and that of the sound equal the speed of light. For a radiative perfect fluid, however, the speed of sound is $1/\sqrt{3}$ of the speed of light and one expects that an initially localized perturbation will be dispersed by the gravitational field. In fact, it is known that when an equation of state different from that of a stiff fluid is imposed, the dynamical equations lead to very complicated behaviour and the formation of shock waves [202]. But as we shall see a soliton structure may be maintained with a spatially anisotropic stress-energy tensor, such that the pressure in the direction of the solitons propagation equals the energy density.

We use the fact that the FLRW cosmological models with a radiative perfect fluid are equivalent to vacuum solutions of Einstein's equations in five dimensions, see for instance ref. [152]. Then by evaluating soliton solutions in five dimensions, which can be interpreted as finite perturbations propagating on a five-dimensional vacuum, one may find perturbations on a radiative FLRW background. This leads us to consider Einstein's equations in more than four dimensions and in particular to discuss the different effective four-dimensional theories that they induce. It is thus useful to comment upon such aspects and the connection with Kaluza–Klein theories [162, 175]. In modern Kaluza–Klein theories the extra space is considered a compact space of the size of the

Planck length whose isometries are the gauge symmetries of some gauge theory [86]. Fourier expanding the N-dimensional metric tensor in terms of the extra coordinates one finds, at the zero mode or the low energy limit, an effective four-dimensional theory for the coupling of gravity with $N - 4$ vector fields (Yang–Mills gauge bosons) and $(N - 4)(N - 3)/2$ scalar fields (presumably Higgs bosons). As described in section 1.5 the vector fields are associated with the mixed metric components of the N-dimensional metric, and the scalar fields are associated with the metric components of the extra space [132, 86].

In order to obtain realistic effective low energy theories the scalar fields play an essential role in these theories. Thus, for instance, in five dimensions the low energy theory is the Jordan–Thiry four-dimensional theory [159, 270, 271] of coupled gravity and electromagnetism with a massless scalar field as in the Brans–Dicke theory. But unlike the Brans–Dicke theory, which contains an arbitrary constant [43, 301], here the constant is fixed, as it should be in a truly unified theory, by the general five-dimensional covariance. However, the theory may be transformed into the source-free Brans–Dicke theory by means of a conformal transformation which involves an arbitrary parameter of the four-dimensional metric or, also, it may be transformed into Einstein's equations coupled to a massless scalar field, [19, 267] which as we have seen in the previous subsection may be equivalent to a stiff perfect fluid. As another example, in six dimensions if one restricts to the scalar sector [198] the Kaluza–Klein ansatz leads to an effective four-dimensional theory which describes the coupling of the Brans–Dicke field with an anisotropic fluid formed by two perfect fluid components [189]. Therefore one may use extra dimensions as a useful auxiliary tool to obtain meaningful four-dimensional theories.

We shall now write down the effective four-dimensional theory from arbitrary dimensions, as in section 1.5, and then we will restrict it to five dimensions. An N-dimensional Kaluza–Klein theory is based on an N-dimensional metric γ_{AB}

$$ds^2_{(N)} = \gamma_{AB}dx^A dx^B = \gamma_{\mu\nu}dx^\mu dx^\nu + 2\gamma_{\mu b}dx^\mu dx^b + \gamma_{ab}dx^a dx^b, \quad (5.54)$$

which we split into the usual four coordinates and $N - 4$ extra coordinates with $A, B, = 0, 1, \ldots, N - 1$, $\mu, \nu = 0, 1, 2, 3$, and, here, $a, b, = 4, \ldots, N - 1$. The general N-dimensional covariance leads to the N-dimensional Einstein field equations in vacuum; in the simplest Kaluza–Klein theories no extra fields are assumed; i.e., $R_{AB} = 0$. Assuming that the extra space is compact and of small size, the zero mode or low energy limit is obtained by the assumption that γ_{AB} has no dependence on the extra coordinates: $\gamma_{AB,a} = 0$. This implements the Kaluza–Klein ansatz. The theory can be written in terms of an effective four-dimensional metric:

$$g_{\mu\nu} = \gamma_{\mu\nu} - \gamma_{ab}A^a_\mu A^b_\nu, \quad (5.55)$$

$$A^a_\mu \equiv \gamma_{b\mu}\hat{\gamma}^{ba}, \quad \hat{\gamma}^{ab}\gamma_{bc} \equiv \delta^a_c. \quad (5.56)$$

One can verify that $g^{\mu\nu} = \gamma^{\mu\nu}$, $\gamma^{a\mu} = -A^{a\mu}$, $\gamma^{ab} = \hat{\gamma}^{ab} + A^{a\mu}A^b_\mu$. The coefficients A^a_μ represent $N - 4$ vector fields on the space of metric $g_{\mu\nu}$ and the coefficients γ_{ab} are $(N - 4) \times (N - 3)/2$ scalar fields. Related to the volume of the extra space there is the scalar field σ defined by $\det(\gamma_{ab}) = \sigma^2$. Note that here our notation differs slightly from that of section 1.5, and also the election of the four-dimensional metric is different since we do not use the Einstein frame.

We wish to consider the scalar sector of the theory only; i.e. we now assume that $A^a_\mu = 0$ ($\gamma_{a\mu} = 0$), which is compatible with the field equations. Now the field equations $R_{AB} = 0$ in terms of the four-dimensional metric $g_{\mu\nu}$ can be written as, see (1.121)–(1.123),

$$\bar{R}_{\mu\nu} = \sigma^{-1}\sigma_{,\mu;\nu} - \sigma^{-2}\sigma_{,\mu}\sigma_{,\nu} - (1/4)\hat{\gamma}^{ab}_{,\mu}\gamma_{ab,\nu}, \tag{5.57}$$

$$(\sigma\gamma_{ab,\mu}\hat{\gamma}^{ac})^{;\mu} = 0, \tag{5.58}$$

$$\sigma_{,\mu}{}^{;\mu} = 0, \tag{5.59}$$

where $\bar{R}_{\mu\nu}$ denotes the Ricci tensor for the metric $g_{\mu\nu}$ and all covariant derivatives are taken in terms of such a four-dimensional metric. Equation (5.57) is a consequence of $R_{\mu\nu} = 0$, (5.58) follows from $R_{ab} = 0$ and (5.59) is just the trace of (5.58). The equations $R_{\mu a} = 0$ are identically verified in view of the assumption $\gamma_{\mu a} = 0$. We now restrict these field equations to the case $N = 5$, then (5.57), (5.58) and (5.59) read

$$\bar{R}_{\mu\nu} = \sigma^{-1}\sigma_{,\mu;\nu}, \quad \sigma_{,\mu}{}^{;\mu} = 0, \tag{5.60}$$

which are Brans–Dicke equations in vacuum with coupling parameter $\omega = 0$ and $\sigma(x^\mu)$ is the massless Brans–Dicke field; see (5.61). Notice that when $N = 6$ (see ref. [198]) one obtains Brans–Dicke equations with coupling parameter $\omega = 1/2$.

Related scalar theories. Let us now discuss related scalar theories, namely Brans–Dicke theory with arbitrary coupling parameter and Einstein equations coupled to a massless scalar field. The Brans–Dicke equations in vacuum are [43, 301]

$$\hat{R}_{\mu\nu} - \frac{1}{2}\hat{g}_{\mu\nu}\hat{R} = \frac{1}{\hat{\sigma}}\hat{\sigma}_{,\mu;\nu} + \frac{\omega}{\hat{\sigma}^2}\left(\hat{\sigma}_{,\mu}\hat{\sigma}_{,\nu} - \frac{1}{2}\hat{g}_{\mu\nu}\hat{\sigma}_{,\rho}\hat{\sigma}^{,\rho}\right), \quad \hat{\sigma}_{,\mu}{}^{;\mu} = 0, \quad (5.61)$$

with metric tensor $\hat{g}_{\mu\nu}$, scalar field $\hat{\sigma}$, and coupling parameter ω. For large ω (i.e. $\omega > 500$) such a theory is compatible with observation [222]. These equations can be obtained from the vacuum equations for the five-dimensional metric (5.54), which we write using (5.55) as

$$ds^2_{(5)} = g_{\mu\nu}(x^\rho)dx^\mu dx^\nu + \sigma^2(x^\rho)(dx^5)^2, \tag{5.62}$$

by defining [19]

$$\hat{g}_{\mu\nu} = \sigma^{1-\Omega} g_{\mu\nu}, \quad \hat{\sigma} = \sigma^{\Omega}, \quad \Omega \equiv (1 + 2\omega/3)^{-1/2}. \tag{5.63}$$

That is, (5.61) are equivalent to (5.60). Consequently a solution of (5.60) can be transformed into a solution of (5.61).

The Einstein equations for a massless scalar field (and stiff fluid, eventually) can be written (see (5.32)–(5.33)) in terms of a four-dimensional metric tensor $\hat{g}_{\mu\nu}$ as

$$\hat{R}_{\mu\nu} = \hat{\sigma}_{,\mu}\hat{\sigma}_{,\nu}, \quad \hat{\sigma}_{,\mu}{}^{;\mu} = 0. \tag{5.64}$$

These equation are equivalent to (5.60), if we define the four-dimensional metric $\hat{g}_{\mu\nu}$ and scalar field $\hat{\sigma}$ in terms of the five-dimensional metric (5.62) by

$$\hat{g}_{\mu\nu} = \sigma g_{\mu\nu}, \quad \hat{\sigma} = (3/2)^{1/2} \ln |\sigma|, \tag{5.65}$$

which corresponds to the Einstein frame. Therefore, a solution of (5.60) can be transformed into a solution of (5.64) by means of the transformation (5.65).

5.4.4 Solitons on radiative, and other, FLRW backgrounds

We shall now consider a few examples of solutions constructed using the Kaluza–Klein ansatz explained in the previous section. This allows us to generate some solutions that can be interpreted as solitons on a radiative FLRW background. From these solutions we can generate other solutions of some scalar field theories like Brans–Dicke, or others representing cosmologies with a perfect fluid source of stiff matter with FLRW limits. We also find some solutions representing solitons on a mixture of stiff and radiative FLRW backgrounds.

Solitons on radiative FLRW backgrounds. Now, let us go back to the cosmological problem of attempting to derive soliton solutions in FLRW backgrounds. The flat radiative FLRW metric, which for convenience we write in cylindrical coordinates

$$ds^2 = t^2(-dt^2 + d\rho^2) + t^2(\rho^2 d\varphi^2 + dz^2), \tag{5.66}$$

with fluid energy density $\epsilon = 3t^{-4}$, pressure $p = \epsilon/3$, four velocity $u^\mu = (t^{-1}, 0, 0, 0)$, is equivalent to the vacuum five-dimensional metric (5.62) with scalar field $\sigma = t^{-1}$,

$$ds_{(5)}^2 = ds^2 + t^{-2}(dx^5)^2. \tag{5.67}$$

This equivalence is also true for the open and closed FLRW models [152]. The connection between the Kaluza–Klein ansatz and anisotropic cosmological evolution has been considered for vacuum in five dimensions [63], and also for classical dust in more than four dimensions [26]. Now, as was explained in section 1.5, we can use (5.67) as the background metric to obtain soliton solutions on such a five-dimensional background and, consequently, soliton solutions on radiative FLRW. To obtain localized waves we take two real opposite pole trajectories (4.49), μ_1^- and μ_2^+. In the coordinates of (5.66), which are not canonical, these pole trajectories are defined by

$$\mu_k^{\pm} = \frac{1}{2\rho}\{-(l_k^2 + \rho^2 + t^2) \pm [(l_k^2 + t^2 + \rho^2)^2 - 4t^2\rho^2]^{1/2}\}, \qquad (5.68)$$

where l_k ($k = 1, 2$) are arbitrary real parameters. As an example we shall consider a diagonal five-dimensional metric with a four-dimensional sector given by,

$$ds^2 = C_1 \left[\frac{(\rho s_1 + t)(\rho s_2 + t)}{\rho(s_1 s_2 - 1)}\right]^2 \left[\frac{(s_1 s_2)^4(s_1 - s_2)}{(s_1^2 - 1)^2(s_2^2 - 1)^2}\right]^{2/3} (-dt^2 + d\rho^2)$$
$$+ t^2(s_1 s_2)^{2/3}(\rho^2 d\varphi^2 + dz^2), \qquad (5.69)$$

and with the scalar field

$$\sigma = t^{-1}(s_1 s_2)^{-2/3}; \quad s_1 = \mu_1^-/t, \quad s_2 = \mu_2^+/t. \qquad (5.70)$$

The regularity condition on the symmetry axis, defined in the lines that follow (5.31), implies that $C_1 = (l_2^2 - l_1^2)^2/l_1^4$.

This metric depends essentially on the factor $s_1 s_2$ and one can see that $s_1 s_2 \approx 1 - (l_1^2 - l_2^2)/t^2$ at time-like infinity ($\rho \ll t \to \infty$), $s_1 s_2 \approx 1 + l_1 l_2/t$ at future null infinity ($\rho \approx t \to \infty$) and $s_1 s_2 \simeq 1 + (l_1^2 - l_2^2)/\rho^2$ at space-like infinity ($t \ll \rho \to \infty$). This indicates that this metric represents cylindrical perturbations on the radiative FLRW background. This is confirmed by analysing the curvature tensor. Choosing the null tetrad: $\vec{n} = (2f)^{-1}(\partial_t + \partial_\rho)$, $\vec{l} = (2f)^{-1}(\partial_t - \partial_\rho)$, $\vec{m} = (2g_{\varphi\varphi})^{-1/2}\partial_\varphi + i(2g_{zz})^{-1/2}\partial_z$ we obtain, from (4.12), $\Psi_0 \sim \Psi_2 \sim 0(t^{-4})$, $\Psi_4 = [(l_2 - l_1)/l_1 l_2]t^{-3}$, at null infinity. Thus the radiative component of the field dominates in this asymptotic region, as usual. The fluid content of the spacetime may now be deduced by identifying the stress-energy tensor $T_{\mu\nu}$ from the right hand side of (5.60). After such an identification one writes the algebraic canonical form [143] of this tensor on an orthonormal basis as

$$T_{ab} = \mathrm{diag}(\mu, p_\rho, p_\varphi, p_z), \qquad (5.71)$$

and it can be proved that these eigenvalues are real and satisfy the strong energy conditions: $\mu + p_i > 0$ ($i = \rho, \varphi, z$) and $\mu + p_\rho + p_\varphi + p_z > 0$ when $l_1 > l_2$ [153]. Note, however, that the hydrodynamical interpretation of this tensor is not known. The asymptotic values at time-like infinity are

$$\mu \sim 3/t^4, \quad p_\rho \sim p_\varphi \sim p_z = 1/t^4, \qquad (5.72)$$

so that the stress-energy tensor becomes that of a radiative perfect fluid; i.e. it gives the background fluid. Similarly, at space-like infinity,

$$\mu \sim 3(l_1^4/l_2^4)/t^4, \quad p_\rho \sim p_\varphi \sim p_z \sim (l_1^4/l_2^4)/t^4, \tag{5.73}$$

so that the stress-energy tensor becomes the background fluid, but with different absolute values. This is because the longitudinal expansion $f(t,\rho)$ at space-like infinity differs from that of the background by a constant. However, at null infinity,

$$\mu \sim p_\rho \sim \left(\frac{5|l_1 - l_2|}{3l_1l_2}\right)^{1/2}\left(\frac{16l_1^5}{(l_1 + l_2)^4 l_2}\right)^{2/3} t^{-7/2} + O(t^{-4}), \quad p_\varphi \sim p_z \sim O(t^{-4}), \tag{5.74}$$

so that it behaves like a null fluid or a directed flux of massless collisionless relativistic particles [314]. Note that although the local energy density decreases in time as the spacetime expands, the ratio of it to the energy density of the background (density contrast), $\mu/\epsilon^{FLRW} \sim t^{1/2}$ grows in time along the direction of propagation of the solitons. This illustrates for an exact metric the growing density modes of the linearized theory of the FLRW perturbations.

The results of (5.72), (5.73) and (5.74) can be understood as the formation of a cylindrical hole (if $l_1 > l_2$) or a halo (if $l_1 < l_2$) with a cylindrical shell whose density grows in time. It shows qualitative agreement with the formation of spherical holes (or halos) in expanding universes [235, 239, 142]. Near the singularity the density contrast is

$$\frac{\mu}{\epsilon^{FLRW}} \sim \frac{l_1^4}{l_2^4}\left(\frac{l_2^2 + \rho^2}{l_1^2 + \rho^2}\right)^{8/3} + O(t^2); \tag{5.75}$$

thus, the energy density becomes singular as $t \to 0$ in the same way as the energy density of the radiative background. In fig. 5.5 the functions (5.71) are shown for several values of t over the background functions of the radiative perfect fluid.

It is interesting to perform a perturbative analysis about the parameter $l_1^2 - l_2^2$ of the solution (5.69). This is reasonable when l_2 is very close to l_1. In this case the perturbations can be analysed using the background coordinates as physical coordinates. One can see that whereas for a metric of type (5.69) the horizon of a particle at ρ_0 grows in time like $\rho - \rho_0 = \pm t$, the pulse width for $t < l_1$ grows like $2\sqrt{2}t$, so that for a particle located at $\rho = 0$ the pulse is, initially, larger that the particle horizon. However, for $t > l_1$ the pulse width grows like $4\sqrt{3}l_1t$, so that for $l_1 > \frac{1}{2}\sqrt{3}$ it is always larger than or equal to the horizon size, but for $l_1 < \frac{1}{2}\sqrt{3}$ it becomes smaller than the horizon. In this last case an observer at a certain radius ρ_0 will see a localized inhomogeneity entering and leaving its horizon on an isotropic and homogeneous background. Thus the parameter l_1 characterizes the time in

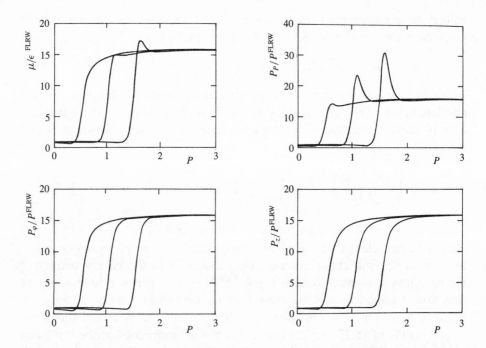

Fig. 5.5. Time evolution of the eigenvalues (5.71) corresponding to metric (5.69) compared with the background values. At space-like infinity the ratios shown approach a constant value different from unity, $l_1^4 l_2^{-4} = 16$, due to the fact that the longitudinal expansion differs from the background by a constant factor. The density contrast and the principal pressure in the radial direction grow in time. This can be interpreted as the formation of a hole with a growing density shell.

which the inhomogeneity becomes smaller than the horizon size. One may also analytically compute the velocities of the metric and of the density perturbations. It is found that such velocities are greater than the speed of light but approach it at null infinity; for instance, the velocity of the maximum of the metric perturbation is $t/\sqrt{t^2 - l_1^2} > 1$, which approaches 1 when $t \to \infty$. This superluminal effect can be understood as the result of the interference produced by the superposition of the two solitons represented by s_1 and s_2. Similarly, the density perturbations cannot be interpreted as lumps of fluid that propagate over the background; instead, different parts of the fluid are perturbed at different points of the spacetime following the metric perturbations. This is similar to the results we found in section 4.6.3 for the general behaviour of gravitational solitons.

Another metric, which is qualitatively similar to (5.69), obtained by the same method is

$$ds^2 = C_2 \frac{1}{\rho^2 (s_1 s_2 - 1)^2} \left[\frac{s_1 s_2 (s_1 - s_2)}{(s_1^2 - 1)^2 (s_2^2 - 1)^2} \right]^{2/3} (-dt^2 + d\rho^2)$$
$$+ t^2 (s_1 s_2)^{-1/3} [(s_1 s_2)^{-1} \rho^2 d\varphi^2 + s_1 s_2 dz^2], \tag{5.76}$$

with the scalar field

$$\sigma = t^{-1} (s_1 s_2)^{1/3}. \tag{5.77}$$

The regularity condition on the symmetry axis requires in this case that $C_2 = (l_2^2 - l_1^2)^2$. The physical properties of this metric are similar to those described for the metric (5.69) but it presents two important differences. In fact, using the conventions at the end of section 4.4.1 this metric has a transverse scale expansion given by $(s_1 s_2)^{-1/3}$, qualitatively similar to metric (5.69), but it also shows a '+' wave polarization with a wave amplitude given by $\ln(s_1 s_2)$ that is absent in the previous metric. Furthermore, the asymptotic value of the longitudinal expansion $f(t, \rho)$ at space-like infinity is exactly that of the background, as it is at time-like infinity. Thus, there is no hole formation and the perturbation is only localized on the cylindrical shell. On the other hand the density contrast grows in time like it does for the metric (5.69).

Related solitons in Brans–Dicke theory. Let us now comment on the application of the solutions such as (5.69)–(5.70) or (5.76)–(5.77) to the related scalar theories discussed above. As we have seen in section 5.4.3 the above solutions can also be considered as solutions of the Brans–Dicke theory with $\omega = 0$ and, by the conformal transformation (5.63), they are related to Brans–Dicke theory with ω arbitrary. The background spacetime and the background scalar field are, respectively,

$$ds^2 = t^{1+\Omega} (-dt^2 + d\rho^2) + t^{1+\Omega} (\rho^2 d\varphi^2 + dz^2), \quad \hat{\sigma} = t^{-\Omega}, \tag{5.78}$$

with Ω given by (5.63). With the appropriate transformation (5.63), solutions (5.69)–(5.70) and (5.76)–(5.77) may be interpreted as representing cylindrical soliton-like perturbations on the background (5.78).

Related soliton solutions with a stiff perfect fluid. In a similar way to transformation (5.65), the above soliton solution may be transformed to a soliton solution on the background,

$$ds^2 = t(-dt^2 + d\rho^2) + t(\rho^2 d\varphi^2 + dz^2), \quad \hat{\sigma} = \sqrt{3/2} \ln t, \tag{5.79}$$

which describes, see (5.39), a stiff fluid on a flat FLRW spacetime with energy density $\epsilon = p = (3/4)/t^3$. The transformed solution may thus be compared to (5.41)–(5.42) but unlike it, here the density of the fluid is also perturbed. A solution of (5.64) can be transformed into a solution of (5.32)–(5.33) by means of (5.66). In particular the soliton solution (5.69)–(5.70) becomes the solution

$$ds^2 = C_1 \frac{[s_1 s_2(\rho s_1 + t)(\rho s_2 + t)]^2}{t\,[\rho(s_1 s_2 - 1)]^2} \left[\frac{s_1 - s_2}{(s_1^2 - 1)^2(s_2^2 - 1)^2} \right]^{2/3} (-dt^2 + d\rho^2)$$
$$+ t(\rho^2 d\varphi^2 + dz^2), \qquad (5.80)$$

with the scalar field

$$\hat{\sigma} = \sqrt{3/2}\,\ln\left[t(s_1 s_2)^{2/3}\right]. \qquad (5.81)$$

Unlike the previous metrics, this metric has no radiative modes; it approaches the flat FLRW background containing stiff fluid (5.79). The source of this solution is a perfect fluid of stiff matter with the fluid four-velocity defined as in (5.34). This solution has no transverse scale expansion or wave polarization because the coefficients $g_{\varphi\varphi}$ and g_{zz} take just the background values. It has been used [153] to illustrate, by means of a finite soliton-like perturbation, the density modes of the linear perturbations of FLRW models; these are the modes relevant for galaxy formation. The density contrast at null infinity on the cylindrical shells, where the soliton wave has a maximum, does not grow in time. In this sense this metric is complementary to (5.41) which contains radiative modes but no independent density modes.

Another solution for stiff matter containing both radiative and density modes can be obtained by transforming the soliton solution (5.69)–(5.70) by means of (5.65). It is

$$ds^2 = C_2 \frac{s_1 s_2}{t\rho^2(s_1 s_2 - 1)^2} \left[\frac{s_1 - s_2}{(s_1^2 - 1)^2(s_2^2 - 1)^2} \right]^{2/3} (-dt^2 + d\rho^2)$$
$$+ t\left[(s_1 s_2)^{-1}\rho^2 d\varphi^2 + s_1 s_2 dz^2 \right], \qquad (5.82)$$

with the scalar field

$$\hat{\sigma} = \sqrt{3/2}\,\ln\left[t(s_1 s_2)^{-1/3}\right]. \qquad (5.83)$$

As for metric (5.76) the longitudinal expansion of this metric exactly approaches the background values at space-like and time-like infinities. Thus, the cylindrical perturbations are located on a cylindrical shell only. Unlike metric (5.80) the spacetime has a '+' wave polarization. On the other hand, like for metric (5.80) the density contrast does not grow in time.

Solitons on FLRW with radiation and stiff fluids. Although we have restricted ourselves to considering perturbations on radiative FLRW backgrounds, it is also possible to extend the above technique to FLRW backgrounds with the equation of state $p = \gamma\epsilon$, where γ is an arbitrary parameter, see Diaz *et al.* [77]. The key to this extension is to consider Einstein's equation in a five-dimensional theory with a massless scalar field χ, instead of vacuum,

$$R_{AB} = \chi_{,A}\chi_{,B}, \tag{5.84}$$

and we recall from subsection 5.4.2, that the ISM can also be applied in this case. The effective equations in four dimensions are now

$$\bar{R}_{\mu\nu} = \sigma^{-1}\sigma_{,\mu;\nu} + \chi_{,\mu}\chi_{,\nu}. \tag{5.85}$$

This corresponds to a formal stress-energy tensor $T_{\mu\nu} = \sigma^{-1}\sigma_{,\mu;\nu} + \chi_{,\mu}\chi_{,\nu} - \frac{1}{2}\chi^{\alpha}\chi^{\alpha}g_{\mu\nu}$, where we have used that the massless scalar field σ satisfies $\sigma^{;\alpha}_{;\alpha} = 0$. Now let us consider the stress-energy tensor for a perfect fluid $T_{\mu\nu} = (\epsilon + p)u_{\mu}u_{\nu} + pg_{\mu\nu}$. This tensor may be decomposed formally as the sum of the stress-energy tensor of a radiative fluid plus the stress-energy tensor of a stiff fluid as follows:

$$T_{\mu\nu} = (\epsilon_r/3)(4u_{\mu}u_{\nu} + g_{\mu\nu}) + \epsilon_s(2u_{\mu}u_{\nu} + g_{\mu\nu}), \tag{5.86}$$

where $\epsilon_r = 3(\epsilon - p)/2$ is the energy density for radiation and $\epsilon_r = (3p-\epsilon)/2$ is the energy density for a stiff fluid, see (5.35). The corresponding fluid pressures are $p_r = \epsilon_r/3$ and $p_s = \epsilon_s$, respectively. Of course such an identification, although formally valid, is not always physically reasonable. The scalar field χ can be interpreted as the velocity potential for a (irrotational) stiff fluid as was done in (5.34)–(5.35). The identification of the traceless part of the tensor, i.e. $\sigma^{-1}\sigma_{,\mu;\nu}$, as the stress-energy tensor of radiation is possible when the spacetime possesses certain symmetries. In particular, this is possible for all isotropic and homogeneous FLRW metrics. In ref. [79] a family of two-soliton metrics that represent cylindrical soliton-like perturbations propagating on the flat FLRW background with a mixture of stiff fluid and radiative fluid was obtained (see ref. [180] for corrected misprints) as

$$ds^2 = C\frac{t^{n-2}(s_1s_2)^{2-p}\left[(\rho s_1 + t)(\rho s_2 + t)\right]^Q (s_1 - s_2)^{u-2}}{\rho^2(s_1s_2 - 1)^2 \left[(1 - s_1^2)(1 - s_2^2)\right]^{u/2}}(-dt^2 + d\rho^2)$$
$$+ t^n[(s_1s_2)^{-p}\rho^2 d\varphi^2 + (s_1s_2)^{-q}dz^2], \tag{5.87}$$

with the two scalar fields

$$\sigma = t^{1-n}(s_1s_2)^{(p+q)/2}, \qquad \chi = \sqrt{3n(2-n)/2}\ln t, \tag{5.88}$$

and where the constant C imposed by the regularity condition on the symmetry axis is $C = (l_2^2 - l_1^2)^2/l_1^{2Q}$. Here the parameters p, q and n are arbitrary constants and $u \equiv 2(q^2 + p^2 + pq)$, $Q \equiv -(q/2)(3n - 2) - (p/2)(3n - 4)$. When $n = 2$ there is only one scalar field and the solution is of the type given before: $p = q = -2/3$ corresponds to (5.69)–(5.70) and $p = 4/3$, $q = -2/3$

corresponds to (5.76)–(5.77). The source of this spacetime is of course an anisotropic fluid, except for some values of the parameters, $p = q = 0$ or $s_1^2 = s_2^2$, when it is a perfect fluid. Note that this family of solutions is of the generalized soliton solutions type, consequently they are not the limit of truly nondiagonal soliton solutions, except for some particular values of the parameters which include the solutions previously studied. Other solutions of this type were deduced in ref. [78] which also generalize (5.69)–(5.70). The properties of the soliton perturbations discussed in detail in ref. [80].

Gravitational waves on FLRW backgrounds. To end this section let us now comment on some solutions which, although not solitonic in general, are closely related to those discussed here and in previous sections, in the sense that they represent gravitational waves (rather than soliton-like waves) propagating on FLRW backgrounds. Griffiths [128] considered a solution which may be interpreted as a gravitational wave propagating on a flat FLRW background of stiff fluid. This solution was then generalized to the case in which two gravitational waves in opposite directions propagate and collide in the same background [129]. These solutions were then extended to open [30] and closed [101] FLRW backgrounds of stiff fluid. The complete metrics for the solutions were not given in these references; a method for obtaining complete solutions was described in ref. [7]. The complete derivation of these metrics as well as a study of the different propagating wavefronts was given in ref. [31].

6

Cylindrical symmetry

Cylindrically symmetric spacetimes also have the symmetries required to generate solutions by the ISM. In this chapter we review, briefly, the soliton solutions in the cylindrical context. The analytic expressions for such solutions can be obtained from the cosmological solutions of chapters 4 and 5 by a simple reinterpretation of the relevant coordinates. For this reason the sections in this chapter are considerably shorter. One of the main interesting features of these spacetimes is that a definition of energy, the so-called C-energy, can be given and, consequently, cylindrically symmetric waves can be understood as waves that carry energy. The study of the C-energy in the soliton solutions will play an important role in the interpretation of the cylindrically symmetric soliton waves. Some general properties are discussed in section 6.1. Diagonal metrics, i.e. one polarization waves, are described in section 6.2; these include all generalized soliton solutions of sections 4.4.1, 4.5 and 4.6 after appropriate transformations. Some attention is paid to solutions which have been used to describe the interaction of a straight cosmic string with gravitational radiation. In section 6.3 solutions with two polarizations are considered and the conversion of one of the modes of polarization into the other is described. This conversion is an effect of the nonlinear interaction between the two modes and is interpreted as the gravitational analogue of the Faraday rotation of electromagnetic waves by a magnetic field and plasma.

6.1 Cylindrically symmetric spacetimes

The main reason for the interest in cylindrically symmetric exact solutions of Einstein's equations is that they are the simplest metrics for which exact gravitational wave solutions are known [90, 179, 213, 214, 272]. For a long time these solutions were the best evidence that general relativity predicts the existence of gravitational waves. Although the asymptotic nonflatness of these spacetimes makes them of little use for the study of any realistic compact source, it is believed that the study of the properties of exact cylindrical gravitational waves will be of help in understanding strong gravitational radiation from some realistic sources.

We have seen that most soliton solutions are obtained by the ISM from a diagonal background solution. This is also the case in the cylindrical context and, in particular, the most extensively used background metric is the static cylindrically symmetric Levi-Cività [201] family,

$$ds^2 = b^2 \rho^{(d^2-1)/2}(d\rho^2 - dt^2) + \rho^{1+d}d\varphi^2 + \rho^{1-d}dz^2, \tag{6.1}$$

where b and d are arbitrary parameters. This is none other than the Kasner metric adapted to cylindrical symmetry. Here ρ is the radial coordinate and φ the polar angle: $0 \leq \varphi \leq 2\pi$. It includes flat spacetime ($d = 1$). In some sense this solution plays, within the context of cylindrical symmetry, the same role that the Schwarzschild solution plays in the spherically symmetry case, although it has a naked singularity on the symmetry axis $\rho = 0$ (except for $d^2 = 1$), and no event horizon hides it. Nevertheless this type of singularity may be understood in most cases by the presence of a source on the axis.

In fact, the following physical interpretation has been suggested for metric (6.1). Levi-Cività [201] was the first to study this solution. In the Newtonian limit he found

$$d \simeq \frac{1+2\lambda}{1-2\lambda}, \tag{6.2}$$

where λ is the (dimensionless) relativistic mass per unit length of a cylinder; reintroducing the physical constants λ becomes $\lambda c^2/G$, where $c^2/G \sim 10^{28}$ g/cm. Thus for $d > 1$ metric (6.1) can be interpreted as the exterior field of an infinite cylinder along the z axis with a uniform mass per unit length, λ. For $d < 1$, Newtonian test particles far from the cylinder suffer a repulsive force [262] and the cylinder cannot be of ordinary matter but of matter with negative mass. For $d \neq 1$ the metric has a naked curvature singularity and we may follow Israel [155] to infer a stress tensor on the axis compatible with the above interpretation. For $\lambda = 0$ ($d = 1$) the space is flat, and d grows without bound when $2\lambda \to 1$. When $d = 1$ one must take $b^2 = 1$ to recover the flat space metric. However, it is convenient to take b arbitrary because in this case we have flat space minus a wedge (if $b^2 > 1$), i.e. a deficit angle which may be interpreted as due to the gravitational field created by a straight gauge cosmic string [291]. Moreover, Marder [213, 214] when studying the matching problem of (6.1) with the interior solution of a physical cylinder, assuming a specific equation of state, concluded that in this case b does depend on d (therefore on the mass density) and cannot be put equal to 1. Thus, in what follows we shall assume that d and b are arbitrary parameters. In section 8.5 we will come back to the interpretation of the Levi-Cività metric in terms of a massive line source along the z axis with a constant linear mass density, i.e. an ILM (infinite line source), using the notation of refs [33, 36].

There is another important difference with the Schwarzschild solution that must be pointed out. The Levi-Cività solution is not the most general exterior solution for a cylinder. From Einstein's equations with (total) cylindrical symmetry one can easily see that gravitational waves can be superimposed on the exterior field of a static cylinder, in contrast with the spherical symmetric case. The most general metric for gravitational waves in a vacuum, with cylindrical symmetry, was written by Kompaneets [176, 160],

$$ds^2 = e^{2(\gamma-\psi)}(d\rho^2 - dt^2) + \rho^2 e^{-2\psi}d\varphi^2 + e^{2\psi}(dz + \omega d\varphi)^2, \qquad (6.3)$$

with the functions γ, ψ and ω depending on t and ρ only. Note that (ρ, t) are canonical coordinates in the sense of section 1.3. The canonical coordinates (α, β) in (1.45) and (1.46) are now $\alpha = \rho$, $\beta = t$; whereas in the cosmological context they were $\alpha = t$, $\beta = z$. Thus we see that exact vacuum cosmological solutions can be interpreted as solutions in the cylindrical context by simply applying the following transformations (together with a change of sign in front of ds^2):

$$t \to \rho, \quad z \to t, \quad x \to i\varphi, \quad y \to iz, \qquad (6.4)$$

where now $0 \le \varphi \le 2\pi$. Gravitational waves with two polarizations are de-scribed by the time-dependent functions $\psi(t, \rho)$ and $\omega(t, \rho)$ which describe the transversal modes. The physical interpretation of the cylindrically symmetric solutions is discussed in ref. [39].

When $\omega(t, \rho) = 0$ the metric reduces to the Einstein–Rosen [90] diagonal form; then when comparing with the expressions $g_{11} = te^\Phi$, $g_{22} = te^{-\Phi}$ of (4.40), we see from (4.41)–(4.42) that $\psi(t, \rho)$ satisfies a linear wave equation, after the substitution

$$\Phi(t, z) \to \ln \rho - 2\psi(t, \rho), \quad \ln f(t, z) \to 2\gamma(t, \rho) - 2\psi(t, \rho). \qquad (6.5)$$

C-energy. As we stressed in section 4.3.3 an important property of a cylindrical spacetime is that gravitational energy may be defined on it. In fact, Thorne [272] was able to introduce a total (cylindrical) energy called C-energy, and a covariant C-energy flux vector P^μ obeying a conservation law $P^\mu{}_{;\mu} = 0$. The energy density is localizable and locally measurable. An observer with four-velocity u^μ measures an energy density $P^\mu u_\mu$ and if X^μ is a space-like vector such that $X_\mu u^\mu = 0$, the C-energy flux is $P^\mu X_\mu$. The four-vector P^μ is derived from a C-energy potential $C(\rho, t)$ defined as

$$C(t, \rho) = \gamma(t, \rho), \qquad (6.6)$$

which is proportional to the total C-energy contained inside a cylinder of radius ρ per unit coordinate length z. It is given by

$$(P^t, P^\rho) = (8\pi\rho)^{-1}e^{-2(\gamma-\psi)}(C_{,\rho}, -C_{,t}), \qquad (6.7)$$

and its components represent the local energy density measured by a local observer with four-velocity $u^\mu = (1, 0, 0, 0)$ and the local flux measured by that observer along the ρ-direction (we take $G = c = 1$). The definition of C-energy was extended by Chandrasekhar [55] in the general nondiagonal case (6.3). Also in analogy with the definition of the three-velocity field defined in section 4.3.3 we may define a *C-velocity* field associated with the C-energy flux as follows. Since the flux vector P^μ is time-like we may define the four-velocity $u^\mu = P^\mu / (-P_\alpha P^\alpha)$, and it is clear that an observer with such a four-velocity will measure no flux of C-energy. The three-velocity of such an observer $v_C^i = u^i / u^0$ ($i = 1, 2, 3$) will be called the C-velocity. In our coordinate system it may be written as,

$$v_C^\rho = -2\psi_{,\rho}\psi_{,t}\left[(\psi_{,\rho})^2 + (\psi_{,t})^2\right]^{-1}. \tag{6.8}$$

The C-energy plays an important role in the study of the physical properties of cylindrical solutions. Thus Marder [214] has calculated the effect of pulse waves on particles moving on geodesics and the change in proper mass per unit length of a particular solid cylinder due to the radiation of such pulses from the cylinder. Thorne [272] found that the decrease of proper mass was equal to the C-energy carried by the waves. It has also been used to characterize the energy properties of soliton solutions [109, 113, 276].

We shall divide the soliton solutions for total cylindrical spacetimes into two classes: those with one mode of polarization (Einstein–Rosen diagonal case) and those with two independent polarizations (nondiagonal metrics). Our emphasis will be on describing solutions which have been shown to have some potential physical relevance.

6.2 Einstein–Rosen soliton metrics

Here we discuss the case of one-polarization waves, i.e. when the two Killing vectors are hypersurface orthogonal. As we know from the cosmological case, these metrics are more easily studied in detail than two-polarization metrics because the computation of the Riemann tensor is relatively simple, see section 4.3. Here the null tetrad (4.11) is replaced by

$$\left.\begin{array}{l} \vec{n} = (1/\sqrt{2f})(\partial_\rho + \partial_t), \quad \vec{l} = (1/\sqrt{2f})(\partial_\rho - \partial_t), \\[2mm] \vec{m} = (1/\sqrt{2g_{zz}})\partial_z - (i/\sqrt{2g_{\varphi\varphi}})\partial_\varphi, \end{array}\right\} \tag{6.9}$$

and the metric is now

$$ds^2 = f(t, \rho)(d\rho^2 - dt^2) + \rho(e^{\Phi(t,\rho)}d\varphi^2 + e^{-\Phi(t,\rho)}dz^2), \tag{6.10}$$

the coefficients of which obey Einstein's equations in vacuum (4.41)–(4.42) after the transformation (6.4). Therefore all generalized vacuum soliton solutions of section 4.4.1 can be reproduced here with the appropriate transformation. These solutions have been considered mainly in connection with cosmic strings [113]. Thus it is useful to recall here the characterization of a cosmic string, its deficit angle and also its connection with the C-energy.

Straight cosmic string. A straight-line cosmic string is characterized by a conical singularity along the axis. As we know, the condition for regularity in the coordinate system (6.10) is that [179] $\lim_{\rho \to 0} X_{,\mu} X^{,\mu}/4X \to 1$, where $X \equiv |\partial_\varphi|^2$. When a metric has no curvature singularity along the axis and it fails to be regular, i.e. the above limit differs from unity, the axis may contain a cosmic string. Equivalently, a cosmic string may also be identified by the deficit angle near the axis which measures the deviation from local flatness around the axis [291]. The deficit angle is defined by

$$\Delta\varphi(\rho) \equiv 2\pi - \left[\int_0^\rho (g_{\rho\rho})^{1/2} d\rho\right]^{-1} \int_0^{2\pi} (g_{\varphi\varphi})^{1/2} d\varphi. \qquad (6.11)$$

In the limit $\rho \to 0$, $\Delta\varphi(0)$ is directly related to the energy density per unit length of the string (string tension) and it is often used in the literature because it has a direct observational manifestation in terms of a light-deflection angle [291, 283]. Moreover, using the field equations (4.41)–(4.42) and (6.6), this last equation reduces to $2C = \ln f + \ln \rho - \Phi$, and it is easily shown that

$$\Delta\varphi = 2\pi [1 - \exp(-C)], \qquad (6.12)$$

on the axis. Thus, for Einstein–Rosen metrics the deficit angle on the axis is related to the C-energy. It turns out that, for the metrics which approach Minkowski space far from the axis, the relation (6.12) between the deficit angle and the C-energy is also valid for large ρ.

As in the cosmological case we distinguish four asymptotic regions: near the symmetry axis ($\rho \to 0$), null infinity ($\rho \sim |t| \to \infty$), space-like infinity ($t \ll \rho \to \infty$) and time-like infinity ($\rho \ll t \to \infty$). We shall restrict ourselves to soliton solutions with complex poles. In this way we avoid the problems of discontinuous first derivatives across the light cones. The light cones are defined by:

$$t_k^2 = \rho^2, \quad t_k = t_k^0 - t, \qquad (6.13)$$

where t_k^0 is the time origin of the k-pole trajectory and the pole trajectories are described by (4.74) after the coordinate change (6.4), so that $z_k^0 - ic_k$ becomes $t_k^0 - ic_k$. Thus we shall consider the generalized soliton solutions (4.79)–(4.82) in the cylindrical context. Besides the coordinate transformation the only difference with the cylindrical case is that the coefficient f is now multiplied by the parameter b^2 of the background (6.1). To study these solutions we thus follow the scheme of section 4.6.

6.2.1 Solutions with one pole

These are the cylindrical versions of (4.84) for composite universes. Also, these solutions can be seen to describe qualitatively, at space-like infinity, the behaviour of the generalized soliton solutions with any number of poles of the same type. The curvature tensor becomes singular on the axis unless $(d + h)^2 = 1$, in which case the axis is quasiregular and the deficit angle is

$$\Delta\varphi(0) = 2\pi(1 - b^{-1}), \tag{6.14}$$

like that of the background metric (6.1). The analysis of this solution and its associated C-energy in the remaining asymptotic regions indicates that these solutions have radiation at null infinity but such radiation cannot be interpreted as being localized and propagating on the static background. Near the axis the metric on $t = $ constant hypersurfaces behaves like a Levi-Città metric with a parameter $d' = d + h$, instead of d. However, at space-like infinity it behaves like a Levi-Città metric with a parameter d. In some sense this metric can be seen as connecting two Levi-Città metrics with different parameters through the light cones which contain gravitational radiation.

Now, from the viewpoint of cosmic strings, only the metrics with the parameter $d' = 1$ (consider the relation $(d + h)^2 = 1$ above) have a string on the axis and its deficit angle is given by (6.14). At space-like infinity, however, the spacetime approaches that of Levi-Città with $d = 1 - h$ ($h \neq 0$) and it is not asymptotically Minkowskian. This may be seen as a consequence of the presence of C-energy surrounding the string. Note the remarkable fact that if we take $0 < h < 1$ the gravitational effect produced by the gravitational energy far from the axis is similar to that of a massive rod with negative energy density. Moreover, we have asymptotically flat solutions ($d = 1$) which have a massive line source with $d' = 1 + h$.

Not all of those solutions admit a physical interpretation because some of them develop singularities at $|t| \to \infty$. Only those satisfying

$$h/|h| = d/|d|, \quad |d| \geq |h| \quad \text{or} \quad (d + h)^2 = 1, \quad h(h + d) \leq 2, \tag{6.15}$$

are free from singularities.

6.2.2 Cosolitons with one pole

These are the cylindrical versions of those of subsection 4.6.2. The Riemann components in the different asymptotic regions are given by (4.92)–(4.93) after using (6.9), (4.12) and the coordinate transformation (6.4). Near the axis the curvature tensor becomes singular, unless $d^2 = 1$, in which case the axis is quasiregular with a deficit angle

$$\Delta\varphi(0) = 2\pi(1 - 2^{h(h-1)/2}b^{-1}). \tag{6.16}$$

At null infinity the metric is radiative and has a finite value of C-energy radiation. At space-like infinity the metric becomes flat when $d^2 + h^2 = 1$, therefore it is not possible to impose that the axis be quasiregular, $d^2 = 1$, together with the requirement that the metric becomes flat far from the axis, unless $h = 0$, i.e. the static case. As for the case of solitons with one pole, this metric has radiation at null infinity but it cannot be interpreted as representing localized radiation propagating on a static background.

6.2.3 Solitons with two opposite poles

These metrics are the cylindrical versions of those of section 4.6.3, i.e. (4.79)–(4.80) with $h_1 = -h_2 = h$, $s = 2$, and the field Φ can be written in terms of the Fourier Bessel integral (4.94), after the coordinate change (6.4). Thus they may be interpreted as representing localized cylindrical gravitational (soliton-like) perturbations 'propagating' on the static Levi-Cività background, coming from past infinity and reflecting off the axis. They have only a curvature singularity on the axis unless $d^2 = 1$ in which case the axis is quasiregular with a deficit angle given by (6.14), like the background metric. At null infinity the radiative Riemann components dominate. The rate of C-energy radiation,

$$C_{,v} = -2h^2(8c_1c_2)^{-1}(c_2^{1/2} - c_1^{1/2})^2[1 + 0(\rho^{-1/2})], \quad v = (t - \rho)/2, \quad (6.17)$$

reaches a finite value.

Perturbative analysis. The localized character of the gravitational perturbation, or pulse wave, is clearly seen by performing a perturbative analysis (as in section 4.6.3) in which one can study analytically how the perturbation propagates on the static background. For this we take $t_1^0 = t_2^0$, define $\delta w \equiv c_2 - c_1$ and call $w \equiv c_1$, $\sigma \equiv \sigma_2$. Since the nonstatic term in (4.79) is now proportional to $\ln(\sigma_1/\sigma_2)$ we may expand this to first order in δw. The function σ_1/σ_2 differs from unity only on a small localized region. Following section 4.6.3 we can write

$$\sigma_1/\sigma_2 \simeq 1 + \delta w \, \partial_w \ln \sigma. \qquad (6.18)$$

The perturbation on the right hand side is too complicated in terms of (ρ, t) to find analytic expressions for its shape and motion directly. Instead we shall introduce new coordinates (R, T) that depends on the parameter w defined, in analogy to (4.86), by

$$\left. \begin{array}{l} \rho = w \cosh(2T) \sinh(2R), \quad t = w \sinh(2T) \cosh(2R), \\ 0 \leq R < \infty, \quad -\infty < T < \infty. \end{array} \right\} \qquad (6.19)$$

In terms of these coordinates $\sigma = \tanh^2 R$. One can now find the maximum in the ρ-direction of the perturbation in (6.18) solving the equation $\partial_\rho \partial_w \ln \sigma = 0$.

This can easily be done after finding $\partial_\rho R$ and $\partial_\rho T$ from the coordinate change (6.19). The solution in (R, T) coordinates is the trajectory

$$3\sinh^2(2T) - \cosh^2(2R) = 0, \tag{6.20}$$

which is the analogue of (4.98). It can now be transformed to (ρ, t) coordinates. The final result, which gives the trajectory of the maximum ρ_m in terms of t, is

$$\left.\begin{array}{ll} \rho_m^2 = (w + t/\sqrt{3})(\sqrt{3}t - w), & t \geq w/\sqrt{3}; \\ \rho_m^2 = (t/\sqrt{3} - w)(\sqrt{3}t + w), & t \leq -w/\sqrt{3}, \end{array}\right\} \tag{6.21}$$

and for $-w/\sqrt{3} \leq t \leq w/\sqrt{3}$ we have $\rho_m = 0$. Thus the pulse has a maximum on the symmetry axis for $-w/\sqrt{3} \leq t \leq w/\sqrt{3}$ and at time $|t| > w/\sqrt{3}$ this maximum propagates on the (ρ, t)-plane. Thus the parameter $w/\sqrt{3}$ characterizes the time of formation of the pulse.

The speed of this pulse in terms of (ρ, t) is now found to be

$$d\rho_m/dt = (w + \sqrt{3}t)(3w + \sqrt{3}t)^{-1/2}(\sqrt{3}t - w)^{-1/2} \quad t \geq w/\sqrt{3}. \tag{6.22}$$

For $t < 0$ the expressions can be trivially obtained from those for $t > 0$ so we shall not discuss them further. From (6.22) we see that the speed of the pulse is infinite at the beginning, $t = w/\sqrt{3}$, and that it tends to the speed of light when $t \to \infty$. The trajectory of the maximum approaches the asymptotes $\rho = t + w/\sqrt{3}$ for the outgoing pulse and $\rho = -t + w/\sqrt{3}$ for the incoming pulse.

The shape of the pulse, its height and width, can now be given analytically. It is found [109] that the height defined as $h_p = |\delta w \partial_w \ln \sigma(\rho_m)|$ is given in the (ρ, t) coordinates by

$$h_p = \frac{1}{2}\delta w \left(\sqrt{3}w^{-1}t^{-1} + 3\sqrt{3}t^{-2}\right)^{1/2}, \tag{6.23}$$

so that it decreases as $t^{-1/2}$ when $t \to \infty$. As $t \to \infty$ the pulse width goes to a constant independent of w [109]. It is also interesting to compute the C-energy flux, from (6.7), or equivalently the C-velocity defined in (6.8) which in this case takes the form [113] (when $d = 1$ which is the case relevant for a cosmic string)

$$v_C^\rho = -\frac{4\sinh(4R)\coth(2T)\,V_1V_2}{4V_2^2 + \sinh^2(4R)\coth^2(2T)V_1^2}, \tag{6.24}$$

where $V_1 = \cosh^2(2R) - 3\sinh^2(2T)$ and $V_2 = 3\cosh^2(2R)\cosh^2(2T) - \cosh^4(2R) - \cosh^2(2T) + 1$. Depending on the sign of this function there are two competing fluxes of C-energy: one of them ingoing (towards the axis) and the other outgoing (from the axis); the two fluxes cancel when $v_C^\rho = 0$. This happens on the world lines $V_1 = 0$ and $V_2 = 0$. The first is the trajectory

(6.20), i.e. the maximum of the perturbation, and the second is the maximum of the perturbation along the t-direction, which is the analogue of (4.96) in the cosmological case.

All this leads to the following interpretation for these solutions. From (6.22) it seems clear that the localized pulse cannot be interpreted as describing the propagation of a causal effect, i.e. a travelling wave. Such a pulse may be understood as a superposition effect produced by the interference of incoming and outgoing fluxes of C-energy. This is similar to what is found in the cosmological case with two opposite poles in section 4.6.3, where the Bel–Robinson superenergy tensor was used, and the solitons could also be understood as a consequence of the interference of two (super)energy fluxes propagating in opposite directions.

To see the effect of the pulse on the axis we can follow Marder [213] who estimated the change in proper mass per unit proper length that a cylinder around the axis suffers when it emits a gravitational wave pulse as $E_{wave} = E_{after} - E_{before}$, where E is given simply in terms of the C-energy potential (6.6) by $E = C/4$. We can evaluate E in our solution for some fixed large radius ρ at $t = 0$, before the passage of the wave, and after at $t \to \infty$. In both cases the metric tends to the static background (6.1) but with a different coefficient b. The explicit result is

$$E_{wave} \sim \frac{1}{2} \ln \left[\frac{4|c_1 c_2|}{(c_1 - c_2)^2} \right]. \qquad (6.25)$$

Thus the passage of the wave has permanently affected the gravitational field: energy has been radiated away.

6.2.4 Cosolitons with two opposite poles

These solutions are the cylindrical version of (4.81)–(4.82) with $s = 2$ and $h_1 = -h_2 = h$. The Riemann components at null infinity are given by (4.93), where one sees that the radiative component dominates. The metric becomes singular on the axis unless $d^2 = 1$, in which case the deficit angle is

$$\Delta\varphi(0) = 2\pi (1 - 2^{h(h-1)} b^{-1}) + O(\rho^2). \qquad (6.26)$$

The rate of C-energy radiation approaches a constant value at null infinity:

$$C_{,v} = -2h^2 (8c_1 c_2)^{-1} (c_2^{1/2} - c_1^{1/2})^2 [1 + 0(\rho^{-1/2})], \qquad (6.27)$$

as in the previous soliton case. A perturbative analysis such as the one performed in the previous case can be carried out. It is found [113] that the maximum of the perturbation is located on the world line

$$\sinh^2 (2T) - 3\cosh^2 (2R) = 0, \qquad (6.28)$$

where (T, R) are defined in (6.19). In terms of the coordinates (ρ, t) such a trajectory is the same as that of (6.20) but for a shift, $2w/\sqrt{3}$, along the t axis ($w \equiv c_1$). Again for $d = 1$, the C-energy flux vanishes along the trajectory (6.28). The interpretation of these metrics is very similar to that of the previous two-soliton solutions. The localized perturbation is the result of the superposition of two competing C-energy fluxes, one incoming towards the axis and the other outgoing.

For $d = 1$, both of these solutions represent the interaction of a static string with incoming and outgoing gravitational radiation localized basically along the light cones. This may be taken as an idealization of gravitational radiation, not necessarily with cylindrical symmetry, surrounding the cosmic string. For $d > 1$ the wave interpretation is similar but now they are on the background of massive line sources.

6.3 Two polarization waves and Faraday rotation

An interesting phenomenon in two polarization cylindrical waves has been described by Piran and Safier [243], who proved that the propagation of these waves displays a reflection of ingoing to outgoing waves and vice versa, combined with a rotation of the polarization vector between the $+$ and $\times$ modes. The rotation of the polarization vector has been interpreted as a consequence of the nonlinear gravitational interaction between the two independent polarizations. This has been described as the gravitational analogue of the electromagnetic Faraday rotation. In fact, although in the electromagnetic theory there is no such interaction, the polarization vector of an incident linearly polarized electromagnetic wave rotates as it propagates through a medium containing a magnetic field and plasma. Tomimatsu [276] has studied this phenomenon in some soliton solutions. Let us introduce the ingoing and outgoing null coordinates $v \equiv (t - \rho)/2$, $u = (t + \rho)/2$, and define, from (6.3),

$$A_+ \equiv 2\psi_{,u}, \quad B_+ \equiv 2\psi_{,v}, \quad A_\times \equiv \frac{1}{\rho}e^{2\psi}\omega_{,u}, \quad B_\times \equiv \frac{1}{\rho}e^{2\psi}\omega_{,v}. \qquad (6.29)$$

Then the metric coefficient $\gamma(t, \rho)$ is determined by these new functions. In fact, from the vacuum Einstein equations for the metric coefficient f (1.40)–(1.41), we have that

$$\gamma_{,\rho} = \frac{\rho}{8}(A_+^2 + B_+^2 + A_\times^2 + B_\times^2), \quad \gamma_{,t} = \frac{\rho}{8}(A_+^2 - B_+^2 + A_\times^2 - B_\times^2). \qquad (6.30)$$

Now one can define the quantities,

$$A \equiv (A_+^2 + A_\times^2)^{1/2}, \quad B \equiv (B_+^2 + B_\times^2)^{1/2}, \qquad (6.31)$$

which may be interpreted as the ingoing and outgoing wave amplitudes, respectively. The indices $+$ and $\times$ denote the different polarizations and the polarization angles are defined by

$$\tan(2\theta_A) \equiv A_\times/A_+, \quad \tan(2\theta_B) \equiv B_\times/B_+. \qquad (6.32)$$

Note that this may be compared to related definitions in the cosmological context by Adams *et al.* [2], see (5.17)–(5.18).

The rotation of the polarization vector between the + and × modes can be understood [243, 242] from an analysis of the equations for A_+, $A_\times$, B_+ and $B_\times$ that follow from the vacuum Einstein equations for the matrix g (1.39):

$$A_{+,u} = \frac{1}{2\rho}(A_+ - B_+) + A_\times B_\times, \tag{6.33}$$

$$B_{+,v} = \frac{1}{2\rho}(A_+ - B_+) + A_\times B_\times, \tag{6.34}$$

$$A_{\times,u} = \frac{1}{2\rho}(A_\times + B_\times) - A_+ B_\times, \tag{6.35}$$

$$B_{\times,v} = -\frac{1}{2\rho}(A_\times + B_\times) - A_\times B_+, \tag{6.36}$$

with the boundary conditions that $A_+ = B_+$ and $A_\times = -B_\times$ at $\rho = 0$. The first terms on the right hand sides of (6.33)–(6.36) couple the ingoing and outgoing waves with the same polarization and this produces the usual cylindrical reflection. The other terms describe a nonlinear interaction between the two polarizations. To understand the nature of this interaction we follow ref. [243] and consider a large value of ρ, so that we can neglect the first terms on the right hand sides of these equations. In this approximation

$$A_\times(u, v) + iA_+(u, v) \simeq \left[A^0_\times(v) + iA^0_+(v)\right] \exp\left[i\int_{u_0}^u B_\times(u', v)du'\right], \tag{6.37}$$

$$B_\times(u, v) + iB_+(u, v) \simeq \left[B^0_\times(u) + iB^0_+(u)\right] \exp\left[i\int_{v_0}^v A_\times(u, v')dv'\right], \tag{6.38}$$

where $A^0_+(v)$, $A^0_\times(v)$, $B^0_+(u)$ and $B^0_\times(u)$ are the initial conditions given on an outgoing, $u = u_0$, and ingoing, $v = v_0$, null hypersurface respectively. When $A_\times$ is present, B_+ and $B_\times$ will oscillate according to (6.38) with a phase difference of $\pi/2$. This means that if the initial outgoing wave is linearly polarized its polarization vector will rotate as it propagates through the $A_\times \neq 0$ region, and the same is true for the incoming waves, see (6.37). For instance, consider the case $B^0_\times = A^0_\times = 0$ and $|B^0_+| \ll |A^0_+|$. Then $A_\times$ remains almost constant, $A_\times \simeq A^0_\times$, B_+ and $B_\times$ oscillate when the B_+ wave crosses the $A_\times \neq 0$ region, $B_+ \simeq B^0_+ \cos[\int^v A_\times(v')dv']$ and $B_\times \simeq -B^0_+ \sin[\int^v A_\times(v')dv']$. The appearance of $B_\times$ causes the conversion of a small part of $A_\times$ to A_+ and if $|B^0_+| \ll 1$ we have

$$A_+(u, v) \simeq -A^0_\times \int_{u_0}^u B^0_+ du' \sin\int_{v_0}^v A^0_\times(v')dv'. \tag{6.39}$$

In general, when both $B^0_\times$ and $A^0_\times$ are large each wave will cause a rotation of the polarization vector of the other. Thus the rotation of the polarization vector can be understood as a consequence of the nonlinear nature of the gravitational interaction. Using the electromagnetic analogy, here the ingoing (for instance) cylindrical waves play the role of both the magnetic field and the plasma as they rotate the polarization vector of the outgoing cylindrical waves propagating through them.

Let us now consider the combined effect of all terms in (6.33)–(6.36). If the initial data contain only the + mode the only effect will be the cylindrical reflection of A_+ to B_+, which is typical of the Einstein–Rosen waves. When the $\times$ mode is present in the initial data there is both conversion of A_+ to B_+ and $A_\times$ to $B_\times$, via the first term in (6.33)–(6.36), and rotations between A_+ and $A_\times$ and B_+ and $B_\times$, which take place simultaneously. For example, an initial $A_\times$ pulse will reflect some $B_\times$ wave. This $B_\times$ wave will, in turn, cause a rotation of some of the initial $A_\times$ pulse into an A_+ pulse and by itself will rotate, due to the presence of $A_\times$, into B_+. This analysis is confirmed by the numerical results of refs [243, 242]. In the latter reference the collision of two cylindrical waves was also numerically analysed and the results are qualitatively analogous to those of the collision between a pulse wave and a soliton wave studied in section 5.3 in the cosmological context where a time shift in the peaks of the waves is also detected after the collision.

6.3.1 One real pole

The first soliton solution by Tomimatsu [276] in the present context is that of one single real pole with the Minkowski background, i.e. (6.1) with $d = 1$. Real poles in the cylindrical case have the same problem as in the cosmological context. Since the pole trajectories are only defined on a region of the canonical coordinate patch and the cylindrical coordinates are defined for all canonical coordinate patches, the soliton solutions with real poles are not defined on the whole relevant spacetime. Thus one may match the soliton metric with the background metric or other metric along some null hypersurface, but this leads to discontinuities and possible shock waves. The pole trajectories for real poles are, from (4.49) and (6.4), defined by

$$\mu_k^\pm = t_k \pm (t_k^2 - \rho^2)^{1/2}, \quad t_k \equiv t_k^0 - t, \tag{6.40}$$

where t_k^0 is the soliton time origin, and now the relevant region is inside the light cone $t_k^2 \geq \rho^2$. Outside this light cone one may match to the background metric. For one single pole, say $k = 1$, there is only one relevant light cone $t_1^2 = \rho^2$, but for several poles one has an intersection similar to that of fig. 4.2 in the cosmological context, with the appropriate coordinate changes. Tomimatsu's solution corresponds to the Belinski and Zakahrov [23] solution (5.1), and has similar problems for a reasonable physical interpretation.

It is perhaps pertinent to recall here an interesting one-pole cylindrically symmetric solution from the Levi-Cività metric derived by Gleiser and Tiglio in the Einstein–Maxwell context [123]. This solution has been interpreted as representing the electrogravitational field interacting with a straight supercon-ducting string.

6.3.2 Two complex poles

The second solution studied by Tomimatsu is the soliton solution with two complex conjugate poles obtained from the Minkowski background. That is, we take

$$\mu_1 = w_1 - t + [(w_1 - t)^2 - \rho^2]^{1/2}, \quad w_1 \equiv t_1^0 - ic_1, \quad \mu_2 = \bar{\mu}_1. \quad (6.41)$$

If we now define $s \equiv -\mu_1/\rho$, choose $t_1^0 = 0$ and $w \equiv c_1$, the transversal metric coefficients are given by [276]

$$e^{2\psi} = |s|^2 F/G, \quad \omega = -\rho H/F, \quad (6.42)$$

where

$$F = \left| \frac{a^2 + 1}{s^2 - 1} \right|^2 - \left(\frac{|a|^2 + 1}{|s|^2 - 1} \right)^2, \quad G = \left| \frac{a^2 + s^2}{s^2 - 1} \right|^2 - \left(\frac{|a|^2 + |s|^2}{|s|^2 - 1} \right)^2,$$

$$H = 2 \operatorname{Re} \left[\frac{\bar{a}}{\bar{s}} \left(\frac{a^2 + 1}{s^2 - 1} - \frac{|a|^2 + 1}{|s|^2 - 1} \right) \right],$$

where a is an arbitrary complex parameter, related to the vector $(m_0)_c^{(1)}$ in (1.80). Both ψ and ω approach zero at space-like infinity ($t \leq \rho \to \infty$), where $|s| \to 1$. Here $|s|$ is the same as $\sqrt{\sigma_1}$ in (4.74) and we can see the asymptotic behaviour in (4.77), where one must transform coordinates according to (6.4). This solution is now defined in the whole coordinate patch (ρ, t).

A particularly simple case is when $a = \bar{a}$. Then the wave amplitudes (6.31) and the polarization angles (6.32) are

$$A = \frac{2(1 - |s|^2)}{\rho|1 + s|^2}, \quad B = \frac{2(1 - |s|^2)}{\rho|1 - s|^2}, \quad \tan \theta_A = -\tan \theta_B = |s|^2/a. \quad (6.43)$$

The function $|s|^2$ runs from zero to unity along an outgoing null ray $v = v_0 > 0$. Therefore if $|a| < 1$, the $+$ mode is completely converted to the $\times$ mode on the time-like world line $|s|^2 = a$, i.e.

$$\rho = \frac{2|a|^{1/2}}{|a| + 1} \left[t^2 + \left(\frac{|a| + 1}{|a| - 1} \right)^2 w^2 \right]^{1/2}. \quad (6.44)$$

Thus, this solution provides an explicit example of the gravitational analogue of the electromagnetic Faraday rotation described above. By analysing the wave amplitudes at the initial time $t = 0$ and at late times $(t \gg |w|)$ in the three asymptotic regions: time-like, space-like and null infinities, one can see [276] that the solution has a local maximum, which corresponds to a gravitational soliton-like pulse propagating at the speed of light. Its polarization angle θ_B is fixed at $\tan \theta_B = -1/a$, although the $+$ mode is dominant when it leaves the axis. The C-energy flux vector (6.7) for this solution shows that the pulse wave transports energy from the axis to null infinity. This solution illustrates a process of generation of gravitational waves from an initial disturbance localized near the axis as well as a full rotation of the polarization vector.

In the context of axisymmetric stationary gravitational fields this solution corresponds to the Kerr–NUT metric, see (8.48), with $a^2 > m^2$, where a and m are the angular momentum and mass, respectively, of the axisymmetric metric, and as such it is of Petrov type D. If we take the background (6.1) with $b > 1$ then the solution represents [87] a soliton-like pulse interacting with a cosmic string on the symmetry axis. It generalizes the Xanthopoulos [303, 304] solution which describes the interaction of gravitational radiation with a cosmic string.

6.3.3 Two double complex poles

This solution and the solution with two real poles are obtained by pole fusion as the limit of four complex or real poles when two of them converge [88]. They correspond in the context of the axisymmetric stationary field to the Kinnersley and Chitre [168, 169] generalization of the $\delta = 2$ Tomimatsu–Sato [277] solution, where δ is the Tomimatsu–Sato distortion parameter, see section 8.6. Real-pole solutions do not have a very clear physical interpretation in the cylindrical context, as we have seen, due to the discontinuities in the first derivatives of the metric across the matching light cones.

7

Plane waves and colliding plane waves

The ISM can also be applied to plane-wave spacetimes as well as to spacetimes describing the collision of two plane waves. In this chapter we shall describe those spacetimes from the point of view of the ISM. In section 7.2 exact gravitational plane waves are defined and the plane-wave soliton solutions are characterized. We illustrate some of the physically more interesting properties of the plane waves with the detailed study of an impulsive plane wave. The more interesting case of solutions describing the head-on collision of plane waves is described in section 7.3. Soliton solutions are seen to describe the interaction region of such a collision since it can be described by a metric in which the transverse coordinates of the incoming plane waves can be ignored. Here again to illustrate the geometry of the colliding waves spacetimes we analyse in some detail a solution representing the head-on collision of two plane waves with collinear polarizations. Soliton solutions are described which include several of the most well known solutions representing the collision of waves with collinear and noncollinear polarizations.

7.1 Overview

Plane waves emerge as a subclass of a larger class of spacetimes: the *pp*-waves. *Plane-fronted gravitational waves with parallel rays* (*pp*-waves) are spacetimes that admit a covariantly constant null Killing vector field l_μ, i.e. $l_{\mu;\nu} = 0$, and were classified by Ehlers and Kundt [89]. They admit a group of isometries G_1 on a null orbit generated by the Killing vector l_μ, all their curvature scalars vanish and therefore are type N in the Petrov classification. Gravitational plane waves are a subclass of *pp*-waves that admit a group G_5 of isometries with an Abelian subgroup G_3 acting on null hypersurfaces; they were first studied by Baldwin and Jeffrey [8]. Plane waves are homogeneous in their wave surfaces and are of infinite extent in all directions in these surfaces, therefore the energy of these waves is infinite. However, they may be considered as realistic approximations to real gravitational waves within finite regions, and they may describe the gravitational field near strong radiating sources [8, 39]. These waves can be purely gravitational, purely electromagnetic or both, depending on the source.

As is well known [184, 241, 32] gravitational plane waves have interesting geometrical properties. One of them is the absence of a space-like Cauchy hypersurface as a consequence of the focusing effect they exert on null rays. The focusing properties of the plane waves are a nonlinear effect of gravity; they are not found in the weak field approximation, which is the relevant approximation for the gravitational field far from isolated sources. The time of focusing is typically inversely proportional to the energy density per unit surface of the waves. These effects, however, may be physically relevant when strong time-dependent gravitational fields are involved such as after the collision of black holes [75], in the decay of a cosmological inhomogenous singularity [308, 309] or when waves are travelling on strongly gravitating cosmic strings [112].

An interesting situation is produced when two of such waves collide. It was soon realized that due to mutual focusing, spacetime singularities could be produced at the focusing points [264, 161]. This type of singularity is a purely gravitational effect, unlike other well known singularities that are produced as the result of the collapse of matter sources. Typically it is found that stronger incoming plane waves have a shorter time between the collision and the focusing singularities. A general study of colliding plane waves was made by Szekeres [265]. It was found that the initial value problem for the construction of colliding wave spacetimes is well posed in the sense that given arbitrary incoming plane waves a unique solution exists in the interaction region. Hauser and Ernst [139, 140, 141] have developed methods for constructing solutions for arbitrary initial data, however, the solutions cannot generally be expressed in closed form. Thus, the standard way [265, 127] to construct colliding wave spacetimes is to obtain a solution in the interaction region and match it to plane-wave solutions: then one interprets the resulting spacetime as describing the collision of the two plane waves.

Khan and Penrose [161] found a solution which describes the collision of two impulsive gravitational plane waves with parallel polarizations, i.e. with collinear polarizations, and later Nutku and Halil [232] were able to describe the collision of two impulsive plane waves which were not parallel polarized, i.e. with noncollinear polarizations. A few years later Chandrasekhar and Ferrari [56] adapted the Ernst formulation of the stationary axisymmetric spacetimes to the colliding wave problem and, as a consequence, many axisymmetric solutions can be used to describe the collision of plane waves [57, 58, 59, 60, 61, 62]. This is so because the interaction region of two colliding plane waves admits two commuting Killing fields, like the axisymmetric spacetimes; although the boundary conditions in the two problems are quite different.

The ISM is a powerful method for obtaining solutions on the interaction region and it has been extensively applied for that purpose [104, 105, 106, 102], together, of course, with other solution-generating techniques such as the Neugebauer and Kramer method [226]. A great deal of attention has been devoted to the nature of the curvature singularities which appear at the focusing hypersurfaces of the colliding plane waves, and new types of singular structures such as the 'fold singularities' have been identified [218]. Some solutions are known in which the singular hypersurfaces are replaced by regular caustics [60, 104, 309, 145, 102, 98]. Such regular caustics, however, are not generic: they are classically unstable against plane-symmetric perturbations [306] or in the presence of an arbitrarily small amount of dust [62]; they are also unstable to quantum effects, in the sense that when a quantum field is coupled to the spacetime the field develops a divergent stress-energy tensor on the caustics [310, 83].

It is not our purpose to give a complete review of colliding plane-wave solutions. An excellent book on colliding plane-wave metrics is now available [127], and for a shorter introduction see ref. [39]. Our main interest will be to review the connection of such solutions with the soliton solutions we have already studied in chapters 4 and 5. We also want to illustrate the physics of the plane waves and of the colliding plane-wave spacetimes with some examples. First, in the next section, we shall concentrate on the plane-wave spacetimes.

7.2 Plane waves

7.2.1 The plane-wave spacetime

Let us begin with the most general *gravitational pp-wave*. In the standard form, [179] its spacetime metric is given by

$$ds^2 = -du\,dV + F(u, X^a)du^2 + \sum_a dX^a dX^a, \tag{7.1}$$

where V is a null coordinate, X^a are space-like transverse coordinates (the Latin indices take only two values, 1 and 2, as usual), and $F(u, X^a)$ is an arbitrary function. The set of coordinates (u, V, X^a), which range over all real values, are called *harmonic coordinates* [117, 114]. From (7.1) it is clear that $\vec{l} = 2\partial_V$ is a covariantly constant null Killing field and that the metric can be written in the form

$$g_{\mu\nu} = \eta_{\mu\nu} + F l_\mu l_\nu, \quad l_\mu = -\delta_{\mu u}, \tag{7.2}$$

where the Greek indices run over all four values of the coordinates (u, V, X^a) and $\eta_{\mu\nu}$ is the Minkowski metric tensor which corresponds here to the case when both coordinates, u and V, are null (i.e. to the metric (7.1) when $F = 0$). Then $g^{\mu\nu} = \eta^{\mu\nu} - F l^\mu l^\nu$ and the Riemann and Ricci tensors take the form

$$R_{\mu\nu\rho\sigma} = 2 l_{[\mu}\partial_{\nu]}\partial_{[\rho} F l_{\sigma]}, \quad R_{\mu\nu} = (1/2)(\eta^{\rho\sigma}\partial_\rho\partial_\sigma F)l_\mu l_\nu. \tag{7.3}$$

From this it is clear that all curvature scalar invariants vanish and the metric is Petrov type N. The Ricci tensor is then proportional to the stress-energy tensor, and from (7.3) we see that the source is in general a null fluid or, if F satisfies $\eta^{\rho\sigma}\partial_\rho\partial_\sigma F = 0$, vacuum.

Gravitational plane waves are the particular case when F is quadratic in X^a, $F(u, X^a) = H_{ab}(u)X^a X^b$, i.e.

$$ds^2 = -du\, dV + H_{ab}(u)X^a X^b du^2 + \sum_a dX^a dX^a, \tag{7.4}$$

where $H_{ab}(u)$ is, without loss of generality, a symmetric matrix. From (7.2) and (7.3) the only nonvanishing component of the Ricci tensor is $R_{uu} = \sum_a H_{aa}$. This Ricci component ($R_{00} = R_{uu}$) implies the following energy density ($\rho = T_{00}$):

$$\rho = -\frac{1}{8\pi G}\sum_a H_{aa}. \tag{7.5}$$

Positivity of the energy density, also known as the weak energy condition [143], implies that $\sum_a H_{aa} \leq 0$. Two special cases of interest are *pure gravitational waves* which correspond to $\sum_a H_{aa} = 0$, i.e. a Ricci flat metric, in which the Weyl and Riemann tensors coincide, and *pure null electromagnetic waves* which correspond to $H_{ab}(u) = H(u)\delta_{ab}$, $H < 0$, in which the Weyl tensor vanishes.

Harmonic coordinates are a convenient set of coordinates because they cover the whole plane-wave spacetime with a single chart and also because direct information on the curvature is contained in a unique metric component. However, they do not display some of the symmetries of the spacetime. It is convenient to introduce the so-called *group coordinates* (u, v, x^a) which range over all real values: u and v are, respectively, retarded and advanced times and x^a are the transverse coordinates adapted to the Killing vectors of the plane symmetry of the wavefronts. In these coordinates the metric (7.4) takes the form

$$ds^2 = -du\, dv + g_{ab}(u)dx^a dx^b. \tag{7.6}$$

The relationship between the two sets of coordinates is given by

$$V = v + \frac{1}{2}\dot{g}_{ab}(u)x^a x^b, \quad X^a = P_b^a(u)x^b, \tag{7.7}$$

where the dot means the derivative with respect to u and the 2×2 matrix P_b^a is determined by substitution of (7.7) into (7.4) and imposing (7.6) as follows. From the coefficient of $dx^a dx^b$ we see that $g_{ab}(u)$ is related to the matrix P_b^a by

$$g_{ab}(u) = \sum_c P_a^c(u) P_b^c(u). \tag{7.8}$$

Imposing that the coefficients of $du\, dx^a$ vanish and using (7.8) we obtain

$$\sum_c \left(\dot{P}_a^c(u) P_b^c(u) - \dot{P}_b^c(u) P_a^c(u) \right) = 0, \tag{7.9}$$

whereas imposing that the coefficient of du^2 vanishes and using the two previous equations we obtain

$$\ddot{P}^a_b(u) = \sum_c H^a_c(u) P^c_b(u),$$ (7.10)

where we have also used that the determinant of the matrix P^a_b should not vanish. Note that the equations of system (7.9)–(7.10) are evolution equations which determine the matrix $P^a_b(u)$. Equations (7.9) are constraint equations for this matrix and are stable against evolution: i.e. once imposed on the initial conditions they hold for all values of u, as can be seen using (7.10) and the fact that $H_{ab}(u)$ is symmetric. Conversely, given a plane wave in group coordinates (7.6), one can find its form in harmonic coordinates (7.4) by solving (7.8) and (7.9) to find the matrix $P^a_b(u)$ and then using (7.10) in order to obtain $H_{ab}(u)$. The nonzero Christoffel symbols and nonzero components of the Riemann and Ricci tensors in the coordinates of (7.6) are

$$\left. \begin{array}{l} \Gamma^a_{bu} = \dfrac{1}{2}g^{ac}\dot{g}_{cb}, \quad \Gamma^v_{ab} = \dot{g}_{ab}, \\[2ex] R^a_{ubu} = -\partial_u\Gamma^a_{bu} - \Gamma^a_{cu}\Gamma^c_{bu}, \quad R_{uu} = -\dfrac{1}{2}g^{ab}\ddot{g}_{ab} - \dfrac{1}{4}\dot{g}^{ab}\dot{g}_{ab}. \end{array} \right\}$$ (7.11)

The group coordinates do not cover the whole spacetime with a single chart, because they become singular for some value of the null coordinate u. It is easy to prove that there is some value of the retarded time $u = u_f$ for which the determinant of the metric (7.6) must vanish. This is the so-called Landau–Raychaudhuri theorem and its proof may be found in ref. [184]. The proof is based on the fact that if we define $\gamma(u) = |\det g_{ab}(u)|^{1/4}$, positivity of the energy density implies that $\ddot{\gamma}/\gamma \leq 0$, which means that $\gamma(u)$ is a convex function (the equality $\ddot{\gamma}/\gamma = 0$ can only hold for flat space), and since γ is positive for some value of u, then it must vanish for at least some other value of $u = u_f$, $\gamma(u_f) = 0$. Note that the explicit use of (7.11) is of help in this proof, see also ref. [114]. Thus at $u = u_f$ we have a coordinate singularity of (7.6). At this null hypersurface geodesics are focused and when two such waves collide curvature singularities usually appear as a result of mutual focusing.

7.2.2 Focusing of geodesics

To understand the geometry and the global properties of the plane-wave space-times it is convenient to study the behaviour of geodesics. We begin with the geodesic equations in harmonic coordinates which as we know cover the whole spacetime. Furthermore, when one considers *sandwich plane waves*, i.e. when $H_{ab}(u)$ is only different from zero on a finite interval of u, the harmonic coordinates are the ordinary Minkowski coordinates outside this range. The geodesic equations can be derived from the Lagrangian

$$L = -\frac{du}{d\lambda}\frac{dV}{d\lambda} + H_{ab}(u)X^a X^b \left(\frac{du}{d\lambda}\right)^2 + \sum_a \frac{dX^a}{d\lambda}\frac{dX^b}{d\lambda}, \qquad (7.12)$$

where we have introduced the affine parameter $\lambda = \tau/m$, τ being the proper time of the particle and m its rest mass. This also includes null geodesics, taking the limit in which both τ and m tend to zero. The Euler–Lagrange equations imply that

$$p_- \equiv -p_V \equiv \frac{1}{2}\frac{du}{d\lambda} = \text{constant}, \qquad (7.13)$$

so u is proportional to the affine parameter, and

$$\ddot{X}^a = H^a_b(u)X^b, \quad \dot{V} = H_{ab}X^a X^b + \sum_c \dot{X}^c \dot{X}^c + \frac{m^2}{4p_-^2}. \qquad (7.14)$$

The first equation of (7.14) indicates that the geodesics suffer a transverse force in these coordinates and the second equation is a constraint equation. As in the previous section overdots indicate derivatives with respect to u. Let ΔX^a be the transverse coordinate separation between two nearby parallel geodesics and then the geodesic deviation equation is simply $\Delta \ddot{X}^a = H^a_b \Delta X^b$, which gives the local tidal forces between the geodesics. Note that when the wave is a pure gravitational wave ($H^a_a = 0$) these geodesics suffer the same pattern of tidal forces as geodesics in a weak gravitational field in the transverse-traceless gauge [222]. On the other hand for a pure null electromagnetic wave, i.e. when $H_{ab} = H(u)\delta_{ab}$, $H < 0$, the geodesics suffer symmetric attractive forces. Only if H_{ab} diverges rapidly near some value of u do the tidal forces become infinite and the geodesics end at a spacetime singularity.

Impulsive plane wave. To illustrate the focusing effects on geodesics we will consider instead of a sandwich wave an *impulsive plane wave*, which is defined by

$$H_{ab}(u) = A_{ab}\delta(u), \qquad (7.15)$$

where A_{ab} is a 2×2 symmetric constant matrix. The main advantage of this spacetime is that the calculations become simpler and the results can be qualitatively extrapolated to the more general sandwich plane waves [114]. The spacetime described by (7.15) may be understood as the matching of two pieces of flat space through the hyperplane $u = 0$, which has an energy density per unit transverse surface, see (7.5), given by $\rho = -(8\pi G)^{-1}A^a_a$. Without loss of generality one can choose A_{ab} to be diagonal by an appropriate rotation of the transverse plane. So let

$$A_{ab} \equiv -\frac{1}{\lambda_a}\delta_{ab}. \qquad (7.16)$$

Let us now rewrite this metric in group coordinates. For this we solve (7.10) taking as initial conditions $P_b^a(u_0) = \delta_b^a$ and $\dot{P}_b^a(u_0) = 0$ for some $u_0 < 0$. Then $P_b^a(u) = \delta_b^a + u\theta(u)A_b^a$, where $\theta(u)$ is the step function and

$$g_{ab}(u) = \delta_{ab}\left(1 - \frac{u}{\lambda_a}\theta(u)\right)^2. \tag{7.17}$$

In these coordinates the interpretation of the spacetime as the matching of two flat regions at $u = 0$ is also clear. In one of the regions ($u < 0$) the metric has the Minkowski form, but in the other ($u > 0$) the transverse coefficients are proportional to $(1 - u/\lambda_a)^2$, which is also flat space but in non-Minkowskian coordinates, as the coordinate transformation (7.7) with $P_b^a(u) = (1 - u/\lambda_a)\delta_b^a$ proves. In these coordinates the metric is continuous but it has discontinuous first derivatives across the null hypersurface $u = 0$. Since the weak energy condition implies that $H_a^a \leq 0$, it is clear from (7.16) that at least one of the λ_a must be positive. Let λ_1 be the minimum positive value of (λ_1, λ_2), then for $u = \lambda_1$ we see that $\det g_{ab}(\lambda_1) = 0$, which is a coordinate singularity of the group coordinates and, as we shall see, $u = \lambda_1$ defines the focusing retarded time. Note that the time of focusing is roughly proportional to the inverse of the energy density per unit surface of the wave $\rho \sim \lambda_a^{-1}$. For a pure gravitational plane wave $\rho = 0$ but then one may use the inverse of the focusing time to define a kind of energy density or 'strength' of the plane wave.

The geodesic equations (7.14) are easily integrated for the impulsive plane wave (7.15)–(7.16). One may easily check that the geodesic trajectories are

$$X^a(u) = b^a + \frac{p^a}{2p_-}u - \frac{b^a}{\lambda_a}u\theta(u), \tag{7.18}$$

$$V(u) = c_0 + \frac{p^a p_a + m^2}{4p_-^2}u - \sum_a \frac{b_a^2}{\lambda_a}\theta(u) + \sum_a \left(\frac{b_a^2}{\lambda_a^2} - \frac{b_a p_a}{\lambda_a p_-}\right)u\theta(u), \tag{7.19}$$

where $b^a \equiv X^a(0)$, $c_0 \equiv V(0)$ are impact parameters and the transverse momenta p_a are the constants of motion associated with the Killing vectors ∂_{x^a} in group coordinates which coincide with ∂_{X^a} in the region $u < 0$. A very important feature of this solution is the shift in $V(u)$ at $u = 0$: $\Delta V = -\sum_a(b_a^2/\lambda_a)$. This is a typical feature of the impulsive nature of the waves [108] (when one considers sandwich plane waves the shift also occurs between the front and the back of the wave but the change in V is continuous). The transverse coordinates, on the other hand, simply change direction as the wave is crossed.

Let us now consider the degenerate case when $\lambda_1 = \lambda_2$ (this is the case of pure electromagnetic plane waves) and assume perpendicular incidence $p_a = 0$. We may introduce polar coordinates $r \equiv (X^a X_a)^{-1/2}$ and $b \equiv (b^a b_a)^{-1/2}$, then

(7.18)–(7.19) read

$$r = b\left(1 - \frac{u}{\lambda_1}\right)\theta(u),\tag{7.20}$$

$$V(u) = c_0 + \frac{m^2}{2p_-}u + \frac{b^2}{\lambda_1}\left(\frac{u}{\lambda_1} - 1\right)\theta(u).\tag{7.21}$$

All the geodesics focus at the same $r(u_f) = 0$, $V(u_f) = c_0 + m^2u_f(2p_-)^{-1/2}$, where $u_f = \lambda_1$, independently of the impact parameter b. In an analogous way we may consider the case of oblique incidence ($p_a \neq 0$). In this case geodesics with the same values of $b^a p_a$ also focus at one point in the degenerate case. In the nondegenerate case when $\lambda_1 \neq \lambda_2$ (for instance, pure gravitational plane waves correspond to $\lambda_1 = -\lambda_2$) and for perpendicular incidence, i.e. $p_a = 0$, all geodesics with the same b_2 meet when $u = u_f = \lambda_1$ at the same point:

$$X^1(u_f) = 0, \qquad X^2(u_f) = b_2\left(1 - \frac{\lambda_1}{\lambda_2}\right),\tag{7.22}$$

$$V = c_0 + \frac{m^2}{2p_-}\lambda_1 - \frac{b_2^2}{\lambda_2}\left(1 - \frac{\lambda_1}{\lambda_2}\right).\tag{7.23}$$

Subfamilies labelled by different values of b_2 meet at different points and the focusing points span a parabola in the $(X^2, V - u)$-plane.

Using this impulsive plane-wave spacetime we may illustrate the focusing of the future null cones of points in front of the wave, a feature which is typical of the plane-wave spacetimes as first shown by Penrose [241]. This is represented in fig. 7.1. Let $X^a(u)$, $V(u)$ be the null geodesics that at some u_0 ($u_0 < 0$) are at the point P:

$$X^a(u_0) = 0, \qquad V(u_0) = 0.\tag{7.24}$$

The trajectories of these null geodesics are obtained from (7.18)–(7.19) by taking $m = 0$. Choosing $b^a = -(2p_-)^{-1}u_0 p^a$ and $c_0 = -(2p_-)^{-2}u_0 p^a p_a$ so that the geodesics satisfy the initial conditions (7.24) these trajectories are, in the region $u < 0$, given by

$$X^a(u) = \frac{p^a}{2p_-}(u - u_0), \qquad V(u) = \frac{p^a p_a}{4p_-^2}(u - u_0).\tag{7.25}$$

Note that there are two rays with $p^a = 0$: one is the ray with perpendicular incidence to the plane wave and the other the ray that moves in the wave direction which has $p_- = 0$ and never crosses the wave. Apart from this single ray any other null ray of (7.25) will cross the plane wave. The family of geodesics (7.25) defines the future null cone of P in front of the wave: $X^a(u)X_a(u) = V(u)(u - u_0)$ ($u_0 \leq u < 0$). Let us now see how this null cone will be focused again behind the plane wave where $u > 0$. For simplicity,

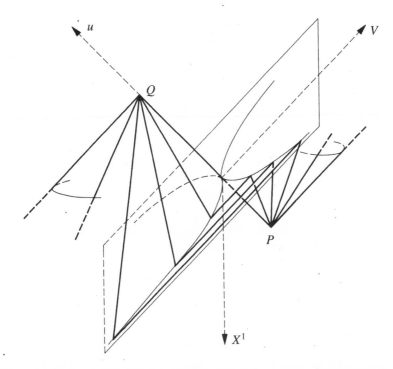

Fig. 7.1.   The future null cone of the point P in front of the impulsive plane wave is focused again at the point Q behind the wave. The null geodesics that form the light cone of P suffer a shift in the coordinate V as they cross the plane wave at $u = 0$: they enter into the upper parabola, which is the intersection of the null cone with the plane $u = 0$ in the figure, and leave from the lower parabola (the intersection of the past null cone of Q with the plane $u = 0$) with the same component of the transverse coordinate X^1. There is no global Cauchy hypersurface in that spacetime: a space-like hypersurface that contains P cannot be extended to space-like infinity behind the wave and be a Cauchy hypersurface.

let us again take the degenerate case, $\lambda_1 = \lambda_2$, then the trajectories of these geodesics in the region $u > 0$ are given by

$$X^a(u) = \frac{p^a}{2p_-}\left(1 + \frac{u_0}{\lambda_1}\right)(u - u_f), \quad V(u) = \frac{p^a p_a}{4p_-^2}\left(1 + \frac{u_0}{\lambda_1}\right)^2 (u - u_f),$$

$$\tag{7.26}$$

where $u_f = \lambda_1(1 + \lambda_1/u_0)^{-1}$. Thus, provided $u_f > 0$ (i.e. when $-\infty \leq u_0 < -\lambda_1$) the future null cone of P focuses again at the point Q: $X^a(u_f) = 0$, $V(u_f) = 0$. This set of geodesics also form the past null cone of the point Q behind the wave: $X^a(u)X_a(u) = V(u)(u - u_f)$ $(0 < u \leq u_f)$. When $u_f < 0$ (i.e. when $-\lambda_1 < u_0 < 0$) there is no refocusing since $u > 0$ behind the wave. The focusing retarded time u_f satisfies that $u_f \geq \lambda_1 > 0$ (recall that $u_0 < 0$)

and it tends to λ_1 when $u_0 \to -\infty$. Thus only the future light cones of points which lie in the range $u_0 \in (-\infty, -\lambda_1)$ will focus again at $u_f \in (\lambda_1, \infty)$, with the rule that $u_0 \to -\infty$ corresponds to $u_f \to \lambda_1$ and $u_0 \to -\lambda_1$ corresponds to $u_f \to \infty$. At the hypersurface $u = \lambda_1$ we have that $\det g_{ab}(\lambda_1) = 0$ which corresponds to the coordinate singularity of the group coordinates.

We can now follow Penrose's argument to show that such a spacetime does not admit a global Cauchy hypersurface. Note, first, that every null or time-like geodesic should intersect a Cauchy hypersurface exactly once. Thus a (space-like) Cauchy hypersurface must lie entirely in the past of the future null cone of P. Now assume that one such space-like hypersurface contains the point P in front of the wave, since the future null cone of P focuses at the point Q, this hypersurface must lie entirely below the past null cone of Q and therefore cannot be extended to spatial infinity behind the wave (note that we can always take points on that hypersurface with arbitrary large values of $|u_0|$). This means that no Cauchy data on an unbound space-like hypersurface in the region $u < 0$ can give information to specify, for instance, a parallel wave that might lie beyond the hypersurface $u = \lambda_1$.

7.2.3 Plane-wave soliton solutions

Let us now go back to the spacetime metric (1.36) and introduce null coordinates $u = t - z$ and $v = t + z$, which are retarded and advanced times, respectively. These are related to the null coordinates (ζ, η) introduced earlier in (1.14) by $u = -2\eta, v = 2\zeta$. In this chapter the coordinates (u, v) are used in order to keep close to the more standard conventions in the literature on this subject. The metric (1.36) can now be written as

$$ds^2 = -\exp[\gamma(u, v)]du\, dv + g_{ab}(u, v)dx^a\, dx^b, \tag{7.27}$$

where $\exp[\gamma] = f$. Clearly the plane-wave metric (7.6) represents a special case of (7.27). Moreover, we will see that with this general form (7.27) one can also describe the interaction region of two colliding plane waves, one of which is coming from $z = -\infty, t = -\infty$ and in this asymptotic region has metric coefficients $\gamma(u), g_{ab}(u)$ and the other of which is coming from $z = +\infty$, $t = -\infty$ with metric coefficients $\gamma(v), g_{ab}(v)$.

In the coordinate system of (7.27), plane-wave solutions are the solutions which have dependence on one null coordinate only, say u, in the matrix g: $g_{ab}(u)$. It is obvious from (1.39)–(1.41) that any arbitrary matrix $g_{ab}(u)$ is a solution of (1.39) and that given $g_{ab}(u)$ the function $\gamma(u, v)$ is determined up to an arbitrary function of v, i.e.

$$\gamma(u, v) = \gamma_1(u) + \gamma_2(v), \tag{7.28}$$

where $\gamma_1(u)$ is a solution of (1.40) and $\gamma_2(v)$ is an arbitrary function. That is,

$$ds^2 = -\exp[\gamma_1(u) + \gamma_2(v)]\, du\, dv + g_{ab}(u)dx^a dx^b, \tag{7.29}$$

which a simple relabelling of u and v puts in the form (7.6) of the group coordinates.

From our discussion of the ISM it should be clear that a plane-wave metric is transformed into another plane wave by the ISM. It is now obvious that canonical coordinates are not appropriate to describe plane-wave spacetimes. In fact, from (1.45) we have that $a(v) =$ constant, and (1.46) implies that α and β are linearly dependent,

$$\alpha = -\beta + \text{constant.} \tag{7.30}$$

This provides a simple way of obtaining plane-wave solutions from the cosmological solutions of chapter 4 (which are described in terms of canonical coordinates): they are the limit in which $\alpha = -\beta +$ constant. Note that when this limit is taken, the arbitrary function $\gamma_2(v)$ in (7.28) is also determined in this way. Some of the interest of the plane-wave solutions is that they can be seen as the limit of some spatially inhomogeneous and homogeneous solutions. Thus Siklos [261] has derived some plane-wave solutions that are the growing modes of some homogenous Bianchi types. Also the Ellis and MacCallum Bianchi solutions (4.62) and their inhomogenous generalizations (4.58)–(4.59) evolve to plane-wave solutions. In fact, the Ellis and MacCallum solutions can be written in the form (4.58)–(4.59) using canonical coordinates ($\alpha \equiv t$, $\beta \equiv z$) and then changing to coordinates adapted to spatial homogeneity (T, Z) by (4.61). After a change in the time direction ($T \to -T$), in the limit $T \to \infty$ ($\alpha \to -\beta +$ constant) the solutions become plane-wave metrics.

Viewed as background metrics for the ISM, Siklos plane-wave metrics are interesting because they are cosmological solutions with horizons and because they are the final evolution stages of several Bianchi types. For such backgrounds the generating matrices $\psi_0(u, v)$ have been obtained by Kitchingham [173]. However, all but the diagonal cases are given in terms of complicated hypergeometric functions, so that explicit calculations soon become cumbersome.

A particularly simple background metric is

$$g_0(u) = \text{diag}(\alpha^{1+d}, \alpha^{1-d}); \quad \gamma_1(u) = -(1 + d^2)au, \quad \gamma_2(v) = -(1 + d^2)av, \tag{7.31}$$

where $\alpha = \exp[-2au]$, and a, d are arbitrary parameters ($a > 0$). This solution is the Siklos plane-wave Bianchi type $\text{VI}_h(\kappa = 0)$ solution. All Bianchi VI_h approach this solution at late times. It is the limit of the Ellis and MacCallum solutions at late times and the limit of the Bianchi $_f\text{VI}_h$ Wainwright and Marshman solutions of class III. It may also be useful as a cosmological model at early times. It describes, for instance, some of Wainwright's spatially homogeneous perfect fluid solutions near the initial singularity [296]. Note that this solution is not the Kasner solution, in spite of the similarity of the transversal coefficients. This solution admits a G_6 group of isometries with V_4

orbits. A G_5 subgroup acts on the wave surfaces $u = $ constant and there is a G_3 subgroup which acts on space-like hypersurfaces of Bianchi type VI_h [260, 261]. The generating matrix ψ_0 for the background (7.31) can easily be calculated by integration of (1.51):

$$\psi_0 = \text{diag} \left((\lambda^2 + 2\alpha\lambda + \alpha^2)^{(1+d)/2}, \ (\lambda^2 + 2\alpha\lambda + \alpha^2)^{(1-d)/2} \right).$$

We remark that in this case the function $a(v)$ in (1.45)–(1.46) is zero and functions α and $-\beta$ coincide. The generalized (diagonal) soliton solutions with this background have been studied [287]. It is found that, generally, the new plane-wave solutions admit a G_5 group of isometries with N_3 orbits. For some values of the parameters they include the background solution which admits a G_6 group and, for some other values, flat spacetime. The soliton components do not introduce new singularities besides the nonscalar curvature singularity of the background; the Weyl scalar Ψ_4 becomes unbounded at a certain null hypersurface $u = $ constant.

7.3 Colliding plane waves

We turn now to the more interesting problem of describing the collision of gravitational plane waves. As first shown by Khan and Penrose [161] the interaction region after the collision of two plane waves propagating in opposite directions is described by a spacetime with two commuting Killing vectors. The usual way to construct these solutions is to consider a particular exact solution to Einstein's equations in the appropriate region and to try to match it to plane-wave solutions and flat spacetime. This must be performed with a suitable matching: the O'Brien and Synge matching conditions [233, 265]. The resulting spacetime is interpreted as describing the collision of the gravitational plane waves.

7.3.1 The matching conditions

The first assumption one makes when trying to construct a spacetime describing the head-on collision of two plane waves propagating in flat spacetime is that the collision does not affect the plane symmetry of the waves. Thus one assumes an Abelian two-parameter group of isometries with Killing vectors ∂_{x^a} $(a = 1, 2)$ along the transverse plane of the wave propagation. One also assumes that at each spacetime point there exist two orthogonal null directions to the above planes, so that we may define two null vectors $\vec{l}$ and $\vec{n}$ along such directions which are proportional to gradients: $l_\mu \propto u_{,\mu}$ and $n_\mu \propto v_{,\mu}$. We may choose u and v as null coordinates and the spacetime will then be described by the coordinates (u, v, x^1, x^2). This spacetime thus admits an orthogonally transitive group of isometries and, consequently, the ISM can be applied to find vacuum solutions. The colliding wave spacetime will be divided in four regions: I, II, III and IV. Region I is the flat region, regions II and III are the incoming plane-wave regions and region IV is the interaction region. The spacetime metric used will be in the form (7.27) which is useful for describing the interaction region (region IV) because, as we saw in section 7.2.3, it can be extended naturally to the initial regions by requiring that the metric coefficients are functions of u only to describe the plane wave that comes from $z \to -\infty$ ($2z = v - u$) (region II), or functions of v only to describe the plane wave that comes from $z \to \infty$ (region III), or constants to describe the flat region between the wavefronts before the collision (region I). We have from (1.44) that $\alpha_{,uv} = 0$ (recall that now we are using (u, v) which are related to (ζ, η) by $u = -2\eta$ and $v = 2\zeta$). From this equation we have the solution (1.45) which we now write as

$$\alpha = F(u) + G(v), \qquad (7.32)$$

where F and G are arbitrary functions. As we have just remarked, α should be a function of u only in region II, of v only in region III, and constant in region I. The wavefronts of the incoming waves will be taken as $u = 0$ and $v = 0$, respectively, and we will impose continuity of α across such null hypersurfaces. Thus we will choose

$$F = \frac{1}{2} \ (u \le 0), \qquad G = \frac{1}{2} \ (v \le 0), \qquad (7.33)$$

so that in the flat region $\alpha = 1$.

To match the different regions suitable matching conditions must be given. All the matching hypersurfaces are null, the appropriate matching conditions in this case were proposed by O'Brien and Synge [233]. Their conditions require that the metric be continuous across the boundaries but discontinuities on certain first derivatives are allowed if they do not produce extra source terms in Einstein's equations. Following ref. [127] we will denote by $x^0 = $ constant a null hypersurface (this corresponds to either $u = 0$ or $v = 0$ in our case), then $g_{00} = 0$ and we will assume that the metric is written in coordinates (x^0, x^1, x^2, x^3), then the O'Brien and Synge conditions reduce to impose that the following functions:

$$g_{\mu\nu}, \quad g^{ij}g_{ij,0}, \quad g^{i0}g_{ij,0} \quad (i, j = 1, 2, 3), \qquad (7.34)$$

are continuous across $x^0 = $ constant. This allows for the impulsive wave defined in (7.15) which has in group coordinates the metric components (7.17).

These conditions imply that the functions F and G should be continuous across $u = 0$ and $v = 0$, respectively, and also that their derivatives should vanish on these hypersurfaces:

$$F(u) = \frac{1}{2}, \quad u \le 0, \quad \dot{F}(0) = 0, \qquad (7.35)$$

$$G(v) = \frac{1}{2}, \quad u \leq 0, \quad \dot{G}(0) = 0, \tag{7.36}$$

so that we have $\alpha = 1$ in region I, $\alpha = 1/2 + F(u)$ in region II, $\alpha = 1/2 + G(v)$ in region III, and $\alpha = F(u) + G(v)$ in region IV with $\dot{F}(0) = 0$ and $\dot{G}(0) = 0$. We have taken F and G in the interaction region IV to be the same functions as in the plane-wave regions II and III, respectively.

Let us now consider in more detail the boundaries between regions I and II, and between I and III. By a coordinate relabelling of u and v it is always possible to take $\gamma = 0$ in (7.27) in these three regions. Equations (1.40)–(1.41) in these regions together with (7.35)–(7.36) imply that F and G are monotonically decreasing functions for positive arguments in regions II and III. Thus since $\alpha = F + G$ eventually $\alpha \to 0$ in the interaction region; this indicates a coordinate singularity of metric (7.27) or, as we shall see, a curvature singularity in most cases. One may then use a relabelling of the coordinates u and v to express F and G in the Szekeres [265] form:

$$F(u) = \frac{1}{2} - (c_1 u)^{n_1} \theta(u), \quad G(v) = \frac{1}{2} - (c_2 v)^{n_2} \theta(v), \tag{7.37}$$

in the four regions, where c_1, c_2, n_1 and n_2 are some real parameters. Here u is a function of the previous u and the same for v (we could use different names, for instance u' and v' instead, but using the same names, as is common practice, should not cause confusion). This parametrization is not always the most convenient, but it may be so in some cases. More generally the leading terms of F and G near the boundaries will be of the form

$$F(u) = \frac{1}{2} - (c_1 u)^{n_1} + \cdots \quad (u \geq 0); \quad G(v) = \frac{1}{2} - (c_2 v)^{n_2} + \cdots \quad (v \geq 0).$$
$$\tag{7.38}$$

In order to satisfy conditions (7.35)–(7.36), i.e. that $\dot{F}(0) = 0$ and $\dot{G}(0) = 0$, the parameters n_a ($a = 1, 2$) are restricted to take $n_a \geq 2$. When $n_a = 2$ the approaching waves usually have impulsive components (an example of this is the metric (7.17) for pure gravity ($\lambda_1 = -\lambda_2$)), when $n_a = 4$ the waves have a step component and when $n_a > 6$ the wavefront is always continuous.

It may be convenient to adopt $F(u)$ and $G(v)$ as coordinates instead of u and v in the interaction region; in fact, many solutions have been given in that form [127]. Then one may write metric (7.27) in the form

$$ds^2 = -\frac{2e^S}{\sqrt{F+G}} dF\, dG + (F+G)[\chi(dx^2)^2 + \chi^{-1}(dx^1 - \omega dx^2)^2], \tag{7.39}$$

where S, χ and ω are functions of F and G. The Einstein equations in vacuum

(1.39)–(1.41) can be written as

$$2\chi_{,FG} = -\frac{1}{F+G}(\chi_{,F} + \chi_{,G}) + \frac{2}{\chi}(\chi_{,F}\chi_{,G} - \omega_{,F}\omega_{,G}), \quad (7.40)$$

$$2\omega_{,FG} = -\frac{1}{F+G}(\omega_{,F} + \omega_{,G}) + \frac{2}{\chi}(\chi_{,F}\omega_{,G} - \chi_{,F}\omega_{,G}), \quad (7.41)$$

and the function S satisfies

$$S_{,F} = -\frac{F+G}{2\chi^2}(\chi_{,F}^2 + \omega_{,F}^2), \quad S_{,G} = -\frac{F+G}{2\chi^2}(\chi_{,G}^2 + \omega_{,G}^2). \quad (7.42)$$

The relationship of S with the function γ of (7.27) is obviously

$$e^\gamma = \frac{2\dot{F}\dot{G}}{\sqrt{F+G}}e^S. \quad (7.43)$$

The continuity of γ across the boundaries where $\dot{F} = 0$ or $\dot{G} = 0$ implies that S should diverge on these boundaries like

$$S = -k_1 \ln(1/2 - F) - k_2 \ln(1/2 - G) + \cdots, \quad (7.44)$$

where $k_a = 1 - 1/n_a$ $(a = 1, 2)$, as one can easily check by substituting the values of (7.38) into (7.44). Naturally this imposes some restrictions on χ and ω via equations (7.42), in the same way that the smoothness of γ imposes restrictions on g_{ab} via equations (1.40)–(1.41).

7.3.2 Collinear polarization waves: generalized soliton solutions

As remarked previously solutions in the interaction region can be found by the ISM. This means, in particular, that we have at our disposal the soliton solutions described in the cosmological context in chapters 4 and 5. There most spacetimes are described in terms of canonical coordinates ($\alpha = t$, $\beta = z$). Here, as we have just seen, such coordinates are not the most appropriate to describe the matching of the interaction region with the plane-wave regions. They are nevertheless useful in the interaction region particularly near $\alpha = 0$, i.e. near the 'cosmological' singularity or 'focusing' hypersurface in the colliding wave problem. Let us now introduce Szekeres prescription (7.37) to write the canonical coordinates (α, β) (see (1.45)–(1.46)) in the interaction region in terms of (u, v) as

$$\alpha = F+G = 1-(c_1 u)^{n_1}-(c_2 v)^{n_2}, \quad \beta = -F+G = (c_1 u)^{n_1}-(c_2 v)^{n_2}, \quad (7.45)$$

where the arbitrary parameters c_a $(a = 1, 2)$ are used for rescaling coordinates. As we have seen in subsection 7.3.1 this prescription gives a suitable matching provided the n_a are restricted to $n_a \geq 2$. The interaction region in the canonical coordinates (α, β) is now compact and is reduced to the triangular region bounded by $\alpha = 0$, $\alpha \pm \beta = 1$. The line element is changed to

$$d\beta^2 - d\alpha^2 = -4n_1 n_2 c_1^{n_1} c_2^{n_2} u^{n_1-1} v^{n_2-1} du\, dv. \qquad (7.46)$$

The spacetime regions defined in terms of the canonical coordinates (α, β) and in terms of the null coordinates (u, v) are shown in fig. 7.2. The triangular region in (a) corresponds to the entire colliding wave spacetime. The interior of the triangle corresponds to the interaction region IV, the point $(\alpha, \beta) = (1, 0)$ to the flat region I. The line $\alpha + \beta = 1$ for $\alpha \in [0, 1)$ corresponds to the plane-wave region III and the line $\alpha - \beta = 1$ for $\alpha \in [0, 1)$ to the plane-wave region II. The $\alpha = 0$ (i.e. $F + G = 0$) for $\beta \in (-1, 1)$ corresponds to the 'focusing' region in IV for $1 = (c_1 u)^{n_1} + (c_2 v)^{n_2}$ and the singular points $(0, 1)$, $(0, -1)$ to the 'focusing' hypersurfaces in the plane-wave regions II and III defined by $(c_1 u)^{n_1} = 1$ and $(c_2 v)^{n_2} = 1$, respectively.

Notice that because of the compact nature of the interaction region in terms of canonical coordinates we can use either real- or complex-pole trajectories (1.68)–(1.69) with $w_k = z_k^0 - ic_k$ to construct soliton solutions, provided we take two-soliton origins at $\beta = z_k^0 = \pm 1$ to satisfy the boundary conditions. Also, now there is no special need to take opposite pole trajectories in order to avoid space-like singularities (since there is no $|\beta| \to \infty$ limit). Nondiagonal solutions in terms of (7.27) correspond to the collision of plane waves with two polarizations, i.e. noncollinear polarizations, whereas diagonal solutions correspond to aligned one-polarization waves, i.e. collinear polarizations. This second case is easier to study and many such solutions have been found and studied in the literature.

Generalized soliton solutions. Here we shall restrict ourselves to diagonal metrics and will consider the generalized soliton solution with real poles in the Kasner background. The diagonal metric (4.6), which we rewrite here in terms of $\alpha \equiv t$ and $\beta \equiv z$ as

$$ds^2 = f(d\beta^2 - d\alpha^2) + \alpha(e^{\Phi} dx^2 + e^{-\Phi} dy^2), \qquad (7.47)$$

where $x = x^1$, $y = x^2$, admits the generalized soliton solution (4.51)–(4.52), which can be rewritten as

$$\Phi = d\ln\alpha + \sum_{k=1}^{s} h_k \ln\left(\frac{\mu_k}{\alpha}\right), \qquad (7.48)$$

$$f = \alpha^{[(d-g)^2-1]/2} \prod_{k=1}^{s} \left[\mu_k^{h_k(h_k+d-g)} (\beta_k^2 - \alpha^2)^{-h_k^2/2} \right] \prod_{k,l=1;k>l}^{s} (\mu_k - \mu_l)^{2h_k h_l}, \qquad (7.49)$$

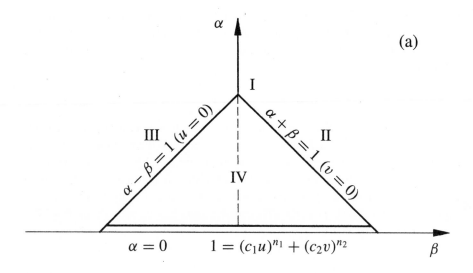

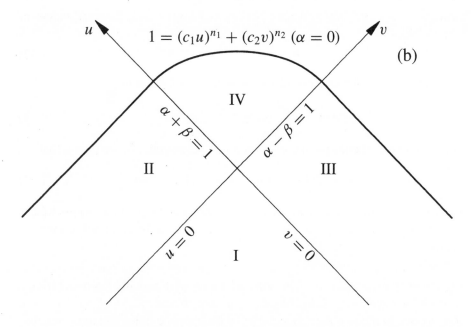

Fig. 7.2. The different relevant regions in the colliding plane-wave problem in terms of (a) canonical coordinates (α, β) and (b) null coordinates (u, v) according to (7.45). Region IV is the interaction region, regions II and III are the plane-wave regions and region I is a portion of flat spacetime.

where $\beta_k \equiv z_k^0 - \beta$, s is the number of pole trajectories, $g \equiv \sum_{k=1}^s h_k$ and z_k^0, d and h_k are arbitrary real parameters. The pole trajectories are (see (4.49)): $\mu_k^{\pm} = z_k^0 - \beta \pm [(z_k^0 - \beta)^2 - \alpha^2]^{1/2}$. The function Φ can be written in an alternative form using

$$\ln\left(\frac{\mu_k^+}{\alpha}\right) = \cosh^{-1}\left(\frac{z_k^0 - \beta}{\alpha}\right). \tag{7.50}$$

As an example we will consider the solutions with two real-pole trajectories, $s = 2$, with $g = h_1 + h_2$ and $z_1^0 = -z_2^0 = -1$. As usual the boundary conditions on the null hypersurface $u = 0$ and $v = 0$ are imposed by working with null coordinates (u, v). From (7.48) and (7.50) it should be clear that there is no problem in matching the transversal metric coefficients. The problem lies in the longitudinal coefficient $f(\alpha, \beta)$. The metric thus has to be written as (7.27). This means that

$$f(d\beta^2 - d\alpha^2) = e^{\gamma(u,v)} du\, dv, \tag{7.51}$$

and using (7.46) to write the left hand side in terms of (u, v) we see that f must diverge as u^{1-n_1} and v^{1-n_2} on these boundaries to ensure a smooth γ. As shown by Feinstein and Ibáñez [102] this divergence in f (see (7.49)) comes precisely from the soliton terms. In fact, consider for instance the first soliton term in (7.48): $h_1 \ln(\mu_1^+/\alpha)$. Its possible divergent contribution in f is in the term $(\beta_1^2 - \alpha^2)^{-h_1^2/2}$, which when $z_1^0 = -1$ and using (7.45) for small values of u and v gives

$$[(1 + \beta)^2 - \alpha^2]^{-h_1^2/2} \sim u^{-n_1 h_1^2/2}. \tag{7.52}$$

The same analysis for the second soliton with $z_2^0 = 1$ leads to

$$[(1 - \beta)^2 - \alpha^2]^{-h_1^2/2} \sim v^{-n_2 h_2^2/2}. \tag{7.53}$$

Imposing that f should go like u^{1-n_1} and v^{1-n_2}, respectively, we obtain that

$$h_1^2 = 2(1 - 1/n_1), \quad h_2^2 = 2(1 - 1/n_2). \tag{7.54}$$

This together with (7.45) means that for a proper matching on the hypersurfaces $u = 0$ and $v = 0$ each pole trajectory is relevant to a different null boundary. Let us consider some examples.

Khan and Penrose solution. The solution found by Khan and Penrose [161] is (7.48)–(7.49) with $s = 2$, $h_1 = -h_2 = 1$, $d = 0$ and $z_1^0 = -z_2^0 = -1$, i.e. it corresponds to the two pole trajectories, with opposite poles and axisymmetric Kasner background. This means that it develops a curvature singularity (related to the cosmological one at $\alpha = 0$), and that according to (7.54) the coordinate transformation (7.45) has $n_1 = n_2 = 2$, which means that the incoming waves have impulsive components.

Szekeres solution. The solution given by Szekeres [264] is (7.48)–(7.49) with $s = 2$, $h_1 = -h_2 = \sqrt{3/2}$, $d = 0$, and $z_1^0 = -z_2^0 = -1$, i.e. it corresponds to the generalized soliton solutions with two opposite poles and axisymmetric Kasner background. Thus it also develops a curvature singularity and, according to (7.54), the transformation (7.45) has $n_1 = n_2 = 4$, which means that the incoming waves have a step component.

We know that not all solutions develop curvature singularities at $\alpha = 0$. For instance, if we take two opposite poles we may choose the flat Kasner background $d = \pm 1$ and $\alpha = 0$ is then a regular hypersurface. More generally, as shown by Feinstein and Ibáñez [102], generalized soliton solutions do not develop curvature singularities at $\alpha = 0$ when $d + g = \pm 1$; this corresponds to the case in which there is no cosmological singularity in the solutions of section 4.5, where several examples have been given. In those cases the interaction region can be extended beyond the focusing hypersurface: $\alpha = 0$. This regular focusing hypersurface is then interpreted as the caustic of the colliding wave spacetime. One example of this type of solution is the generalized soliton solution with two opposite poles, $h_1 = -h_2$, $d = \pm 1$ ($z_1^0 = -z_2^0 = -1$), i.e. with the Minkowski background. This solution, which was first described in the soliton context by Ferrari and Ibáñez [104], is locally isometric to a region inside the horizon of the Schwarzschild metric [60, 307, 309]. A detailed analysis of the geometry of this solution can be found in refs [145, 82]; its main features, however, will be discussed in section 7.3.3 to illustrate the geometry of the colliding wave spacetimes. Another example of this type is a solution with two, not opposite, pole trajectories $g = h_1 + h_2 = \pm 1$ and $d = 0$ [102], i.e. with the axisymmetric Kasner background.

Instead of canonical coordinates (α, β), or the previous null coordinates (u, v), the collision wave problem has often been discussed in another convenient set of compact nonnull coordinates: ϕ, θ. These are defined by

$$\alpha = \sin\phi \sin\theta, \quad \beta = -\cos\phi \cos\theta; \quad 0 \le \phi \le \pi/2, \ 0 \le \theta \le \pi, \quad (7.55)$$

which for the corresponding line element imply

$$d\beta^2 - d\alpha^2 = (\cos^2\phi - \cos^2\theta)(d\theta^2 - d\phi^2). \quad (7.56)$$

In the new coordinates ϕ is the time and θ is a space-like variable. We choose the region in which ϕ and θ change to be the triangle in the (ϕ, θ)-plane with the vertices $(\phi = 0, \theta = 0)$, $(\phi = 0, \theta = \pi)$ and $(\phi = \pi/2, \theta = \pi/2)$. In this region the Jacobian of the transformation (7.55) is positive and there is a one-to-one mapping between the points of this region and the points of the triangle in the (α, β)-plane with vertices $(\alpha = 0, \beta = -1)$, $(\alpha = 0, \beta = 1)$ and $(\alpha = 1, \beta = 0)$. Inside the (ϕ, θ) triangle we have $\cos\phi \pm \cos\theta \ge 0$. We choose $z_1^0 = -z_2^0 = 1$, in which case we have $\beta_1 > 0$, $\beta_2 < 0$ and prescription (4.49) gives for the real-pole trajectories the following simple form:

$$\mu_1^- = (1 - \cos\phi)(1 - \cos\theta), \quad \mu_2^- = -(1 - \cos\phi)(1 + \cos\theta); \quad (7.57)$$

note that μ_k^+ ($k = 1, 2$) follows from the relations $\mu_1^- \mu_1^+ = \mu_2^- \mu_2^+ = \sin^2 \phi \sin^2 \theta$. Then, for instance, the generalized soliton solution with two opposite real poles ($s = 2$), i.e. (7.48)–(7.49) with $\mu_1 = \mu_1^-$, $\mu_2 = \mu_2^-$ but with $h_1 = -h_2 \equiv h$, takes the simple form

$$ds^2 = (\sin \phi \sin \theta)^{(d^2-1)/2} (1 + \cos \theta)^{h^2-hd} (1 - \cos \theta)^{h^2+hd}$$
$$\times (\cos^2 \phi - \cos^2 \theta)^{1-h^2} (d\theta^2 - d\phi^2)$$
$$+ (\sin \phi \sin \theta)^{1+d} \left(\frac{1 - \cos \theta}{1 + \cos \theta}\right)^h dx^2$$
$$+ (\sin \phi \sin \theta)^{1-d} \left(\frac{1 + \cos \theta}{1 - \cos \theta}\right)^h dy^2, \tag{7.58}$$

which includes most of the solutions just described.

7.3.3 Geometry of the colliding waves spacetime

In this subsection we will use a particular solution to describe the geometry of the head-on collision of two aligned one-polarization waves [82] (collinear polarizations). The general properties are found in this example, the main difference with other solutions is that the focusing hypersurface is a nonsingular caustic here, whereas most of the solutions develop curvature singularities. As we have mentioned, regular caustics are not generic. We will consider the solution (7.58) when $d = h = 1$ (i.e. one soliton with Minkowski background). This case is thus a truly solitonic solution in the sense that it may be considered as the limit of a nondiagonal soliton solution.

For convenience, instead of the coordinates of (7.58) we will use some dimensionless null coordinates, which we also denote as u and v, defined by

$$u = \pi + \frac{1}{2}(\theta - \phi), \qquad v = \frac{\pi}{2} + \frac{1}{2}(\theta + \phi). \tag{7.59}$$

In this case the metric in the interaction region can be written as,

$$ds_{IV}^2 = -4L_1 L_2[1 + \sin(u + v)]^2 du\, dv + \frac{1 - \sin(u + v)}{1 + \sin(u + v)} dx^2$$
$$+ [1 + \sin(u + v)]^2 \cos^2(u - v) dy^2, \tag{7.60}$$

where L_1 and L_2 are positive arbitrary parameters that we have introduced for convenience (this can always be done) and we have interchanged x and y. Given (7.60), we can always restrict the range of (u, v) to between 0 and $\pi/2$, thus we will define the interaction region IV ($0 \leq u < \pi/2$, $0 \leq v < \pi/2$) with boundaries $u = 0$, $v = 0$ and $u + v = \pi/2$. Note that these coordinates do not satisfy Szekeres prescription (7.37), although conditions (7.38) are obviously satisfied for $n_1 = n_2 = 2$. The hypersurface $u + v = \pi/2$, as we shall see, is a Cauchy horizon (a Killing–Cauchy horizon) and these coordinates become singular there. The extension of this metric to regions II ($0 \leq u < \pi/2$, $v \leq 0$) and III ($0 \leq v < \pi/2$, $u \leq 0$) through the boundaries ($v = 0$, $0 \leq u < \pi/2$) and ($u = 0$, $0 \leq v < \pi/2$), respectively, consists of substituting $u \to u\theta(u)$ and $v \to v\theta(v)$ in (7.60). The resulting metrics in the different regions are

$$ds_{II}^2 = -4L_1L_2(1 + \sin u)^2 du\, dv + \frac{1 - \sin u}{1 + \sin u} dx^2 + (1 + \sin u)^2 \cos^2 u\, dy^2,$$

$$(7.61)$$

$$ds_{III}^2 = -4L_1L_2(1 + \sin v)^2 du\, dv + \frac{1 - \sin v}{1 + \sin v} dx^2 + (1 + \sin v)^2 \cos^2 v\, dy^2,$$

$$(7.62)$$

$$ds_I^2 = -4L_1L_2 du\, dv + dx^2 + dy^2. \qquad (7.63)$$

The parameters L_1 and L_2 have dimensions of length; in the flat region I ($u < 0$, $v < 0$) the physical retarded and advanced times are $u_{phys} = 2L_1 u$ and $v_{phys} = 2L_2 v$, respectively. Naturally by construction (see section 7.3.1) the O'Brien and Synge matching conditions (7.34) are satisfied. It is interesting to see this explicitly, for instance at $v = 0$: $g_{\mu\nu,v}^{IV} \neq g_{\mu\nu,v}^{II}$ but $g_{\mu\nu,u}^{IV} = g_{\mu\nu,u}^{II}$. From this it follows that the Ricci tensor is zero at the boundaries (it is a vacuum solution everywhere), however, the Weyl tensor has δ-function components at the boundaries, otherwise it is regular and nonzero (except in region I where it is zero) [104]. This is interpreted as wavefronts of impulsive (shock) pure gravitational waves. The resulting spacetime is then interpreted as representing the collision of two pure gravitational shock waves followed by trailing gravitational radiation. From our discussion of the meaning of the focusing retarded time u_f, see the discussion following (7.17) for example, the parameters L_1 and L_2 represent the inverse of the strength of the waves. To see this consider the solution in region II and use $v_{phys} = 2L_2 v$ and $du_{phys} = 2L_1(1 + \sin u)^2 du$, then the focusing retarded time $u_f = \pi/2$ corresponds to $u_{phys} = 2L_1(2 + 3\pi/4)$ and the inverse of this physical retarded time represents the strength (energy density) of the u-plane wave.

The interaction region. This is the region bounded by the hypersurfaces $u = 0$, $v = 0$ and $u + v = \pi/2$. Let us now see that this region is locally isometric to a part of the interior Schwarzschild metric. Just consider the coordinate change:

$$\left. \begin{aligned} t &= x, & r &= M[1 + \sin(u + v)], \\ \varphi &= 1 + y/M, & \theta &= \pi/2 - (u - v), \end{aligned} \right\} \qquad (7.64)$$

where we have defined $M = \sqrt{L_1 L_2}$. Metric (7.60) then becomes

$$ds_{IV}^2 = -\left(\frac{2M}{r} - 1\right)^{-1} dr^2 + \left(\frac{2M}{r} - 1\right) dt^2 + r^2(d\theta^2 + \sin^2\theta\, d\varphi^2), \quad (7.65)$$

which is the interior of the Schwarzschild metric ($r < 2M$) where the Killing field ∂_t is space-like. The hypersurface $u + v = \pi/2$ (the caustic) corresponds to the black hole event horizon $r = 2M$. The boundary $v = 0$ corresponds to $r = M(1 + \cos\theta)$ and $u = 0$ corresponds to $r = M(1 - \cos\theta)$, these are the boundaries of the plane waves. These boundaries meet at $r = M$ (spacetime point of the collision) and they meet the caustic at $\theta = 0$ and $\theta = \pi$. This region of the Schwarzschild interior does not contain the singularity $r = 0$ and thus the interaction region has no curvature singularity. Moreover, the caustic $u + v = \pi/2$, which corresponds to the event horizon, is a Cauchy horizon. The above local isometry is not global, however: the coordinates θ and φ are cyclic in the black hole case but in the plane-wave case $-\infty < y < \infty$ and $-\infty < v - u < \infty$. Note, however, that a possible extension of the interaction region through the Cauchy horizon is achieved by making y cyclic and matching to the past event horizon of the Schwarzschild metric in the Kruskal–Szekeres coordinates [105, 145, 82]. But since regular horizons do not seem generic in the physics of colliding waves we shall not insist on this point.

From this discussion we may note that since the Schwarzschild coordinates are not adapted to describe the event horizon, the group coordinates are also not adapted to describe the caustic. For this reason it is convenient to introduce a set of Kruskal–Szekeres-like coordinates to describe the interaction region. First we introduce dimensionless time and space coordinates (ξ, η); $\xi = u + v$, $\eta = v - u$, with ranges $0 \le \xi < \pi/2$ and $-\pi/2 \le \eta < \pi/2$. Then we define a new time coordinate

$$\xi^* = 2M \ln\left(\frac{1 + \sin\xi}{2\cos^2\xi}\right) - M(\sin\xi - 1), \quad (7.66)$$

and a new set of null coordinates $\tilde{U} = \xi^* - x$, $\tilde{V} = \xi^* + x$. Finally we introduce

$$U' = -2M \exp\left(\frac{-\tilde{U}}{4M}\right) \le 0, \quad V' = -2M \exp\left(\frac{-\tilde{V}}{4M}\right) \le 0, \quad (7.67)$$

and the metric in the interaction region then reads:

$$ds_{IV}^2 = -\frac{2\exp[(1 - \sin\xi)/2]}{1 + \sin\xi} dU' dV' + M^2(1 + \sin\xi)^2 d\eta^2$$
$$+ (1 + \sin\xi)^2 \cos^2\eta\, dy^2, \quad (7.68)$$

where ξ and η are functions of U' and V'. From the previous coordinate changes we have that

$$U'V' = \frac{8M^2 \cos^2 \xi}{1 + \sin \xi} \exp\left(\frac{\sin \xi - 1}{2}\right), \qquad \frac{U'}{V'} = \exp\left(\frac{x}{2M}\right). \qquad (7.69)$$

These last equations show that the curves $\xi = $ constant and $x = $ constant are, respectively, hyperbolas and straight lines through the origin of coordinates $U' = V' = 0$. The Cauchy horizon (the caustic) corresponds to the limit of the hyperbolas when $\xi \to \pi/2$, which are the lines $U' = 0$ or $V' = 0$. Note that the transverse coordinate x is involved in the above coordinate change: it is not a good coordinate at the caustic because the whole range of x collapses into the point $U' = V' = 0$; whereas the lines $U' = 0$ and $V' = 0$ represent $x \to -\infty$ and $x \to \infty$, respectively.

To see the structures of the different regions it is illustrative to represent the boundaries between the plane waves and the interaction region as in fig. 7.3. The boundary of region IV with regions II and III, $\mathcal{S}_3$ $\{(v = 0, 0 \le u < \pi/2) \cup (u = 0, 0 \le v < \pi/2)\}$ is shown on the Kruskal–Szekeres-like coordinates. On the other hand the boundary of region II with the interaction region, $\mathcal{S}_2$ $(v = 0, 0 \le u < \pi/2)$ and the boundary of region III with the interaction region, $\mathcal{S}'_2$ $(v = 0, u = 0, 0 \le v < \pi/2)$ are shown in the harmonic coordinates (7.7). Boundaries $\mathcal{S}_2$ and $\mathcal{S}'_2$ must be properly identified with $\mathcal{S}_3$. We can see that $\mathcal{L}$ $(u = \pi/2)$, in region II, is a line rather than a surface contrary to what fig. 7.2 seems to indicate, and represents the focusing effect of the u-plane wave. This line splits into two lines, $\mathcal{M}_1$ and $\mathcal{M}_2$, on the boundary of region IV. In the same way the line $\mathcal{L}'$ $(v = \pi/2)$ in region III splits into two lines, $\mathcal{M}'_1$ and $\mathcal{M}'_2$, on the boundary region IV. To clarify further the global topology of the colliding wave spacetime it is useful to analyse the spacetime null geodesics.

Analysis of null geodesics. We have just seen that in the interaction region the caustic at $u + v = \pi/2$ is not singular and we have introduced a convenient set of coordinates to describe it. In the plane-wave regions II and III, the focusing lines $u = \pi/2$ and $v = \pi/2$, respectively, present coordinate singularities (the metric coefficients of dx^2 and dy^2 vanish). For a single plane wave we have seen in subsection 7.2.1 that this is only due to a bad choice of coordinate patch in this region. In fact, using harmonic coordinates the single plane-wave spacetimes can be continued beyond u_f ($u_f = \pi/2$ for the u-plane wave (7.61)). However, such an extension is not possible in the colliding wave spacetime as we shall see in what follows. Also the colliding wave spacetimes admit global Cauchy hypersurfaces, unlike the single plane-wave spacetimes.

Thus let us now concentrate on the plane-wave region II (a similar analysis can be carried out in region III). Let us introduce harmonic coordinates,

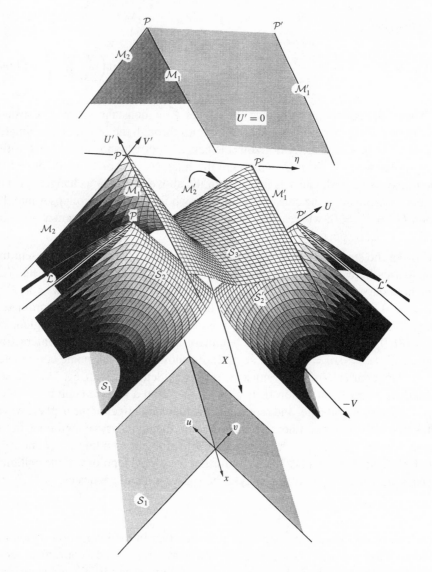

Fig. 7.3.   This is the three-dimensional plot of the colliding plane-wave spacetime in which the boundaries of the four regions I, II, III and IV are shown. The surface $\mathcal{S}_1$ is the boundary between the flat region I and the plane-wave regions II and III. The surface $\mathcal{S}_2$ and the line $\mathcal{L}$ are the boundaries of the plane-wave region II with region IV and the surface $\mathcal{S}_2'$ and the line $\mathcal{L}'$ the boundaries of the plane-wave region III with region IV. The surface $\mathcal{S}_3$ and the lines $\mathcal{M}_1$, $\mathcal{M}_2$, $\mathcal{M}_1'$ and $\mathcal{M}_2'$, are the boundaries of region IV and the two plane-wave regions II and III. The lines $\mathcal{M}_1$ and $\mathcal{M}_2$ of region IV have to be identified with the line $\mathcal{L}$ of region II and lines $\mathcal{M}_1'$ and $\mathcal{M}_2'$ of region IV have to be identified with line $\mathcal{L}'$ of region III. The Cauchy horizon is the 'roof' $\{U' = 0, V' < 0\} \cup \{V' = 0, U' < 0\}$. Causality is globally preserved in this figure: region I stays below the surface $\mathcal{S}_1$, region II stays between the surfaces $\mathcal{S}_2$ and $\mathcal{S}_1$, region III stays between $\mathcal{S}_2'$ and $\mathcal{S}_1$ and region IV stays between the surface $\mathcal{S}_3$ and the roof $\{U' = 0, V' < 0\} \cup \{V' = 0, U' < 0\}$.

$(U, V, X \equiv X^1, Y \equiv X^2)$, according to (7.7):

$$\left.\begin{array}{ll} U = u', & V = v' + \dfrac{-X^2 + (1 - 2\sin u)Y^2}{2L_1(1 + \sin u)^2 \cos u}, \\[3mm] X = \dfrac{\cos u}{1 + \sin u}x, & Y = [(1 + \sin u)\cos u]y, \end{array}\right\} \tag{7.70}$$

where u' is found from $du' = 2L_1(1 + \sin u)^2 du$, $v' = 2L_2 v$ and (7.61) can be written as

$$ds_{II}^2 = -dU\, dV + \frac{3}{4L_1^2(1 + \sin u)^5}(X^2 - Y^2)dU^2 + dX^2 + dY^2, \tag{7.71}$$

which is of the form (7.4). The null geodesics are easily found in the coordinates (u, v, x, y): there are three conserved momenta p_x, p_y and p_v associated with the Killing fields ∂_x, ∂_y and ∂_v, respectively, in region II. Moreover $p^\mu p_\mu = 0$, where p^μ is the four-momentum of a massless particle. Thus, after a simple integration we have:

$$\left.\begin{array}{l} y(u) = y(0) - L_1\dfrac{p_y}{p_v}\tan u, \\[5mm] x(u) = x(0) - L_1\dfrac{p_x}{p_v}\left[\dfrac{(1 + \sin u)^2}{2\cos u}(9 - \sin u) + \dfrac{15}{2}\cos u - \dfrac{15}{2}u - 12\right], \\[5mm] v(u) = v(0) + \dfrac{L_1 p_x^2}{4L_2 p_v^2}\left[\dfrac{(1 + \sin u)^2}{2\cos u}(9 - \sin u) + \dfrac{15}{2}\cos u - \dfrac{15}{2}u - 12\right] \\[5mm] \qquad\quad + \dfrac{L_1 p_y^2}{4L_2 p_v^2}\tan u. \end{array}\right\} \tag{7.72}$$

We now see that almost all geodesics entering region II cross the surface ($v = 0, 0 \le u < \pi/2$) to enter the interaction region IV. The geodesics with $p_v = 0$ travel in the same sense as the wave and do not collide with it, thus they are not of interest here. Geodesics with $p_x \ne 0$ or $p_y \ne 0$ always cross ($v = 0, 0 \le u < \pi/2$) because given any finite value of negative $v(0)$ ($v(0) < 0$), there will always be a value of $u < \pi/2$ for which the terms with either $p_x^2/\cos u$ or $p_y^2 \tan u$ will make $v(u) = 0$. Null geodesics with a perpendicular incidence, i.e. $p_x = p_y = 0$ (x and y are constants) have $v(u) = v(0)$ and intersect $\mathcal{L}$ ($u = \pi/2$). Note that any family of null geodesics with $p_x \ne 0$ or $p_y \ne 0$ will enter region IV but if the parameters p_x and p_y are small they will enter region IV close to $\pi/2$. This is true for any values of p_x and p_y close to zero, but in the limit $(p_x, p_y) = (0, 0)$, the geodesics reach the line $\mathcal{L}$; see fig. 7.4. This behaviour suggests that we can identify the line $\mathcal{L}$ with the point $\mathcal{P}$, i.e. when a geodesic reaches $\mathcal{L}$ it is immediately sent to the point $\mathcal{P}$. This is called a 'fold singularity' [218]. The line $\mathcal{L}$ concentrates null geodesics with perpendicular incidence and sends them to $\mathcal{P}$. This means that we cannot extend the spacetime through $\mathcal{L}$, since beyond $\mathcal{L}$ lies the interaction region. The point $\mathcal{P}$ ($u = \pi/2$, $v = 0$) of fig. 7.3 is an accumulation point of all geodesics. The possible confusion regarding such points that may arise from figs 7.2(b) or 7.4 is due to the fact that the projection of the spacetime into the (u, v)-plane does not preserve the causality globally. The accumulation of geodesics near points $\mathcal{P}$ and $\mathcal{P}'$ in fig. 7.4 is related to the infinite blueshift produced at massless modes propagating on this spacetime [82], as one can easily see by noting that the geometrical optics approximation is exact in the plane-wave regions.

All this may be further clarified by writing the boundaries $\mathcal{S}_1$ ($u = 0$, $v < 0$), $\mathcal{S}_2$ ($v = 0$, $u < 0$) and $\mathcal{L}$ ($u = \pi/2$) of region II in harmonic coordinates: $\mathcal{S}_1$: $\{U = 0, \ V = v' + (Y^2 - X^2)(2L_1)^{-1}\}$; $\mathcal{S}_2$: $\{0 \leq U < 2L_1(2 + 3\pi/4), \ V = (-X^2 + (1 - 2\sin u)Y^2)[2L_1(+\sin u)^2 \cos u]^{-1}\}$; $\mathcal{L}$: $\{U = 2L_1(2 + 3\pi/4), \ V = v \leq 0$, if $X = Y = 0$, $V = -\infty$ otherwise$\}$. Note that the surface $\mathcal{S}_2$ is formed by all null geodesics with $p_x = p_y = v(0) = 0$. In fact, these geodesics in harmonic coordinates are

$$
\left.
\begin{aligned}
&X = x(0)\frac{\cos u}{1 + \sin u}, \qquad Y = y(0)(1 + \sin u)\cos u, \\[2mm]
&V = \frac{\cos u}{L_1(1 + \sin u)^2}\left[\frac{-x^2(0)}{(1 + \sin u)^2} + y^2(0)(1 - 2\sin u)(1 + \sin u)^2\right].
\end{aligned}
\right\} \tag{7.73}
$$

The null geodesics $x = $ constant and $y = $ constant, which form $\mathcal{S}_2$, all end at the focal point $\mathcal{P}$: $U = 2L_1(2 + 3\pi/4)$, $V = X = Y = 0$ and the line $\mathcal{L}$ corresponds to a limit curve of these geodesics as $x^2 + y^2 \to 0$; $\mathcal{L}$ does not belong to $\mathcal{S}_2$, it is in the closure of $\mathcal{S}_2$. However, $\mathcal{L}$ does not belong to region II. To show this it is convenient to consider the family of hypersurfaces $v = $ constant < 0. Each of these surfaces is formed by the family of hypersurfaces with $p_x = p_y = 0$ and $v < 0$, and all of these geodesics end at $X = Y = 0$, $V = v$, which is one point of $\mathcal{L}$. Thus, $\mathcal{L}$ and $\mathcal{P}$ are topological singularities and must be excluded from the spacetime. Note that only a null measure geodesics set reaches these points since all time-like and almost all null geodesics avoid $\mathcal{L}$ and $\mathcal{P}$.

7.3.4 Noncollinear polarization waves: nondiagonal metrics

In this subsection we consider soliton solutions that have been used to describe the head-on collision of two gravitational plane waves with noncollinear polarizations propagating on a Minkowski spacetime. The most general solution of this type is the two-parameter solution obtained by Ferrari *et al.* [106, 107] as a two-soliton solution with two real-pole trajectories on a Kasner background. The solution includes the Nutku and Halil solution [232] representing the collision of pure impulsive gravitational plane waves with noncollinear polarizations, which generalizes the Khan and Penrose solution for impulsive waves with collinear polarizations. The Ferrari *et al.* solution also includes the

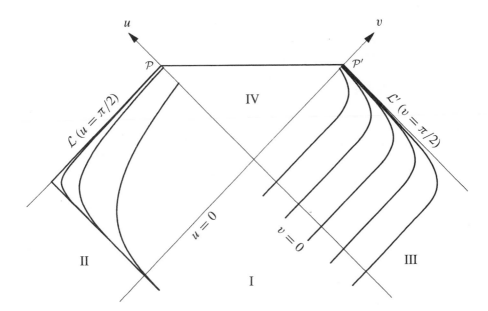

Fig. 7.4.   Some null geodesics are projected onto the (u, v)-plane. Only null geodesics with perpendicular incidence, $p_x = p_y = 0$ reach the lines $\mathcal{L}$ or $\mathcal{L}'$, but if either $p_x \neq 0$ or $p_y \neq 0$ the geodesics reach region IV. In the limit $p_x \to 0$ and $p_y \to 0$, the geodesics reach region IV at points closer and closer to $\mathcal{P}$ or $\mathcal{P}'$. One can interpret the null geodesic with $p_x = p_y = 0$ as being sent to the lines $\mathcal{L}$ and $\mathcal{L}'$ and then immediately sent to the points $\mathcal{P}$ and $\mathcal{P}'$, respectively. On the left hand side of the figure null geodesics, which cross the boundary between regions I and II at the same point v_0, with several values of the parameter p_x and with $p_y = 0$ are represented (the parameter p_x changes from one geodesic to the next by a factor of 10). After entering region II, as the impact parameter v_0 grows the geodesics reach the horizon at points closer and closer to $\mathcal{P}$. On the right hand side of the figure we represent null geodesics which cross the boundary between regions I and III at regularly spaced points on the u axis and with the same values for the parameters p_x and p_y. This figure suggests that the points $\mathcal{P}$ and $\mathcal{P}'$ are accumulation points of null geodesics.

generalization by Ernst *et al.* [93] of the Nutku and Halil solution which was obtained by means of an Ehlers transformation applied to the Ernst potential of the Nutku and Halil metric. The solution depends on two parameters: one represents the angle between the direction of polarization of the two waves and the other is the Kasner parameter. This solution has been further generalized by a three-parameter solution obtained by Ernst *et al.* [94].

Following ref. [106] we take two real-pole trajectories (4.49) μ_1^- and μ_2^+, with origins $z_1^0 = 1$ and $z_2^0 = -1$. In canonical coordinates ($\alpha = t$, $\beta = z$) these are: $\mu_1^- = 1 - \beta - \sqrt{(1-\beta)^2 - \alpha^2}$ and $\mu_1^+ = -1 - \beta + \sqrt{(1+\beta)^2 - \alpha^2}$. As shown in subsection 7.3.2 the interaction region in the colliding wave problem is the triangle formed by the intersection of the exterior light cones $(1 - \beta)^2 = \alpha^2$ and $(1 + \beta)^2 = \alpha^2$, see fig. 7.2. As in subsection 7.3.2 we also use the Kasner metric as the background solution. In the conventions of ref. [106], the solution is given by (1.87), (1.100) and (1.110) with $n = 2$, where the vectors $m_a^{(k)}$ which make the matrix Γ_{kl} $(k, l = 1, 2)$ are defined in (4.36), with the real constant vectors $m_{0a}^{(k)}$ restricted by

$$\left. \begin{array}{ll} m_{02}^{(1)} m_{02}^{(2)} + m_{01}^{(1)} m_{01}^{(2)} = p, & m_{02}^{(1)} m_{02}^{(2)} - m_{01}^{(1)} m_{01}^{(2)} = 1, \\[2mm] m_{02}^{(1)} m_{01}^{(2)} - m_{01}^{(1)} m_{02}^{(2)} = q, & m_{02}^{(1)} m_{01}^{(2)} + m_{01}^{(1)} m_{02}^{(2)} = 0, \end{array} \right\} \quad (7.74)$$

where p and q are real parameters. From (7.74) it follows that these parameters are related by

$$p^2 + q^2 = 1. \quad (7.75)$$

Note that this should be compared with analogous expressions (8.44)–(8.45) or (8.51), in the axisymmetric context. In such a context one introduces the three parameters of the Kerr–NUT solution: the mass parameter m, the parameter a, and the Taub–NUT parameter b. In the Ferrari et al. solution the counterpart of the Kerr parameter is zero; it corresponds to the zero in the last equation of (7.74). Thus a three-parameter solution, which is related to the solution of ref. [94], can easily be obtained; see also ref. [127]. We may use the two-soliton solution derived in section 2.2 to write the Ferrari et al. solution in the interaction region in canonical coordinates. In fact, the explicit form of this solution is given by (2.20)–(2.25) with $u_0 = d \ln \alpha$ and $\rho_k = d \ln(\mu_k / \alpha) + C_k$, according to (2.27). The parameters C_k can be determined from (7.74) via (4.37): it is easy to see that C_1 and C_2 are both linear in $\ln \sqrt{(p + 1)/(p - 1)}$.

Ferrari et al. chose to give the explicit form of the metric in the dimensionless coordinates (ϕ, θ) defined by

$$\alpha = \sin \phi \sin \theta, \quad \beta = \cos \phi \cos \theta; \quad 0 \le \pi, \theta \le \pi. \quad (7.76)$$

The solution in the interaction region is then defined inside the circle $\cos^2 \phi + \cos^2 \theta = 1$ and the pole trajectories are given by $\mu_1^- = (1 - \cos \phi)(1 + \cos \theta)$ and $\mu_2^+ = -(1 - \cos \phi)(1 - \cos \theta)$. We refer the reader to ref. [106] for the complete expression. The resulting metric reduces to the Nutku and Halil solution [232] when $d = 0$ and to the Khan and Penrose solution when $d = 0$ and $q = 0$. Ferrari et al. computed the Weyl scalar Ψ_2 for this metric and found that this function is always singular on the focusing hypersurface with the exception of the degenerate solutions $d = \pm 1$, which correspond to the Minkowski background.

After a solution in the interaction region has been identified, the next step is to extend it to the plane-wave regions. For this we need to write the solution in terms of the null coordinates (u, v) and, as in fig. 7.2, define region I as the points $\{u < 0, v < 0\}$, region II as $\{u > 0, v < 0\}$, and region III as $\{u < 0, v > 0\}$. Writing the metric in the form (7.27) the extension to these regions across the null boundaries $u = 0$ and $v = 0$ is made, as usual, by the substitutions $u \rightarrow u\theta(u)$ and $v \rightarrow v\theta(v)$. These coordinates may be defined from the canonical ones using Szekeres prescription (7.45) with $c_1 = c_2 = 1$ and $n_1 = n_2 = 2$ so that

$$\alpha = 1 - u^2 - v^2, \qquad \beta = u^2 - v^2. \tag{7.77}$$

In these coordinates the focusing hypersurface $\alpha = 0$ is defined by $u^2 + v^2 = 1$. The relationship between the null coordinates (u, v) and the previous (ϕ, θ) is

$$\cos\phi = u\sqrt{1 - v^2} + v\sqrt{1 - u^2}, \qquad \cos\theta = u\sqrt{1 - v^2} - v\sqrt{1 - u^2}. \tag{7.78}$$

Ferrari [103] has given a physical interpretation of the parameters of the complete solution. First, we should note that we have taken the soliton origins $z_1^0 = -z_2^0$ equal to unity. This is mathematically convenient and common practice, but it makes the coordinates (α, β) dimensionless. Thus for a physical interpretation we should restore z_1^0 as an arbitrary length parameter (this is similar to the introduction of the length parameters L_1 and L_2 in subsection 7.3.3). We have also the two parameters p and q which are related by (7.75) and, finally, there is the Kasner parameter d. By analysing the Weyl scalar Ψ_0 in the plane-wave region II, Ferrari found that $(z_1^0)^{-2}$ gives the amplitude of the impulsive component of the wave, i.e. the term proportional to $\delta(u)$. It is also seen that z_1^0 determines the time of focusing of the waves, and its inverse is proportional to the strength of the waves. This is consistent with our analysis in section 7.3.3 and also with a result on the focusing of colliding graviton beams [289], where the scattering gravitons converge to a focus in a time which is proportional to the energy density per unit area of the beam. Our discussion in subsection 7.2.2 suggests that the energy density of a pure gravitational plane wave should in fact be proportional to the inverse of the focusing time. The parameter q measures the angle between the polarization directions of the two waves: $q = 0$ corresponds to collinear polarizations. The only effect of the polarization parameter q in the focusing time is to delay the time by a factor $\sqrt{1 + q^2}$. The Kasner parameter, on the other hand, determines the power in the power law divergence of the Weyl scalars on the focusing singularity.

The degenerate solution. This corresponds to the solution with the Minkowski background $d = \pm 1$. This case is interesting in its own right and was studied in some detail in ref. [105]. It may be written in the coordinates of (7.76) as

$$ds^2 = -C(1 + \cos^2\phi + 2p\cos\phi)(d\phi^2 - d\theta^2)$$
$$+ \left(\frac{1 - \cos^2\phi}{1 + \cos^2\phi + 2p\cos\phi}\right)(dx - 2q\cos\theta\,dy)^2$$
$$+ \sin^2\theta(1 + \cos^2\phi + 2p\cos\phi)dy^2, \tag{7.79}$$

where p and q are related by (7.75) and C is an arbitrary constant. Let us now see that this solution is isometric to a region of the Taub–NUT metric. For this we introduce new parameters (σ, m, n) by $p = m/\sigma$ and $q = -b/\sigma$, thus $\sigma = \sqrt{m^2 + b^2}$; then define new coordinates (t, r, φ) by

$$\cos\phi = \frac{r - m}{\sigma}, \quad x = t, \quad y = \sigma\varphi, \tag{7.80}$$

and take $C = \sigma^2$. Metric (7.79) then becomes

$$ds^2 = \frac{r^2 + b^2}{r^2 - 2mr - b^2}dr^2 - \frac{r^2 - 2mr - b^2}{r^2 + b^2}(dt + 2b\cos\theta\, d\varphi)^2$$
$$+ (r^2 + b^2)(d\theta^2 + \sin^2\theta\, d\varphi^2). \tag{7.81}$$

This is the Taub–NUT solution [268, 230]; it is given by (8.48) when $C_f = -1$ and $a = 0$, and is of Petrov type D. Since under the previous change $1 - \cos^2\phi = (-r^2 + 2mr + b^2)(m^2 + b^2)^{-1}$, the interaction region described by (7.79) is isometric to the Taub region $r^2 - 2mr + b^2 < 0$, where the two Killing fields ∂_t and ∂_φ are both space-like. Note the similarity with (7.65). The focusing hypersurface $u^2 + v^2 = 1$ corresponds to the hypersurface $r_+ = m + \sigma$ which makes the nonsingular event horizon $r^2 - 2mr + b^2 = 0$. Also, it is easy to see using (7.78) that the null hypersurfaces $u = 0, 0 < v < 1$ and $v = 0$, $0 < u < 1$, which match to the two plane-wave regions, correspond to $r = m - \sigma\cos\theta$ and $r = m + \sigma\cos\theta$, respectively. Thus, as happened for the case of the solution considered in subsection 7.3.3, this solution is not singular at the focusing hypersurface $u^2 + v^2 = 1$, which can now be seen as the colliding wave caustic. If the caustic is regular, the metric can be extended across this surface by matching to the horizon of the Taub–NUT metric [105]. This is very similar to the analysis carried out in subsection 7.3.3.

8
Axial symmetry

In the previous four chapters we discussed metrics which admit two commuting space-like Killing vector fields. In this chapter we deal with stationary axisymmetric spacetimes where one of the two Killing fields is time-like. These spacetimes have been investigated for a long time due to the possibility of describing the gravitational fields of compact astrophysical sources. The field equations for the relevant metric tensor components are now elliptic rather than hyperbolic as in the nonstationary case but the solutions can be formally related via complex coordinate transformations. In section 8.1 we again formulate the ISM, but in this case, because of the different ranges of the coordinates, some of the previous expressions become much simpler. In section 8.2 the general n-soliton solution is explicitly constructed in this axisymmetric context. In section 8.3 the Kerr, Schwarzschild and Kerr–NUT solutions are constructed as simple two-soliton solutions on the Minkowski background. The asymptotic flatness of the general n-soliton solution is discussed in section 8.4 and we show that asymptotic flatness can always be imposed by certain restrictions on the soliton parameters; the resulting spacetimes can be interpreted as a superposition of Kerr black holes on the symmetry axis. In section 8.5 we discuss the diagonal metrics (static Weyl class). In this case the soliton metrics contain many well known static solutions and some generalized soliton solutions can be constructed as in the previous chapters; a few particularly interesting solutions are considered in some detail. Finally, in section 8.6 we show how the well known Tomimatsu–Sato solution can be obtained from an n-soliton solution by pole fusion, and by the same procedure we obtain the Neugebauer superposition of Kerr–NUT metrics.

8.1 The integration scheme

Let us write the metric for a stationary and axisymmetric gravitational field in the form:

$$ds^2 = f(d\rho^2 + dz^2) + g_{ab}dx^a dx^b, \tag{8.1}$$

where the metric coefficients f and g_{ab} are functions of ρ and z only. For the stationary case throughout this book we use the notation $(x^0, x^1, x^2, x^3) = (t, \varphi, \rho, z)$, and the Latin indices $a, b, c, \ldots$ take the values 0 and 1, which correspond to the coordinates t and φ. The metric coefficient f in (8.1) is nonnegative. This block diagonal form is guaranteed in vacuum by Papapetrou's theorem [236], assuming a regular symmetry axis and the presence of the two commuting Killing vector fields, one of them time-like (in our coordinates ∂_t) and the other space-like (in our coordinates ∂_φ). The latter defines closed orbits around the symmetry axis and vanishes on it. The nonvacuum version of this theorem that guarantees that the orbits of the isometry group admit orthogonally transitive surfaces in the stationary axisymmetric spacetimes can be found in refs [182, 179].

Using the remaining freedom in the choice of the coordinates ρ and z we can, without loss of generality, impose on the 2×2 matrix g (with components g_{ab}) the following supplementary condition

$$\det g = -\rho^2. \tag{8.2}$$

Note that we did not use a choice analogous to (8.2) in chapter 1 because for nonstationary metrics $\det g$ can be both time-like and space-like. Thus to cover both possibilities in the calculations simultaneously it was better to keep $\det g = \alpha^2$ without specifying the function $\alpha(\zeta, \eta)$. In the stationary case for the spacetime metric (8.1) the $\det g$ can have space-like character only. Consequently, it is better to use the simplification (8.2) from the beginning, which means that we are using canonical coordinates. This is analogous to what we did in section 4.4 in the cosmological context.

It is now easy to show that the full system of Einstein equations in vacuum for the metric (8.1)–(8.2) separates into two groups. The first determines the matrix g and has the form

$$\left(\rho g_{,\rho} g^{-1}\right)_{,\rho} + \left(\rho g_{,z} g^{-1}\right)_{,z} = 0. \tag{8.3}$$

The second group of equations determines the metric coefficient f for a given solution of (8.3) and can be written in the form

$$(\ln f)_{,\rho} = -\frac{1}{\rho} + \frac{1}{4\rho} \text{Tr}(U^2 - V^2), \quad (\ln f)_{,z} = \frac{1}{2\rho} \text{Tr}(UV), \tag{8.4}$$

where the 2×2 matrices U and V are defined as

$$U = \rho g_{,\rho} g^{-1}, \quad V = \rho g_{,z} g^{-1}. \tag{8.5}$$

It is easy to see that after the formal transformations $z = \zeta + \eta$, $\rho = -i(\zeta - \eta)$, $\rho = -i\alpha$, $z = \beta$, $U = -(A+B)/2$, $V = i(A-B)/2$ from the variables ρ, z and the matrices U, V to the variables and matrices used in chapter 1, metric (8.1) and (8.2)–(8.5) will be formally reduced to the equations we studied previously; see also section 4.4. For this reason all the formal aspects of the integration scheme can be obtained from the results of chapter 1. We shall discuss here only the basic points that are necessary for a complete exposition and we shall not go into the details of the proofs. For more details see refs [23, 24].

Using the results of section 1.3 we can easily find the 'L–A pair', or spectral equations, for the matrix equation (8.3) in the variables ρ and z:

$$D_1\psi = \frac{\rho V - \lambda U}{\lambda^2 + \rho^2}\psi, \quad D_2\psi = \frac{\rho U + \lambda V}{\lambda^2 + \rho^2}\psi, \tag{8.6}$$

where the commuting differential operators D_1 and D_2 are given by

$$D_1 = \partial_z - \frac{2\lambda^2}{\lambda^2 + \rho^2}\partial_\lambda, \quad D_2 = \partial_\rho + \frac{2\lambda\rho}{\lambda^2 + \rho^2}\partial_\lambda, \tag{8.7}$$

and where λ is a complex spectral parameter independent of the coordinates ρ and z. It is not hard to verify that the conditions of compatibility of (8.6) for the generating matrix function $\psi(\lambda, \rho, z)$ are identical to the original (8.3) and (8.4), if we rewrite them, and also the conditions for their compatibility, in terms of the matrices U and V in the same way as was done in section 1.3. The required matrix g is the value of the matrix $\psi(\lambda, \rho, z)$ at $\lambda = 0$:

$$g(\rho, z) = \psi(0, \rho, z). \tag{8.8}$$

The procedure for the integration of (8.6) assumes knowledge of some particular solution. Let the matrices g_0, U_0 and V_0 be some particular solution of (8.3) and (8.5), and let $\psi_0(\lambda, \rho, z)$ be the solution of (8.6) with such matrices. We then seek the solution for ψ of the form

$$\psi = \chi\psi_0, \tag{8.9}$$

and we get, from (8.6), the following equations for the dressing matrix $\chi(\lambda, \rho, z)$:

$$D_1\chi = \frac{\rho V - \lambda U}{\lambda^2 + \rho^2}\chi - \chi\frac{\rho V_0 - \lambda U_0}{\lambda^2 + \rho^2}, \quad D_2\chi = \frac{\rho U + \lambda V}{\lambda^2 + \rho^2}\chi - \chi\frac{\rho U_0 + \lambda V_0}{\lambda^2 + \rho^2}. \tag{8.10}$$

Now, as before, it can be shown that to ensure that the matrix g is real and symmetric, supplementary conditions have to be imposed on the solutions of (8.10). For the reality of g we need that

$$\bar{\chi}(\bar{\lambda}) = \chi(\lambda), \quad \bar{\psi}(\bar{\lambda}) = \psi(\lambda), \tag{8.11}$$

where a bar denotes complex conjugation, and for g to be symmetric we need that

$$g = \chi(\lambda)g_0\tilde{\chi}(-\rho^2/\lambda), \tag{8.12}$$

where a tilde indicates transposition. Besides this, compatibility of (8.12) with (8.9) requires that

$$\chi(\infty) = I, \tag{8.13}$$

where I is the unit matrix; here and often in this chapter we omit the arguments ρ and z of some functions for simplicity. We can now turn to the construction of the n-soliton solution for the axisymmetric gravitational field.

We should remark that the relationship between the ISM and other solution-generating techniques in the stationary axisymmetric context such as Kinnersley–Chitre transformations [167, 168, 169], the Hauser–Ernst formalism [137, 138], Harrison's Bäcklund transformation [135, 136] or Neugebauer's Bäcklund transformation [224] were given and studied by Cosgrove [64, 65, 66]. These relationships were adapted to the hyperbolic time-dependent context by Kitchingham [172, 173, 174].

8.2 General n-soliton solution

The soliton solutions for the matrix g correspond to the presence of pole singularities of the dressing matrix $\chi(\lambda, \rho, z)$ in the complex plane of the spectral parameter λ. Let us consider the general case in which the matrix χ has n such poles, which we assume to be simple. The dressing matrix $\chi(\lambda, \rho, z)$ can then be represented in the form

$$\chi = I + \sum_{k=1}^{n} \frac{R_k}{\lambda - \mu_k}, \tag{8.14}$$

where the matrices R_k and the numerical functions μ_k now depend on the variables ρ and z only.

Substitution of (8.14) into (8.10), using the supplementary condition (8.12), completely determines the pole trajectories $\mu_k(\rho, z)$ and the matrices $R_k(\rho, z)$. The functions μ_k are determined from the requirement that on the left hand sides of (8.10) there are no poles of second order at the points $\lambda = \mu_k$. The result is that each function $\mu_k(\rho, z)$ (with each index $k = 1, 2, \ldots, n$) satisfies a pair of differential equations:

$$\mu_{k,z} = \frac{-2\mu_k^2}{\mu_k^2 + \rho^2}, \qquad \mu_{k,\rho} = \frac{2\rho\mu_k}{\mu_k^2 + \rho^2}, \tag{8.15}$$

whose solutions are the roots of the quadratic algebraic equation,

$$\mu_k^2 + 2(z - w_k)\mu_k - \rho^2 = 0, \tag{8.16}$$

where w_k are arbitrary, generally complex, constants.

Accordingly, for each index k (i.e. for each pole) we have an arbitrary constant w_k that determines two possible solutions for the pole trajectory $\mu_k(\rho, z)$:

$$\mu_k = w_k - z \pm [(w_k - z)^2 + \rho^2]^{1/2}, \tag{8.17}$$

with the appropriate definition of the square root. This formula shows the essential difference between stationary solitons and solitonic gravitational waves. Here the solutions with real poles (i.e. with w_k real) will have no discontinuities and unperturbed regions because the quantity under the square root will be nonnegative through the whole spacetime.

The matrices R_k are degenerate and have the form

$$(R_k)_{ab} = n_a^{(k)} m_b^{(k)}; \tag{8.18}$$

the two-component vectors $m_a^{(k)}$ are found directly from (8.10) by requiring that these equations be satisfied at the poles $\lambda = \mu_k$, and the vectors $n_a^{(k)}$ are then determined from condition (8.12). The vectors $m_a^{(k)}$ can be expressed in terms of the given partial solution for the background generating matrix $\psi_0(\lambda, \rho, z)$ taken at the value μ_k of the argument λ. These vectors are of the following form:

$$m_a^{(k)} = m_{0b}^{(k)} [\psi_0^{-1}(\mu_k, \rho, z)]_{ba}, \tag{8.19}$$

where here, and from now on, *summation is understood to be over repeated indices a, b, c, d, f, which take the values 0 and 1; whereas summation over other indices occurs only when explicitly indicated*. In (8.19) the $m_{0b}^{(k)}$ are arbitrary constants.

The vectors $n_a^{(k)}$ can then be determined from the following nth order system of algebraic equations:

$$\sum_{l=1}^{n} \Gamma_{kl} n_a^{(l)} = \mu_k^{-1} m_c^{(k)} (g_0)_{ca}, \quad k, l = 1, 2, \ldots, n, \tag{8.20}$$

where the matrix Γ_{kl} is symmetric with matrix elements

$$\Gamma_{kl} = m_c^{(k)} (g_0)_{cb} m_b^{(l)} (\rho^2 + \mu_k \mu_l)^{-1}. \tag{8.21}$$

If we introduce the symmetric matrix D_{kl}, inverse to the Γ_{kl},

$$\sum_{p=1}^{n} D_{kp} \Gamma_{pl} = \delta_{kl}, \tag{8.22}$$

then we get from (8.20) for the vectors $n_a^{(k)}$

$$n_a^{(k)} = \sum_{l=1}^{n} D_{lk} \mu_l^{-1} L_a^{(l)}, \tag{8.23}$$

where

$$L_a^{(k)} = m_c^{(k)}(g_0)_{ca}. \tag{8.24}$$

According to (8.8)–(8.9), and (8.14), the required matrix g is

$$g = \psi(0) = \chi(0)\psi_0(0) = \chi(0)g_0 = \left(I - \sum_{k=1}^{n} R_k \mu_k^{-1}\right) g_0. \tag{8.25}$$

Now, using (8.18), (8.23) and (8.24), we get the metric components g_{ab}:

$$g_{ab} = (g_0)_{ab} - \sum_{k,l=1}^{n} D_{kl} \mu_k^{-1} \mu_l^{-1} L_a^{(k)} L_b^{(l)}. \tag{8.26}$$

This expression shows that the matrix g is symmetric. The question of it being real can be treated in the same way as in section 1.4 for the nonstationary case. The result is the same, namely, to ensure the reality of g it is necessary to choose the arbitrary constants w_k in (8.16) and $m_{0b}^{(k)}$ in (8.19) to be real for each real-pole trajectory μ_k (the vectors $m_a^{(k)}$ are real) or, alternatively, to choose these constants to be complex-conjugate for each pair of complex conjugate trajectories μ_k and $\mu_{k+1} = \overline{\mu}_k$, i.e. $w_{k+1} = \overline{w}_k$ and $m_{0b}^{(k+1)} = \overline{m}_{0b}^{(k)}$ (in this case the vectors $m_a^{(k+1)}$ and $m_a^{(k)}$ are also complex conjugate to each other).

Now we need to make sure that condition (8.2) for the matrix g is satisfied. Using an approach analogous to that of section 1.4, we can calculate the determinant of the matrix g with components (8.26). The result is

$$\det g = (-1)^n \rho^{2n} \left(\prod_{k=1}^{n} \mu_k^{-2}\right) \det g_0. \tag{8.27}$$

If we take the particular solution g_0, which by definition satisfies $\det g_0 = -\rho^2$, it follows from (8.27) that the number of solitons, n, must always be even, since an odd number would change the sign of $\det g$ and lead to an unphysical metric signature. Therefore, in contrast to the nonstationary case, on a physical background all stationary axisymmetric solitons (even those which correspond to real poles $\lambda = \mu_k$) can only appear in pairs forming bound two-soliton states. Nevertheless, we can obtain physical solutions with an odd number of solitons, but for this it is necessary to take a background solution with a nonphysical signature, $\det g_0 = \rho^2$. The first examples of solutions of this kind were obtained and investigated in ref. [284].

In order to obtain a physical n-soliton solution $g^{(ph)}$ that satisfies not only equation (8.3) but also the supplementary condition (8.2) we remark that for any solution g of equation (8.3) det g satisfies the equation

$$\rho^{-1}[\rho(\ln \det g),_\rho],_\rho + (\ln \det g),_{zz} = 0. \tag{8.28}$$

Then it is easy to verify that the matrix

$$g^{(ph)} = \pm\rho(-\det g)^{-1/2}g, \tag{8.29}$$

(with any sign on the right hand side of this formula) also satisfies (8.3) and the condition $\det g^{(ph)} = -\rho^2$. Thus we get from (8.27) and (8.29) the final expression for the metric tensor,

$$g^{(ph)} = \pm\rho^{-n}\left(\prod_{k=1}^{n}\mu_k\right)g, \quad \det g^{(ph)} = -\rho^2, \tag{8.30}$$

where the sign in the expression for $g^{(ph)}$ should be appropriately chosen in each individual case to ensure the correct metric signature.

The computation of the metric coefficient f can be made in direct analogy with the computation in the nonstationary case described in section 1.4 (see ref. [24] for details). Substituting the nonphysical solution g given by (8.26) into (8.4) we get for the nonphysical value of f

$$f = C_n f_0 \rho^n\left(\prod_{k=1}^{n}\mu_k^2\right)\left(\prod_{k=1}^{n}(\mu_k^2 + \rho^2)\right)^{-1}\det\Gamma_{kl}, \tag{8.31}$$

where C_n are arbitrary constants, f_0 is the background value of the coefficient f corresponding to the background solution g_0 and the matrix Γ_{kl} is given by (8.21).

From the definition (8.5) of the matrices U and V and the definition (8.29) we get the obvious expressions

$$U^{(ph)} = \rho g^{(ph)}_{,\rho}(g^{(ph)})^{-1} = U + \left[1 - \frac{1}{2}\rho(\ln \det g),_\rho\right]I,$$

$$V^{(ph)} = \rho g^{(ph)}_{,z}(g^{(ph)})^{-1} = V - \frac{1}{2}\rho(\ln \det g),_z I. \tag{8.32}$$

When we now substitute the matrices $U^{(ph)}$ and $V^{(ph)}$ instead of U and V into (8.4), we find that the physical coefficient $f^{(ph)}$ is given by the formula

$$f^{(ph)} = 16C_f f_0 \rho^{-n^2/2}\left(\prod_{k=1}^{n}\mu_k\right)^{n+1}\left[\prod_{k>l=1}^{n}(\mu_k - \mu_l)^{-2}\right]\det\Gamma_{kl}, \tag{8.33}$$

where C_f is an arbitrary constant (but with a sign that ensures the condition $f^{(ph)} \geq 0$), and the structure of the product $\prod(\mu_k - \mu_l)^{-2}$ was explained in section 1.4 after the analogous formula (1.110). The factor 16 in (8.33) is introduced just for future convenience.

Consequently, the final form of the vacuum stationary n-soliton solution is

$$ds^2 = f^{(ph)}(d\rho^2 + dz^2) + g_{ab}^{(ph)} dx^a dx^b, \tag{8.34}$$

where $f^{(ph)}$ is given by (8.33) and the matrix elements $g_{ab}^{(ph)}$ are determined by (8.30) and (8.26).

If the background metric (g_0, f_0) is flat and given (in cylindrical coordinates) by the interval

$$ds^2 = -dt^2 + \rho^2 d\varphi^2 + d\rho^2 + dz^2, \tag{8.35}$$

then we have $f_0 = 1$ and $g_0 = \text{diag}(-1, \, \rho^2)$ with the obvious property that $\det g_0 = -\rho^2$. The matrix V_0 is equal to zero, and the matrix U_0 is: $U_0 = \text{diag}(0, \, 2)$. From (8.6) we get the corresponding solution for the generating matrix $\psi_0(\lambda, \rho, z)$:

$$\psi_0 = \text{diag}(-1, \, \rho^2 - 2z\lambda - \lambda^2), \tag{8.36}$$

which satisfies the requirement that $\psi(\lambda = 0) = g_0$. From this and (8.19), using (8.16) we find the components of the vectors $m_a^{(k)}$:

$$m_0^{(k)} = C_0^{(k)}, \quad m_1^{(k)} = C_1^{(k)} \mu_k^{-1}, \tag{8.37}$$

where $C_0^{(k)}$ and $C_1^{(k)}$ are arbitrary constants. Then from (8.21) we get the elements of the matrix Γ_{kl}:

$$\Gamma_{kl} = (-C_0^{(k)} C_0^{(l)} + C_1^{(k)} C_1^{(l)} \mu_k^{-1} \mu_l^{-1} \rho^2)(\rho^2 + \mu_k \mu_l)^{-1}, \tag{8.38}$$

and from (8.24) we get the components of the vectors $L_a^{(k)}$:

$$L_0^{(k)} = -C_0^{(k)}, \quad L_1^{(k)} = C_1^{(k)} \mu_k^{-1} \rho^2. \tag{8.39}$$

Formulas (8.37)–(8.39) together with (8.17) for the functions μ_k give all we need for the construction of the n-soliton solution on a flat space background.

8.3 The Kerr and Schwarzschild metrics

As has already been stated in the previous section, stationary solutions on a physical background (with either complex or real poles) can appear only in pairs. Consequently, the simplest case will be a two-soliton solution, i.e. two poles $\lambda = \mu_1$ and $\lambda = \mu_2$ on the flat space background (8.35). In this section we show that a double stationary soliton on a flat background, corresponding to a pair of complex conjugate poles $\mu_2 = \bar{\mu}_1$, gives a Kerr–NUT [230] solution with an 'anomalous large' angular momentum (i.e. a solution without horizons and with a naked singularity). On the other hand, if both functions μ_1 and μ_2 are real, the solutions correspond to the 'normal' situation, with the singularity hidden from an outside observer by event horizons [23, 24].

These assertions can be verified by direct calculation of the metric for the two-soliton case. Let us represent the constants w_1 and w_2 that appear in (8.16)–(8.17) for $k = 1, 2$ in the forms

$$w_1 = \tilde{z}_1 + \sigma, \quad w_2 = \tilde{z}_1 - \sigma, \tag{8.40}$$

where $\tilde{z}_1$ and σ are new arbitrary constants. The constant $\tilde{z}_1$ is always real, and σ is either real (for real-pole trajectories μ_1 and μ_2) or pure imaginary (for complex conjugate trajectories $\mu_2 = \bar{\mu}_1$). We now introduce instead of ρ and z the new coordinates r and θ:

$$\rho = [(r - m)^2 - \sigma^2]^{1/2} \sin\theta, \quad z = \tilde{z}_1 + (r - m)\cos\theta, \tag{8.41}$$

where m is an arbitrary constant whose value will be specified later. The square root in (8.41) is defined in such a way that the leading terms of these expressions in the asymptotic region $r \to \infty$ (for r real and positive) give the usual transformation between cylindrical and spherical coordinates. From (8.40)–(8.41) it follows that we can define the quantities $[(z - w_k)^2 + \rho^2]^{1/2}$, for $k = 1, 2$, in (8.17) by the formulas

$$[(z - w_1)^2 + \rho^2]^{1/2} = r - m - \sigma\cos\theta, \quad [(z - w_2)^2 + \rho^2]^{1/2} = r - m + \sigma\cos\theta. \tag{8.42}$$

In order to get the expressions for the function μ_1 and μ_2 we need to choose in (8.17) either the same sign for μ_1 and μ_2, or opposite signs. Both cases lead to the same metric (within a choice of sign for the arbitrary constant C_f in the metric component $f^{(ph)}$).

Let us consider first the case in which the signs are the same. If we choose the plus sign in (8.17) for both μ_1 and μ_2, then substituting (8.40)–(8.42) we get

$$\mu_1 = 2(r - m + \sigma)\sin^2(\theta/2), \quad \mu_2 = 2(r - m - \sigma)\sin^2(\theta/2). \tag{8.43}$$

Now, without loss of generality we can impose the following two conditions on the arbitrary constants $C_0^{(k)}$ and $C_1^{(k)}$, for $k = 1, 2$, that appear in (8.37) for the vectors $m_a^{(k)}$:

$$C_1^{(1)}C_0^{(2)} - C_0^{(1)}C_1^{(2)} = \sigma, \quad C_1^{(1)}C_0^{(2)} + C_0^{(1)}C_1^{(2)} = -m. \tag{8.44}$$

The first equation is possible because the constants $C_0^{(k)}$ and $C_1^{(k)}$ contain some nonphysical arbitrariness due to the normalization freedom: $C_a^{(k)} \to \zeta^{(k)}C_a^{(k)}$. In fact, after such a transformation the metric components g_{ab} will not change, i.e. all constants $\zeta^{(k)}$ will disappear from the final result of the matrix g. The first relation of (8.44) just fixes, partially, such nonphysical arbitrariness. The second equation, on the other hand, is simply the definition of the constant m. We then introduce two new arbitrary constants a and b, defined by

$$C_1^{(1)}C_1^{(2)} - C_0^{(1)}C_0^{(2)} = -b, \quad C_1^{(1)}C_1^{(2)} + C_0^{(1)}C_0^{(2)} = a. \tag{8.45}$$

From (8.44)–(8.45) it follows that

$$\sigma^2 = m^2 - a^2 + b^2. \tag{8.46}$$

Now, using (8.41) and (8.43) to express ρ, μ_1 and μ_2 in terms of the variables r, θ, we can calculate from (8.26), (8.30), (8.33), (8.38) and (8.39) for $k, l = 1, 2$ the metric coefficients $g_{ab}^{(ph)}$ and $f^{(ph)}$. The resulting expressions for the metric contain only those combinations of the constants $C_0^{(k)}$, $C_1^{(k)}$ which are expressible through the three independent arbitrary parameters m, a and b according to (8.44)–(8.46). If we substitute these results into the interval (8.34) with the simultaneous transformation of the line element $d\rho^2 + dz^2$ to the r, θ variables according to (8.41), namely,

$$d\rho^2 + dz^2 = [(r - m)^2 - \sigma^2 \cos^2 \theta] \left\{ [(r - m)^2 - \sigma^2]^{-1} dr^2 + d\theta^2 \right\}, \tag{8.47}$$

we get the following final form for the physical metric:

$$\begin{aligned}
ds^2 = &-C_f \omega (\Delta^{-1} dr^2 + d\theta^2) - \omega^{-1}(\Delta - a^2 \sin^2 \theta)(dt + 2ad\varphi)^2 \\
&+ \omega^{-1}[4\Delta b \cos \theta - 4a \sin^2 \theta (mr + b^2)](dt + 2ad\varphi)d\varphi \\
&- \omega^{-1}[\Delta(a \sin^2 \theta + 2b \cos \theta)^2 - \sin^2 \theta (r^2 + b^2 + a^2)^2]d\varphi^2, \tag{8.48}
\end{aligned}$$

where ω and Δ are defined as

$$\omega = r^2 + (b - a \cos \theta)^2, \quad \Delta = r^2 - 2mr + a^2 - b^2. \tag{8.49}$$

These formulas are the standard expression for the Kerr–NUT solution in the Boyer–Lindquist coordinates if we take $C_f = -1$ for the arbitrary constant C_f, and adopt $t + 2a\varphi$ as the time variable.

It can be seen from this that the Kerr–NUT solution with horizons corresponds to real poles $\lambda = \mu_1$ and $\lambda = \mu_2$, since in this case the constant σ is real $(m^2 + b^2 > a^2)$, and the constants w_1 and w_2 and the pole trajectories μ_1 and μ_2 are real along with σ. If σ is pure imaginary $(m^2 + b^2 < a^2)$, then the constants w_1 and w_2 and the pole trajectories μ_1 and μ_2 are complex and conjugate to each other. This case corresponds to a solution without horizons. Furthermore, metric (8.48) and the constants m, a, b are, of course, still real, but the original constants $C_a^{(k)}$, as (8.44)–(8.45) show, must be taken complex and related by $C_a^{(2)} = \overline{C}_a^{(1)}$ which, as we see from (8.37), means that $m_a^{(2)} = \overline{m}_a^{(1)}$. This agrees with the rule for choosing real solutions with a complex conjugate pair of poles that was formulated in section 8.2.

Let us now look at the second possibility we have for the solutions μ_1 and μ_2 of (8.16), namely, the possibility of using different signs in (8.17). Choosing the plus sign for μ_1 and the minus sign for μ_2, we get

$$\mu_1 = 2(r - m + \sigma) \sin^2(\theta/2), \quad \mu_2 = -2(r - m + \sigma) \cos^2(\theta/2). \quad (8.50)$$

Direct calculation shows that in this case we have the same expression (8.48) for the metric but with a plus sign in front of the constant C_f. Thus to get the standard expression of the Kerr–NUT metric we should take $C_f = 1$. In addition, the expressions for the parameters m, a and b through the original arbitrary constants $C_a^{(k)}$ are essentially different:

$$\left.\begin{array}{ll} C_0^{(1)} C_0^{(2)} + C_1^{(1)} C_1^{(2)} = \sigma, & C_0^{(1)} C_0^{(2)} - C_1^{(1)} C_1^{(2)} = m, \\ C_0^{(1)} C_1^{(2)} - C_1^{(1)} C_0^{(2)} = -a, & C_0^{(1)} C_1^{(2)} + C_1^{(1)} C_0^{(2)} = -b, \end{array}\right\} \quad (8.51)$$

but relation (8.46) between σ and the parameters m, a, b is still valid.

In conclusion we point out that the only actual physical solution is that of Kerr [165] , which corresponds to $b = 0$, since the presence of the NUT parameter b makes the metric no longer asymptotically flat and produces a number of nonphysical properties of the solution. A first physical analysis of the Kerr–NUT metric was given by Misner [220]. For a more recent review of possible physical interpretations of this metric, as well as of other stationary axisymmetric solutions, see ref. [35]. It is remarkable that such well known solutions as the Kerr metric ($b = 0$), the Taub–NUT metric ($a = 0$) [268, 230] and the Schwarzschild metric [258] ($b = 0$, $a = 0$) are, in fact, solutions representing double soliton states on the flat Minkowski background. We hope that future developments in gravitational theory will clarify the real significance of this fact.

The rotating disc solution. An important related solution was obtained by Neugebauer and Meinel [227, 228, 229] who formulated the problem of a uniformly rotating stationary and axisymmetric disc of dust particles as a boundary value problem of the Ernst equation and solved it by the ISM. The solution is given in terms of two linear integral equations and depends on two parameters, namely, the angular velocity of the disc and the relative redshift from the centre of the disc. The analytic solution of one of the equations leads to explicit expressions of the Ernst potential on the symmetry axis, the disc metric, and the surface mass density. It turns out that the complete solution of the problem may be represented, up to quadratures, in terms of ultraelliptic functions. A remarkable feature of the exterior solution is that in the limit of infinite redshift it approaches exactly the extreme Kerr solution.

8.4 Asymptotic flatness of the solution

In this section we consider some general properties of the n-soliton solutions, confining ourselves to one of their possible types. We shall assume that on the background of a flat space with the metric (8.35) an even number n of solitons are introduced, corresponding to the poles $\lambda = \mu_1$, $\lambda = \mu_2$, ..., $\lambda = \mu_n$. We divide the functions μ_k ($k = 1, \ldots, n$) into pairs and introduce the Greek index γ, which will take only the odd values $\gamma = 1, 3, \ldots, n - 1$, to enumerate such pairs. We thus have $n/2$ pairs of pole trajectories $(\mu_\gamma, \mu_{\gamma+1})$.

To understand the physical meaning of the solution it is helpful to examine first a special case which corresponds to a diagonal matrix g, i.e. to the static field remaining after the rotation has been turned off (metrics of the Weyl class). As in subsection 4.4.1, to obtain such a special case we set all the arbitrary constants $C_0^{(k)}$ in (8.37) equal to zero and all the $m_0^{(k)}$ also equal to zero. It then follows from (8.20) that all the $n_0^{(k)} = 0$ and the matrices R_k take the form

$$R_k = \begin{pmatrix} 0 & 0 \\ 0 & n_1^{(k)} m_1^{(k)} \end{pmatrix}. \tag{8.52}$$

The nonphysical matrix g, in (8.25), can be represented (see details in ref. [24]) in the form

$$g = \left(\prod_{k=1}^{n} [I - \mu_k^{-2}(\mu_k^2 + \rho^2) P'_k] \right) g_0, \tag{8.53}$$

which is the stationary analogue of (1.97), where the matrices P'_k satisfy (1.98). Then (8.52) implies that the matrices P'_k in (8.53) take the form

$$P'_k = \begin{pmatrix} 0 & 0 \\ 0 & 1 \end{pmatrix}, \tag{8.54}$$

and from (8.53) and (8.30) we get the following solution for the diagonal case under consideration:

$$g_{00}^{(ph)} = \rho^{-n} \prod_{k=1}^{n} \mu_k, \quad g_{01}^{(ph)} = 0, \quad g_{11}^{(ph)} = -\rho^2/g_{00}^{(ph)}. \tag{8.55}$$

The metric coefficient $f_n^{(ph)}$ can be found from (8.33), by computing the determinant of the matrix Γ_{kl} with $C_0^{(k)} = 0$. It is simpler, however, to determine $f_n^{(ph)}$ directly from (8.4), since for the solution (8.55) such equations are simple and easy to integrate. The result is

$$f_n^{(ph)} = \text{constant}\, \rho^{(n^2+2n)/2} \left(\prod_{k=1}^{n} \mu_k \right)^{1-n} \left[\prod_{k=1}^{n} (\mu_k^2 + \rho^2) \right]^{-1} \left[\prod_{k>l=1}^{n} (\mu_k - \mu_l)^2 \right]. \tag{8.56}$$

We now determine from (8.16) and (8.17) the function μ_k, which we have arranged in the pairs $(\mu_\gamma, \mu_{\gamma+1})$. Confining our treatment to the case in which the signs in (8.17) are chosen differently for the functions of each pair, we have

$$\mu_\gamma = w_\gamma - z + [(w_\gamma - z)^2 + \rho^2]^{1/2}, \quad \mu_{\gamma+1} = w_{\gamma+1} - z - [(w_{\gamma+1} - z)^2 + \rho^2]^{1/2}.$$
(8.57)

Instead of the pairs of arbitrary constants w_γ and $w_{\gamma+1}$, it is convenient to introduce new constants $\tilde{z}_\gamma$ and m_γ:

$$w_\gamma = \tilde{z}_\gamma - m_\gamma, \quad w_{\gamma+1} = \tilde{z}_\gamma + m_\gamma.$$
(8.58)

If we now introduce $n/2$ pairs of functions $r_\gamma(\rho, z)$ and $\theta_\gamma(\rho, z)$, giving to each pair of poles their own 'radial' and coordinates, through the relations

$$\rho = [r_\gamma(r_\gamma - 2m_\gamma)]^{1/2} \sin \theta_\gamma, \quad z - \tilde{z}_\gamma = (r_\gamma - m_\gamma) \cos \theta_\gamma,$$
(8.59)

we get from (8.57)

$$\mu_\gamma = 2(r_\gamma - 2m_\gamma) \sin^2(\theta_\gamma/2), \quad \mu_{\gamma+1} = -2(r_\gamma - 2m_\gamma) \cos^2(\theta_\gamma/2).$$
(8.60)

Using these expressions for ρ and μ_k we get from (8.55) the component $g_{00}^{(ph)}$ in terms of a product of $n/2$ factors

$$g_{00}^{(ph)} = - \left(1 - \frac{2m_1}{r_1}\right) \left(1 - \frac{2m_3}{r_3}\right) \cdots \left(1 - \frac{2m_{n-1}}{r_{n-1}}\right).$$
(8.61)

For the case of the two-soliton solution (8.61) will have only one factor which is the Schwarzschild expression for the coefficient $g_{00}^{(ph)}$. Computing from (8.56) the coefficient $f_2^{(ph)}$ for this case and writing out the interval, we indeed get the standard expression for the Schwarzschild metric with radial coordinate r_1 and polar angle θ_1. Of course, this result also follows from the general form of the two-soliton Kerr–NUT solution, given in the preceding section, i.e. (8.50) and (8.51) with $C_0^{(1)} = C_0^{(2)} = 0$.

To interpret the static solution with the 'potential' (8.61) we must choose a suitable radial variable. Any one of the functions $r_\gamma(\rho, z)$ could now be used as a radial coordinate, but it is more natural to define the radial variable in such a way that the dipole moment relative to it vanishes in the expansion at infinity of the Newtonian potential, Φ_N, of the system in question. As is well known, the Newtonian potential here is $2\Phi_N = 1 + g_{00}^{(ph)}$, and from (8.61) we have

$$2\Phi_N = 1 - \left(1 - \frac{2m_1}{r_1}\right) \left(1 - \frac{2m_3}{r_3}\right) \cdots \left(1 - \frac{2m_{n-1}}{r_{n-1}}\right).$$
(8.62)

Let us try to define the 'true' radial coordinate r and polar angle θ by relations of the same form as (8.59):

$$\rho = [r(r - 2m)]^{1/2} \sin \theta, \quad z - z_0 = (r - m) \cos \theta,$$
(8.63)

but with new constants m and z_0, to be defined. From (8.63) and (8.59) we can find functions $r_\gamma(r, \theta)$ and $\theta_\gamma(r, \theta)$ and obtain their asymptotic expansions for $r \to \infty$ (in the first approximation we simply have for $r \to \infty$: $r_\gamma = r$ and $\theta_\gamma = \theta$). Substituting these expansions into (8.62) we find the expansion of the potential Φ_N, and from the condition that it must contain no dipole term we can determine the constants m and z_0. We get

$$m = \sum_{\gamma=1}^{n-1} m_\gamma, \quad z_0 = \left(\sum_{\gamma=1}^{n-1} m_\gamma \tilde{z}_\gamma\right)\left(\sum_{\gamma=1}^{n-1} m_\gamma\right)^{-1}, \tag{8.64}$$

and then the expansion for Φ_N takes the form

$$2\Phi_N = \frac{2m}{r} + q\frac{3\cos^2\theta - 1}{r^3} + \cdots, \tag{8.65}$$

where q is the quadrupole moment of the system. For instance, in the case of the four-soliton solution, where the index γ takes only the values 1 and 3, we have

$$q = m_1 m_3[(\tilde{z}_1 - \tilde{z}_3)^2 - m^2](m_1 + m_3)^{-1}. \tag{8.66}$$

These results show that the static solution is a localized perturbation in an asymptotically flat space. For a sufficiently remote observer such a field can be regarded as the external gravitational field produced by $n/2$ localized axially symmetric structures, each of which has its own mass m_γ and its centre of mass lying on the axis of symmetry at the point with coordinate $\tilde{z}_\gamma$. Equations (8.64) show that the total mass of these $n/2$ objects equals the sum of their masses, and the coordinate z_0 of their centre of mass is given by the usual expression in particle mechanics. All the multipole moments of the field can also be expressed in definite ways in terms of the constants m_γ and $\tilde{z}_\gamma$.

If we now suppose that rotational motion around the axis of symmetry appears in this system the resulting case will correspond to a nondiagonal metric with $g_{01}^{(ph)} \neq 0$. In the special case of the two-soliton system considered in the preceding section, this change corresponds to the change from the Schwarzschild solution to that of Kerr. Just as in that special case we must also make sure that the solution with n solitons is asymptotically flat. In the two-soliton case it was necessary to set the NUT parameter to zero. This means that the off-diagonal component $g_{01}^{(ph)}$ of the metric must decrease like r^{-1} as $r \to \infty$; note that in the Kerr–NUT solution, $g_{01}^{(ph)} \sim b\cos\theta + O(r^{-1})$ for $r \to \infty$. Then the coefficient of r^{-1} in $g_{01}^{(ph)}$ gives the total angular momentum of the system.

It is not hard to find the behaviour of the components $g_{ab}^{(ph)}$ for $r \to \infty$ in the general case of the n-soliton metric. As in the two-soliton case, we must introduce the notation (8.40) for each pair of constants w_γ, $w_{\gamma+1}$ and for each pair of functions μ_γ, $\mu_{\gamma+1}$ we must introduce the new pairs of 'coordinates' r_γ, θ_γ by the formulas (8.41). After this we get from (8.57) expressions for μ_γ and $\mu_{\gamma+1}$ of the form (8.50). At infinity all the variables r_γ, θ_γ coincide, so that if we are concerned only with the first terms in the expansion for $r \to \infty$ it is irrelevant which pair we take as spherical coordinates r and θ.

Now from (8.37) we get the asymptotic form of the vectors $m_a^{(k)}$, and from (8.38) and (8.20) that of the vectors $n_a^{(k)}$. From these it is easy to find the behaviour of the components $g_{ab}^{(ph)}$. The results show that the asymptotic behaviour of the metric coefficients $g_{ab}^{(ph)}$ for $r \to \infty$ is exactly the same as in the two-soliton case:

$$g_{00}^{(ph)} \to -1, \quad g_{11}^{(ph)} \to r^2 \sin^2 \theta, \quad g_{01}^{(ph)} \to b_1 \cos \theta + b_2 + O(r^{-1}), \quad (8.67)$$

where b_1 and b_2 are constants which can be expressed in terms of $C_0^{(k)}$ and $C_1^{(k)}$. For the metric to be asymptotically flat at $r \to \infty$ the parameter b_1 must be zero, and this gives a supplementary condition to the constants $C_a^{(k)}$:

$$b_1(C_0^{(k)}, C_1^{(k)}) = 0. \quad (8.68)$$

The constant b_2 can then be eliminated from the asymptotic form of the metric coefficient $g_{01}^{(ph)}$ with a linear transformation of the form $t = t' + b_2\varphi$.

This analysis shows that the technique developed in the previous sections ensures asymptotic flatness to the n-soliton solution almost automatically. For this we have to impose only condition (8.68) to the arbitrary constants.

8.5 Generalized soliton solutions of the Weyl class

Up to this point we have considered only the Minkowski background when generating soliton solutions in the axisymmetric context, as we have seen this background produces physically interesting solutions. Here we will consider the Levi-Città metric as the background metric; this is the analogue of the Kasner solution in the cosmological context that we considered in chapter 4. We saw in sections 4.4–4.6 that when we restrict ourselves to diagonal metrics we may use the linearity of the field equations together with the ISM to derive a set of generalized solitons solutions, which describe physically interesting cosmological models in a unified way.

In the axisymmetric context when the two Killing vectors are hypersurface orthogonal the spacetime is static (because it admits a hypersurface orthogonal time-like Killing vector) and the metric expressed in appropriate coordinates (e.g. Weyl coordinates) becomes diagonal. The solutions in this case are classified as Weyl class. In this section we classify all the generalized soliton solutions in the Weyl class obtained from the Levi-Città background. Axisymmetric soliton solutions obtained from an arbitrary Weyl class background were considered by Letelier [193], who also paid special attention to the Levi-Città background and the diagonal solutions.

To facilitate the interpretation and identification of the solutions we give the Ernst potentials explicitly, since these potentials have been used traditionally to characterize stationary and axisymmetric solutions. Furthermore, the Ernst potential may be regarded as a complexified nonlinear generalization of the Newtonian potential [170], thus helping us to understand the physical meaning of the solutions. A review on the physical interpretation of solutions of the Weyl class can be found in refs [36, 35].

In the case of real-pole trajectories, the generalized soliton solutions associated with a number of such poles gives rise to well known solutions. Among these there are uniformly accelerated metrics, SILM (semi-infinite line mass) metrics, the C-metric, the Curzon–Chazy metrics, the γ-metrics or Voorhees–Zipoy metrics [292] and their superpositions [64, 278, 193], which include the superpositions of Schwarzschild black holes as discussed in section 8.4. The generalized soliton solutions arising from the real and imaginary parts were studied in ref. [49] and are dealt with in a similar way to the generalized solutions in the cosmological context, which we studied in section 4.6, or in the cylindrical context, as studied in section 6.2.

Using Weyl coordinates the general static metric with axial symmetry may be written as (8.1), where

$$g_{\phi\phi} = \rho^2 \exp[-2U(\rho, z)], \quad g_{tt} = -\exp[2U(\rho, z)], \quad g_{t\phi} = 0. \tag{8.69}$$

The Ernst potential for such solutions is real and is given by

$$\mathcal{E} = \exp(2U). \tag{8.70}$$

The generalized soliton solutions can be read off from the solutions in sections 4.4.1, 4.5 and 4.6, after the appropriate coordinate changes discussed after (8.5). As we did in subsection 4.4.1 the potential U will be written as

$$U = U_0 + U_s, \quad U_0 = \frac{1}{2}(d+1)\ln\rho, \tag{8.71}$$

where U_0 is the potential of the background solution (d is the Levi-Città parameter) and U_s corresponds to the generalized soliton part. Note that here, in order to follow the traditional Weyl form of the metric, U differs slightly from Φ in (4.41) (apart from the change $t \to i\rho$), the exact correspondence is: $\Phi \to 2U - \ln\rho$.

Levi-Città background. The background metric in Weyl coordinates coordinates is

$$ds^2 = \rho^{(d^2-1)/2}(d\rho^2 + dz^2) + \rho^{1-d}d\phi^2 - \rho^{1+d}dt^2, \tag{8.72}$$

which is equivalent to metric (6.1) (with $b = 1$). Although we will use the same symbol, the Levi-Città parameter d here should not be confused with the Kasner parameter d of (6.1): the new parameter d in terms of the old one is $(d^2 - 3)/2$. Let us now return to the interpretation of these metrics [36]. The first observation is that the Newtonian potential U_N created by a massive line source along the z axis with a linear mass density λ, in units where $G = c = 1$, is

$$U_N = 2\lambda \ln \rho. \tag{8.73}$$

Thus comparing with U_0 in (8.71) the first naive interpretation is that such a metric could represent the gravitational field of an ILM (infinite line mass) with a linear mass density $\lambda = (d + 1)/4$. Replacing the physical constants, λ is changed to $\lambda c^2/G$ and $\lambda = 1$ corresponds to 10^{28} g/cm. Of course, this interpretation does not always hold. When $d = -1$ ($\lambda = 0$) metric (8.72) is obviously a flat spacetime in polar coordinates. Also for $d = 1$, which corresponds to $\lambda = 1/2$, the metric is flat but it can be interpreted as an accelerated metric. In fact, with the coordinate change $T = \rho \sinh t$, $X = \rho \cosh t$, $Y = \phi$ and $Z = z$, this metric becomes

$$ds^2 = dX^2 + dY^2 + dZ^2 - dT^2,$$

for $X^2 \geq T^2$ ($X \geq 0$). Thus the Levi-Città metric for $d = 1$ corresponds to the Rindler wedge of Minkowski spacetime. The congruences of ∂_ρ define trajectories of particles with a constant proper acceleration, they correspond to a family of observers with hyperbolic motion. For this reason this metric is sometimes known as a *uniformly accelerated metric* [250, 251].

Metrics with $d = 0$ and $d = \pm 3$ are Petrov type D and admit four Killing vectors, otherwise they are Petrov type I, with the obvious exceptions of $d = \pm 1$, which are flat. The *ILM metric* interpretation seems to hold for the values $-1 < d \leq 0$ ($0 < \lambda \leq 1/4$) and also for the values $-3 < d < -1$ ($-1/2 < \lambda < 0$), which correspond to a negative mass source. The case $d = 0$ is one of Kinnersley's type D metrics [166]. In all other cases the ILM interpretation does not hold. The case $d = 3$ ($\lambda = 1$) has been interpreted as the metric of a cosmic string [246], and $d = -3$ is one of Taub's plane-symmetric metrics [268]. This last metric can also be obtained as an infinite mass limit of the Schwarzschild solution [252, 116] and can be seen to describe the interaction region of the head-on collision of two impulsive gravitational plane waves of null matter [85].

The generalized soliton solutions obtained from the Levi-Città background will be classified as follows. Generalized one-soliton solutions (one-soliton solutions and their superposition): (a1) real-pole trajectories, and (a2) complex-pole trajectories. Generalized two-soliton solutions (soliton–antisoliton solutions and their superpositions): (b1) real-pole trajectories, and (b2) complex-pole trajectories.

8.5.1 Generalized one-soliton solutions

(a1) Real-pole trajectories: We start with case (a1), i.e. *real-pole trajectories.*
Then the parameters w_i are real and, following (4.51), we can write

$$U_s = \frac{1}{2} \sum_{k=1}^{s} h_k \ln \left(\frac{\mu_k^+}{\rho} \right) = \frac{1}{2} \sum_{k=1}^{s} h_k \sinh^{-1} \left(\frac{w_k - z}{\rho} \right), \tag{8.74}$$

where the real parameters h_k play the role of the degeneracy of the k-pole when
they are integers. The explicit form of the coefficient $f(\rho, z)$ can be read off
from (4.52), where again one defines $g = \sum_{k=1}^{s} h_k$.

The Ernst potential is given by

$$\mathcal{E} = \rho^{d+1} \prod_{k=1}^{s} \left(\frac{\mu_k}{\rho} \right)^{h_k}. \tag{8.75}$$

Some insight into these solutions is gained if we consider just the one-pole
case, namely $s = 1$. This solution is singular along the symmetry axis $\rho = 0$
unless $d = \pm 1$ (Minkowski background) and at $z \to \pm \infty$ [284]; however, the
singularity at $z \to \pm \infty$ is not present in the case $h_1^2 = d^2 + 3$, thus for $d = \pm 1$
and $h_1^2 = 4$ the metric is asymptotically flat.

Newtonian potential of SILM. To help the interpretation of these metrics it is
useful to note that $\ln \mu_k$ has a simple Newtonian interpretation [36]. Let us
consider an SILM with a uniform linear mass density λ along the z axis that goes
from $z = w_k$ to $z \to \infty$. The Newtonian potential dU_N created by a line element
dz' at a certain point (ρ, z) may be written as $dU_N(\rho, z) = -\lambda[(z - z')^2 + \rho^2]^{-1/2} dz'$. With this prescription the potential vanishes at infinity, but since
the line mass goes to infinity this is not a convenient choice for the potential
origin and leads to divergences. To deal with this problem we may subtract,
for instance, $dU_N(\rho_1, z_1)$, i.e. the same potential at an arbitrary point (ρ_1, z_1).
Integration of this difference along the z axis leads to the difference between the
Newtonian potentials created by the SILM at the point of interest (ρ, z) and at
the point (ρ_1, z_1). This difference is free of divergences and if we then fix the
arbitrary potential origin in a convenient way we can write

$$U_N(\rho, z) = \lambda \ln \mu_k^+(\rho, z). \tag{8.76}$$

Analogously if the line source goes from $z = w_k$ to $z \to -\infty$ the Newtonian
potential is given by $U_N(\rho, z) = \lambda \ln(-\mu_k^-(\rho, z))$, which taking into account
that $\mu_k^+ \mu_k^- = -\rho^2$ may be written as

$$U_N(\rho, z) = 2\lambda \ln \rho - \lambda \ln \mu_k^+(\rho, z). \tag{8.77}$$

Superposing these two potentials we get the Newtonian potential of an ILM.
The potential of a finite rod with the same linear mass of length $w_2 - w_1$, whose

ends are at $z = w_1$ and $z = w_2$ ($w_2 > w_1$), is given by

$$U_N = \lambda \ln \left(\frac{\mu_1^+}{\mu_2^+} \right). \tag{8.78}$$

Thus the presence of the real-pole trajectories μ_k in the potential U suggests the interpretation of some of these metrics in terms of an SILM with a uniform linear mass density $\lambda = h_k/2$. Of course this interpretation does not always hold. Let us now consider some particular solutions of (8.75), and recall that $U = \frac{1}{2} \ln \mathcal{E}$; in the following we will also assume that μ_k stands for μ_k^+.

Uniformly accelerated metric. This corresponds to $s = 1$, $d = 0$ and $h_1 = 1$ in (8.75). If we define new coordinates (Z, r) by

$$\rho = Zr, \quad 2(w_1 - z) = Z^2 - \rho^2, \tag{8.79}$$

then $\mu_1 = Z^2$ and the metric becomes

$$ds^2 = dZ^2 + dr^2 + r^2 d\phi^2 - Z^2 dt^2, \tag{8.80}$$

which is the Rindler metric [250] in polar coordinates. In fact, the further coordinate change $T = Z \sinh t$, $\tilde{Z} = Z \cosh t$, transforms this metric into the Rindler wedge of Minkowski spacetime: $ds^2 = d\tilde{Z}^2 + dr^2 + r^2 d\phi^2 - dT^2$, for $\tilde{Z}^2 \geq T^2$ ($\tilde{Z} \geq 0$). Here the congruence of ∂_Z defines trajectories of particles with uniform proper acceleration in the Rindler wedge. Note that the two flat metrics interpreted as accelerated metrics correspond to the Ernst potential being $\mathcal{E} = \ln \rho$ and $\mathcal{E} = \frac{1}{2} \ln \mu_1$, respectively.

The SILM metric. This corresponds to $s = 1$ and $d = h_1 - 1$ in (8.75). In this case the metric can be written as

$$ds^2 = \mu_1^{-h_1} \left[\frac{\mu_1}{2\sqrt{(w_1 - z)^2 + \rho^2}} \right]^{h_1^2} (dz^2 + d\rho^2) + \mu_1^{-h_1} \rho^2 d\phi^2 - \mu_1^{h_1} dt^2. \tag{8.81}$$

When $h_1 \leq 2$ it can be interpreted [36] as the spacetime created by an SILM which lies on the z axis from $z = w_1$ to $z \to \infty$, and has a uniform linear mass density $\lambda = h_1/2$. For $\lambda = -1/2$ ($h_1 = -1$) the coordinate change (8.79) followed by a change from polar coordinates (r, ϕ) to Cartesian coordinates (x, y), together with a redefinition $\tilde{Z} = Z^4$ puts the metric in the form

$$ds^2 = \tilde{Z}(dx^2 + dy^2) + \tilde{Z}^{-1/2}(d\tilde{Z}^2 - dt^2), \tag{8.82}$$

which is one of Taub's plane-symmetric vacuum solutions [268, 179]. Thus there is also an alternative interpretation in this case. It can be shown that for

$\lambda > 1$, at $t = $ constant a radial line that goes from some point z_0 ($z_0 < w_1$) on the z axis to infinity has a finite proper length, and that also the proper length of the semi-infinite axis from z_0 to $z \to -\infty$ is finite; thus the SILM interpretation does not hold. For the limiting case $\lambda = 1$ the metric admits four Killing vectors, instead of the two that these SILM metrics generally have. Metric (8.81) has only a (naked) curvature singularity at the position of the SILM.

The C-metric. This corresponds to $s = 3$, $d = -1$ and $h_1 = h_2 = h_3 = 1$ in (8.75), where the parameters w_1, w_2 and w_3 are three roots of the cubic equation [158, 193]

$$2a^4 w^3 - a^2 w^2 + m^2 = 0. \tag{8.83}$$

This metric is supposed to describe the field of accelerated particles [124, 34]; see also refs [171, 96] for more details on the C-metric. The parameters m and a are identified as the particle mass and acceleration, respectively.

(a2) Complex-pole trajectories: We turn now to the case (a2) of *complex-pole trajectories.* In this case we follow section 4.6, and the parameters w_k are complex, according to (4.73): $w_k = z_k - i c_k$, where z_k^0 and c_k are real parameters. In analogy with (4.74) we can write the pole trajectories in the form

$$\mu_k/\rho = \sqrt{\sigma_k} e^{i\gamma_k}, \tag{8.84}$$

where $\sqrt{\sigma_k}$ is understood as a positive quantity and the explicit forms of the functions $\sigma_k(\rho, z)$ and $\gamma_k(\rho, z)$ are

$$\left. \begin{array}{l} \sigma_k^{\pm} = L_k \pm (L_k^2 - 1)^{1/2}, \\ L_k \equiv (z_k^2 + c_k^2)\rho^{-2} + [1 + 2(z_k^2 - c_k^2)\rho^{-2} + (z_k^2 + c_k^2)^2 \rho^{-4}]^{1/2}, \end{array} \right\} \tag{8.85}$$

$$\gamma_k = \cos^{-1}\left(\frac{2 z_k \sqrt{\sigma_k}}{\rho(1 + \sigma_k)} \right), \tag{8.86}$$

where, as usual, $z_k \equiv z_k^0 - z$ and the minus and plus signs stand for the *in* and *out* pole labels, respectively; also, $\sigma^+ = (\sigma^-)^{-1}$ and $0 < \sigma_k^- < 1$. The asymptotic values of σ^- can also be read off in (4.77) by simply changing $t \to i\rho$.

Now, as in section 4.6, we get two classes of solution corresponding to the real and imaginary parts of $\ln(\mu_k/\rho)$, which is the main ingredient of the linear potential U. These will be called class I and class II, respectively.

Class I. These solutions correspond to (4.79):

$$U_s = \frac{1}{2} \sum_{k=1}^{s} h_k \ln \sqrt{\sigma_k}. \tag{8.87}$$

The explicit form of the metric coefficient $f(\rho, z)$ can be read from (4.80) after the change $t \rightarrow i\rho$ and recalling that an arbitrary parameter can always multiply f.

In the limit $c_k \rightarrow 0$ this family of solutions behaves like the previous real-pole solutions (a1). For a single pole and when $c_k \neq 0$, it is possible to define a set of new coordinates such that σ_k takes a simpler form than in the Weyl coordinates. In fact, define

$$\rho = c_k \cosh(2a_k \hat{R}_k) \sin \hat{\theta}_k, \quad z_k = c_k \sinh(2a_k \hat{R}_k) \cos \hat{\theta}_k, \tag{8.88}$$

where a_k are arbitrary real parameters, then the functions σ_k^- read

$$\sigma_k^- = \frac{1 - \cos \hat{\theta}_k}{1 + \cos \hat{\theta}_k}. \tag{8.89}$$

In these new coordinates the line element $d\rho^2 + dz^2$ still keeps its diagonal structure and transforms into

$$d\rho^2 + dz^2 = c_k^2 [\sinh^2(2a_k \hat{R}_k) + \cos^2 \hat{\theta}_k](4a_k^2 d\hat{R}_k^2 + d\hat{\theta}_k^2). \tag{8.90}$$

The geometrical meaning of these coordinates can be seen if we consider their asymptotic behaviour: assuming (at large distances) a relationship between Weyl and spherical coordinates of the type,

$$\rho \sim r \sin \theta, \quad z_k \sim r \cos \theta \tag{8.91}$$

the new coordinates then behave as

$$\tan \hat{\theta}_k \sim \frac{\rho}{z_k}, \quad \exp(4a_k \hat{R}_k) \sim \frac{\rho^2 + z_k^2}{c_k}, \tag{8.92}$$

i.e. $\hat{\theta}_k$ behaves as a spherical angular coordinate and $\hat{R}_k \sim \ln r$.

Class II. These solutions correspond to (4.81):

$$U_s = \frac{1}{2} \sum_{k=1}^{s} h_k \gamma_k, \tag{8.93}$$

and the explicit form of the $f(\rho, z)$ coefficient can be read from (4.82) after the change $t \rightarrow i\rho$. The behaviour of these solutions is quite different from that of the class I solutions. In the limit $c_k \rightarrow 0$ (real limit) we get $\gamma_k \rightarrow 0$, which does not produce any new solution.

In the case of a single pole and $c_k \neq 0$ we can again use the coordinates defined in (8.88) to simplify the expression of function γ_k:

$$\gamma_k^\pm = \cos^{-1}[\pm \tanh(2a_k \hat{R}_k)], \quad \gamma_k^- = \pi \pm \gamma_k^+. \tag{8.94}$$

To see the physical meaning of these solutions we again take the single-pole case and expand its associated Ernst potential (8.70) asymptotically in a power series of r^{-1} [170]. Then at large distances we have

$$\mathcal{E} = -\frac{1 - (1 + 4c_k^2 \cos^2\theta)^{1/2}}{\sqrt{2}\cos\theta}\frac{1}{r} + O\left(\frac{1}{r^2}\right), \qquad (8.95)$$

where we have assumed the same relationship between Weyl and spherical coordinates as in (8.91)–(8.92). This last expression shows that these solutions are asymptotically flat, provided the background is flat. However, no classical interpretation is possible here, since even at the first order in r^{-1} an angular dependence is present. For this reason the real parameters c_k are not simply related to the mass or momenta of the source of the gravitational field, in spite of their occurrence at first order.

8.5.2 *Generalized two-soliton solutions*

An interesting family of solutions is obtained by the superposition of soliton–antisoliton pairs, since the divergences which appear in some of the previous solutions will no longer be present. Some of the solutions obtained by such superpositions will be asymptotically flat and their parameters can be given a physical interpretation after an asymptotic expansion is made.

(b1) Real-pole trajectories: Thus, we start with case (b1), i.e. *real-pole trajectories* (w_γ real). From (4.51) the potential U made by the superposition of pairs of opposite poles is

$$U_s = \frac{1}{2}\sum_{\gamma=1}^{n-1} h_\gamma [\ln(\mu_{\gamma+1}/\rho) - \ln(\mu_\gamma/\rho)], \qquad (8.96)$$

where for definiteness we will assume that μ_γ here means μ_γ^+, see (4.49), and we use the convention introduced in section 8.4 of assuming that the Greek index γ takes only odd values: $\gamma = 1, 3, \ldots, 2n - 1$. The Ernst potential (8.70) for these solutions is given by

$$\mathcal{E} = \rho^{d+1}\prod_{\gamma=1}^{n-1}\left(\frac{\mu_{\gamma+1}}{\mu_\gamma}\right)^{h_\gamma}. \qquad (8.97)$$

Note that this potential may be seen as a particular case of (8.75), where $h_k = -h_{k+1}$ and $s = n/2$. Comparing $U = \frac{1}{2}\ln\mathcal{E}$ with the Newtonian potential (8.78) for a rod on the symmetry axis, it seems that metrics with $d = -1$ corresponding to the Minkowski background could be interpreted as the field of a superposition of finite rods along the axis with linear mass densities $h_\gamma/2$ and with ends at $(w_\gamma, w_{\gamma+1})$, where $w_\gamma > w_{\gamma+1}$. Let us now consider some interesting particular cases.

The γ-metric. This corresponds to $n = 2$, $d = -1$, $h_1 = -m/a$ and $w_1 = -w_2 = -a$, where m is the mass of the rod and $2a$ its length. Since $\mu_k^+ \mu_k^- = -\rho^2$ we have that $\mu_1^+/\mu_2^+ = (\mu_1^+ - \mu_2^-)/(\mu_2^+ - \mu_1^-)$ and the potential U can be written in the form

$$U = \frac{m}{2a} \ln \left[\frac{R_1 + R_2 - 2a}{R_1 + R_2 + 2a} \right], \tag{8.98}$$

where $R_1 = \sqrt{(a + z)^2 + \rho^2}$ and $R_2 = \sqrt{(a - z)^2 + \rho^2}$ are the distances from the ends of the rod to the point (ρ, z). This metric was discovered by Darmois and is also known as the Voorhees–Zipoy metric [316, 292], see ref. [179]. It has a directional singularity if $m > 2a$ but not for $m < 2a$ [115, 35]. At infinity the metric represents an isolated body with higher mass moments. For $m = a$ it is the Schwarzschild solution, i.e. a rod of mass $\lambda = 1/2$ and length $2m$, in Weyl coordinates. However, while the Voorhees–Zipoy metrics generally have an Abelian symmetry group G_2, the Schwarzschild metric has a larger symmetry group G_4 (four Killing vectors); in particular, it has spherical symmetry and the rod interpretation does not hold.

The Curzon–Chazy metric. This metric [179] can be obtained as a limit of the previous one. In fact, taking the limit of vanishing rod length, $a \to 0$, but keeping m finite in (8.98) we get

$$U = -\frac{m}{\sqrt{z^2 + \rho^2}}, \tag{8.99}$$

which is the Newtonian potential for a spherical particle. Of course, it is different from Schwarzschild because this metric has no spherical symmetry. Its far field is that of a mass at the origin with multipoles, it has no horizon and it has a directional naked curvature singularity at the origin. For instance, the scalar invariant I defined in (4.13) vanishes if we approach the origin along the symmetry axis, otherwise it diverges [115, 35]. One may superpose two Curzon–Chazy particles at different points on the symmetry axis. In this case there is a singularity along the z axis between the particles. This singularity represents a stress which holds the particles apart; this stress has zero active gravitational mass and does not exert a gravitational field [156, 155]. This is how one gets a static solution with two massive particles.

Accelerated particles metric. Another interesting solution which may be interpreted as the gravitational field of two uniformly accelerated particles was constructed by Bonnor and Swaminarayan [38]. It is obtained by superposing two Curzon–Chazy particles on the accelerating metric (8.80), i.e. (8.75) with $s = 1$, $d = 0$ and $h_1 = 1$. The potential U for this metric is thus

$$U = \ln \mu_1 - \frac{m_2}{\sqrt{(a - z)^2 + \rho^2}} - \frac{m_4}{\sqrt{(b - z)^2 + \rho^2}}. \tag{8.100}$$

This metric may be obtained from (8.75) with $s = 5$, $d = -1$, $h_2 = -h_3 = m_2/\epsilon$, $h_4 = -h_5 = m_4/\epsilon$, $w_2 = a - \epsilon$, $w_3 = a + \epsilon$, $w_4 = b - \epsilon$, $w_5 = b + \epsilon$, and then taking the limit $\epsilon \to 0$. We recall that the coordinate change (8.79) makes $\mu_1 = Z^2$ and the congruences of ∂_Z define the hyperbolic motion in the Rindler wedge. In Weyl coordinates the particles are located on the symmetry axis at $(\rho = 0, z = a)$ and $(\rho = 0, z = b)$. In the new coordinates (r, ϕ, Z, t) defined by (8.79) the location of these two particles is at $(r = 0, Z = \sqrt{2(w_1 - a)})$ and $(r = 0, Z = \sqrt{2(w_1 - b)})$, respectively. But these are hyperbolic trajectories in the Rindler wedge plane $(T, \tilde{Z})$ defined by $T = Z \sinh t$, $\tilde{Z} = Z \cosh t$. For that reason this solution is interpreted as the gravitational field of two Curzon–Chazy particles moving with uniform acceleration along the symmetry axis. See refs [179, 158, 193] for more details and generalizations of this metric.

The Ernst potential (8.70) for the generalized two-soliton solutions can be written in another form, familiar in the literature on axisymmetric solutions, by changing to prolate spheroidal coordinates or to Boyer–Lindquist coordinates. Prolate spheroidal coordinates x_γ, y_γ are defined by

$$\rho^2 = m_\gamma^2 (x_\gamma^2 - 1)(1 - y_\gamma), \quad z - \tilde{z}_\gamma = -m_\gamma x_\gamma y_\gamma, \tag{8.101}$$

where m_γ and $\tilde{z}_\gamma$, are defined in (8.58). Then the Ernst potential can be written as

$$\mathcal{E} = -\rho^{d+1} \prod_{\gamma=1}^{n-1} \left(\frac{x_\gamma - 1}{x_\gamma + 1} \right)^{h_\gamma}. \tag{8.102}$$

As remarked before this family of solutions includes the γ-metric or the Voorhees–Zipoy metric ($n = 2, d = -1$) with the Voorhees–Zipoy parameter $\delta = h_1$ [179]. In these coordinates an interpretation of the γ-metric as the field of a circular disc in 3-space with Euclidean topology is also possible [37].

Using the Boyer–Lindquist coordinates $(r_\gamma, \theta_\gamma)$ defined in (8.59) with the parameters m_γ and $\tilde{z}_\gamma$ defined in (8.58), i.e. $m_\gamma = (w_{\gamma+1} - w_\gamma)/2$, $\tilde{z}_\gamma = (w_{\gamma+1} + w_\gamma)/2$, the Ernst potential takes the form

$$\mathcal{E} = -\rho^{d+1} \prod_{\gamma=1}^{n-1} \left(1 - \frac{2m_\gamma}{r_\gamma} \right)^{h_\gamma}, \tag{8.103}$$

which represents a generalization of (8.62). For $n > 2$ and $h_\gamma = 1$ we get a superposition of static Schwarzschild black holes along the z axis to the Levi-Cività source, which is generally an ILM when $-3 < d \leq 0$. If $d = -1$ the solution is, of course, asymptotically flat. Like in the superposition of Curzon–Chazy particles there is a singularity along the z axis between the particles.

(b2) Complex-pole trajectories: We turn now to case (b2) which corresponds to the superposition of *pairs of complex opposite poles*. Again we classify the solutions as class I, if they come from the real part of $\ln(\mu_\gamma/\rho)$, and as class

II if they come from the imaginary part. Class I solutions are obtained from (4.79)–(4.80) and class II ones from (4.81)–(4.82), after the change $t \to i\rho$.

To study the asymptotic behaviour of these solutions it is useful to again use the Boyer–Lindquist coordinates for each pair of poles defined in (8.59). At large distances, these coordinates become simply $\rho \sim r_\gamma \sin\theta_\gamma$ and $z - \tilde{z}_\gamma \sim r_\gamma \cos\theta_\gamma$, and thus $z_\gamma \sim m_\gamma - r_\gamma \cos\theta_\gamma$ and $z_{\gamma+1} \sim -(m_\gamma + r_\gamma \cos\theta_\gamma)$.

Class I. The Ernst potential for this class takes the form

$$\mathcal{E} = \rho^{d+1} \prod_{\gamma=1}^{n-1} \left(\frac{\sigma_{\gamma+1}}{\sigma_\gamma}\right)^{h_\gamma}, \tag{8.104}$$

where for definiteness we take σ_γ as σ_γ^+. This expression reduces to (8.103) when $c_\gamma \to 0$ ($w_\gamma = z_\gamma^0 - ic_\gamma$). To see the physical meaning of the parameters m_γ and c_γ that appear in these solutions, let us restrict ourselves to the case of a single pair of poles and expand $\sigma_{\gamma+1}/\sigma_\gamma$ at large distances in a power series of r_γ^{-1}. We get

$$\frac{\sigma_{\gamma+1}}{\sigma_\gamma} = 1 - \frac{4m_\gamma}{r_\gamma} + \frac{8m_\gamma^2 + (c_{\gamma+1}^2 - c_\gamma^2)\cos\theta_\gamma}{r_\gamma^2} + \cdots, \tag{8.105}$$

which shows the asymptotic flatness of the solution (if the background is flat). We also see that to first order in r_γ^{-1}, it reproduces the Newtonian potential of a mass $2m_\gamma$. The second order contains the parameters c_γ in a way that suggests that these parameters are related to the dipolar moment of the source. In fact, if we shift the origin of distances by an amount $2m_\gamma$ (i.e. $r_\gamma = r'_\gamma - 2m_\gamma$) then the term $8m_\gamma^2$ in (8.105) cancels out in an expansion in r'^{-1}_γ, and we get,

$$\frac{\sigma_{\gamma+1}}{\sigma_\gamma} = 1 - \frac{4m_\gamma}{r'_\gamma} + \frac{(c_{\gamma+1}^2 - c_\gamma^2)\cos\theta_\gamma}{r'^2_\gamma} + \cdots. \tag{8.106}$$

Therefore $c_{\gamma+1}^2 - c_\gamma^2$ is exactly the dipolar moment. For references about higher degree multipoles see, for instance, refs [64, 149].

Class II. The Ernst potential for this class of solutions can be written as

$$\mathcal{E} = \rho^{d+1} \exp\left[\sum_{\gamma=1}^{n-1}(\gamma_\gamma - \gamma_{\gamma+1})\right]. \tag{8.107}$$

Again we note that when $c_\gamma \to 0$, the $\gamma_\gamma \to 0$ and no new solution is obtained, apart from the background solution.

The asymptotic flatness of solutions (8.107) follows immediately from the asymptotic behaviour of (8.94). As in the previous case we consider a single pair of poles and expand the Ernst potential in powers of r_γ^{-1} to obtain

$$\mathcal{E} = -\frac{1 - (1 + 4c_\gamma^2 \cos^2 \theta_\gamma)^{1/2}}{\sqrt{2} \cos \theta_\gamma} + \frac{1 - (1 + 4c_{\gamma+1}^2 \cos^2 \theta_\gamma)^{1/2}}{r_\gamma} + \cdots. \quad (8.108)$$

These solutions do not reproduce the Newtonian potential of localized sources and they are clearly different from those of class I. The same remarks as in the case of a single pole apply here, in the sense that no classical interpretation is possible because of the angular dependence at first order. Therefore, no interpretation of the parameters c_γ in terms of the mass and momenta of the source of the gravitational field is possible either, despite the fact that they appear at first order.

8.6 Tomimatso–Sato solution

In subsection 2.1.1 we outlined the pole fusion procedure that corresponds to the multiple pole structure of the dressing matrix χ. One interesting example of the solution which can be constructed in this way is the well known Tomimatso–Sato metric [277]. In the stationary case, as we know, the number of solitons that one can introduce on a given background is even. Let us assume that it is $2N$, where N is a positive integer, then we can divide the poles into two equal sets and fuse them into two poles only, each with multiplicity N. When we use for this procedure the flat background metric (8.35) the resulting solution will be the so-called extended version of the Tomimatso–Sato solution. The original Tomimatso–Sato solution can be obtained from this last one by adding some constraints on the arbitrary constants of the extended version of the Tomimatso–Sato metric. This was discovered by Tomimatso and Sato [275, 278], see also ref. [6]; we will follow these references closely.

The scheme to derive the Tomimatso–Sato solution is the following. One starts from the background metric (8.35) and constructs the $2N$-soliton solution as explained at the end of section 8.2; see (8.36)–(8.39). It is convenient to parametrize the arbitrary constants $C_0^{(k)}$ and $C_1^{(k)}$ as

$$C_0^{(k)} = c_k \sin \eta_k, \qquad C_1^{(k)} = c_k \cos \eta_k, \quad (8.109)$$

where the new constants c_k and η_k are assumed to be real. The pole trajectories μ_k are also assumed to be real: they are solutions of the quadratic equation (8.16) with real constant parameters w_k. As in section 8.4, the sets η_k, w_k and μ_k may be divided into the pairs

$$(\eta_\gamma, \eta_{\gamma+1}), \quad (w_\gamma, w_{\gamma+1}), \quad (\mu_\gamma, \mu_{\gamma+1}), \quad (8.110)$$

where γ runs over odd values only: $\gamma = 1, 3, 5, \ldots, 2N - 1$. Each pair of the constants w_k is represented in a way analogous to (8.58), i.e.

$$w_\gamma = \tilde{z}_\gamma - \sigma_\gamma, \qquad w_{\gamma+1} = \tilde{z}_\gamma + \sigma_\gamma, \tag{8.111}$$

where $\tilde{z}_\gamma$ and σ_γ are some new arbitrary constants. Instead of the 'polar coordinates' r_γ, θ_γ for each pair of poles that we used in section 8.4, Tomimatsu and Sato used the prolate spheroidal coordinates x_γ, y_γ which were defined in (8.101):

$$\rho^2 = \sigma_\gamma^2 (x_\gamma^2 - 1)(1 - y_\gamma^2), \qquad z - \tilde{z}_\gamma = \sigma_\gamma x_\gamma y_\gamma. \tag{8.112}$$

For the pole trajectories $\mu_\gamma, \mu_{\gamma+1}$ they chose the following roots of the quadratic equation (8.16):

$$\mu_\gamma = \sigma_\gamma (x_\gamma - 1)(1 - y_\gamma), \qquad \mu_{\gamma+1} = -\sigma_\gamma (x_\gamma - 1)(1 + y_\gamma). \tag{8.113}$$

As we have shown in section 8.3 the simplest case, $N = 1$, gives just the Kerr–NUT metric. A simple examination of the relation between coordinates x_1, y_1 and r, θ, which was used in section 8.3, shows that the choice (8.113) for μ_1 and μ_2 corresponds to the choice (8.50) (with $\sigma_1 = -\sigma$). Consequently, from (8.51) and (8.109) we have for the NUT parameter b,

$$b = -C_0^{(1)} C_1^{(2)} - C_1^{(1)} C_0^{(2)} = -c_1 c_2 \sin(\eta_1 + \eta_2). \tag{8.114}$$

To ensure asymptotic flatness we should adopt the constraint

$$\eta_1 + \eta_2 = 0, \tag{8.115}$$

and we arrive at the Kerr metric.

When $N = 2$ we have a four-soliton solution (double Kerr solution). The index γ now takes the values 1 and 3 and there are two pairs of pole trajectories $(\mu_\gamma, \mu_{\gamma+1})$ and two pairs of constants $(w_\gamma, w_{\gamma+1})$ and $(\eta_\gamma, \eta_{\gamma+1})$. We can now fuse pole μ_3 with μ_1 and pole μ_4 with μ_2 to get a two-pole solution with multiplicity 2 at each pole. For this we perform the following limiting procedure (see subsection 2.1.1):

$$w_3 \to w_1, \quad \mu_3 \to \mu_1, \quad \eta_3 \to \eta_1; \quad w_4 \to w_2, \quad \mu_4 \to \mu_2, \quad \eta_4 \to \eta_2, \tag{8.116}$$

keeping the ratios ξ_3 and ξ_4:

$$\xi_3 = \frac{\eta_3 - \eta_1}{w_3 - w_1}, \quad \xi_4 = \frac{\eta_4 - \eta_2}{w_4 - w_2}, \tag{8.117}$$

finite and arbitrary. In this limit we get the extended Tomimatso–Sato metric with five arbitrary parameters, namely $\sigma_1, \eta_1, \eta_2, \xi_3$ and ξ_4, which was obtained by Kinnersley and Chitre [167, 168, 169]. In addition, if we adopt the constraints

$$\xi_3 = \xi_4 = 0, \tag{8.118}$$

and the asymptotic flatness condition (8.115), we get a solution with two arbitrary parameters σ_1, η_1 which is just the original Tomimatso–Sato solution with the distortion parameter $\delta = 2$. This limiting case corresponds to the fusion of two Kerr black holes in such a way that together with the vanishing of the distance between the centres of the two Kerr metrics ($\tilde{z}_3 - \tilde{z}_1 \to 0$), all other differences between these two objects disappear. In this limit we have an overlap of two identical Kerr black holes centred at the same point.

The same limiting procedure can be used for any number N in the Kerr–NUT solutions. When N is arbitrary the $2N$-soliton solution describes the nonlinear Neugebauer superposition of the Kerr–NUT metrics [225]. Each such metric (for each value of γ) can be characterized by the essential quantities (8.110). First we take the limits

$$w_\gamma \to w_1, \quad \mu_\gamma \to \mu_1, \quad \eta_\gamma \to \eta_1; \quad w_{\gamma+1} \to w_2, \quad \mu_{\gamma+1} \to \mu_2, \quad \eta_{\gamma+1} \to \eta_2, \tag{8.119}$$

and get the metric with $2N + 1$ parameters σ_1, η_1, η_2, ξ_3, ξ_4, ..., ξ_{2N}, where the $2N - 2$ parameters ξ_3, ..., ξ_{2N} are

$$\xi_\gamma = \frac{\eta_\gamma - \eta_1}{w_\gamma - w_1}, \quad \xi_{\gamma+1} = \frac{\eta_{\gamma+1} - \eta_2}{w_{\gamma+1} - w_2}, \quad \gamma = 3, 5, \ldots, 2N - 1, \tag{8.120}$$

which are kept finite under the limiting procedure. If we adopt the additional condition (8.115), then the solution will be the asymptotically flat extended Tomimatso–Sato solution. Finally, if we choose the particular case

$$\xi_\gamma = \xi_{\gamma+1} = 0, \quad \gamma = 3, 5, \ldots, 2N - 1, \tag{8.121}$$

we recover the original Tomimatso–Sato solution with the distortion parameter $\delta = N$. Such a solution describes an overlap of N identical Kerr black holes centred at the same point. This solitonic interpretation of the Tomimatso–Sato solution gives a natural explanation of the fact (which was rather mysterious in the beginning) that the distortion parameter δ can take positive integer values only, since it just represents the number of soliton pairs of the solution. Note that in the diagonal limit (static limit) this restriction does not apply to the generalized soliton solutions.

Bibliography

[1] D.J. Adams, R.W. Hellings and R.L. Zimmerman, *Astrophys. J.* **288**, 14 (1985). *See* §4.1, 5.4.

[2] D.J. Adams, R.W. Hellings, R.L. Zimmerman, H. Farhoosh, D.I. Levine and Z. Zeldich, *Astrophys. J.* **253**, 1 (1982). *See* §4.1, 4.4.1, 4.3.3, 4.6.3, 5.4, 6.3.

[3] A. Albrecht and P.J. Steinhardt, *Phys. Rev. Lett.* **48**, 120 (1982). *See* §4.1.

[4] G.A. Alekseev, *JETP Lett.*, **32** 277 (1981). *See* §3.1, 3.3.3.

[5] G.A. Alekseev, *Proceedings of the Steklov Institute of Mathematics* (Providence, RI: American Mathematical Society) **3**, 215 (1988). *See* §3.1, 3.2, 3.3.3, 3.4, 3.5.

[6] G.A. Alekseev and V.A. Belinski, *Sov. Phys. JETP* **51**, 655 (1981). *See* §8.6.

[7] G.A. Alekseev and J.B. Griffiths, *Phys. Rev.* **D52**, 4497 (1995). *See* §5.4.4.

[8] D.R. Baldwin and G.B. Jeffrey, *Proc. R. Soc. London* **A111**, 95 (1926). *See* §7.1.

[9] J.D. Barrow, *Nature* **272**, 211 (1977). *See* §5.4.2.

[10] L. Bel, *C.R. Acad. Sc. Paris* **247**, 1094 (1958). *See* §4.3.3.

[11] L. Bel, *C.R. Acad. Sc. Paris* **248**, 1297 (1959). *See* §4.3.3.

[12] V.A. Belinski, *Sov. Phys. JETP* **50**, 623 (1979). *See* §5.4.2.

[13] V.A. Belinski, *JETP Lett.* **30**, 28 (1979). *See* §3.1.

[14] V.A. Belinski, *Phys. Rev.* **D44**, 3109 (1991). *See* §2.3, 4.6.3.

[15] V.A. Belinski and D. Fargion, *Nuovo Cimento* **59B**, 143 (1980). *See* §5.1.2.

[16] V.A. Belinski and M. Francaviglia, *Gen. Rel. Grav.* **14**, 213 (1982). *See* §4.5, 5.2.

[17] V.A. Belinski and M. Francaviglia, *Gen. Rel. Grav.* **16**, 1189 (1984). *See* §5.2.

[18] V.A. Belinski and I.M. Khalatnikov, *Sov. Phys. JETP* **29**, 911 (1969). *See* §4.1.

[19] V.A. Belinski and I.M. Khalatnikov, *Sov. Phys. JETP* **36**, 591 (1973). *See* §5.4.3.

[20] V.A. Belinski, I.M. Khalatnikov and E.M. Lifshitz, *Adv. Phys.* **19**, 525 (1970). *See* §4.1, 4.2, 4.4.1, 5.4.

[21] V.A. Belinski, E.M. Lifshitz and I.M. Khalatnikov, *Adv. Phys.* **31**, 639 (1982). *See* §4.2.

[22] V.A. Belinski and R. Ruffini, *Phys. Lett.* **89B**, 195 (1980). *See* §1.5.

[23] V.A. Belinski and V.E. Zakharov, *Sov. Phys. JETP* **48**, 985 (1978). *See* §1.2, 1.3, 1.5, 5.1.1, 6.3.1, 8.1, 8.3.

[24] V.A. Belinski and V.E. Zakharov, *Sov. Phys. JETP* **50**, 1 (1979). *See* §1.2, 1.4.2, 1.5, 8.1, 8.2, 8.3, 8.4.

[25] M. Berg and M. Bradley, *Physica Scripta* **62**, 17 (2000). *See* §5.2.

[26] R. Bergamini and C.A. Orzalesi, *Phys. Lett.* **B135**, 38 (1984). *See* §5.4.4.

[27] B. Bertotti, *Astrophys. Lett* **14**, 51 (1973). *See* §4.1.

[28] B. Bertotti and B.J. Carr, *Astrophys. J.* **236**, 1000 (1980). *See* §4.1.

[29] B. Bertotti, B.J. Carr and M.J. Rees, *Mon. Not. R. Astron. Soc.* **203**, 945 (1983). *See* §4.1.

[30] J. Bičák and J.B. Griffiths, *Phys. Rev.* **D49**, 900 (1994). *See* §5.4.4.

[31] J. Bičák and J.B. Griffiths, *Ann. Phys.* **252**, 180 (1996). *See* §5.4.4.

[32] H. Bondi and F.A.E. Pirani, *Proc. R. Soc. London* **A421**, 395 (1989). *See* §7.1.

[33] W.B. Bonnor, *J. Phys.* **A12**, 847 (1979). *See* §6.1.

[34] W.B. Bonnor, *Gen. Rel. Grav.* **15**, 535 (1983). *See* §8.5.1.

[35] W.B. Bonnor, *Gen. Rel. Grav.* **24**, 551 (1992). *See* §8.3, 8.5, 8.5.2.

[36] W.B. Bonnor and M.A.P. Martins, *Class. Quantum Grav.* **8**, 727 (1991). *See* §6.1, 8.5, 8.5.1.

[37] W.B. Bonnor and A. Sackfield, *Commun. Math. Phys.* **8**, 338 (1968). *See* §8.5.2.

[38] W.B. Bonnor and Swaminarayan, *Z. Phys.* **177**, 240 (1964). *See* §8.5.2.

[39] W.B. Bonnor, J.B. Griffiths and M.A.H. MacCallum *Gen. Rel. Grav.* **26**, 687 (1994). *See* §4.1, 6.1, 7.1.

[40] P.T. Boyd, J.M. Centrella and S.A. Klasky, *Phys. Rev.* **D43**, 379 (1991). *See* §4.6.3, 5.1.2.

[41] M. Bradley and A. Curir, *Gen. Rel. Grav.* **25**, 539 (1993). *See* §5.2.

[42] M. Bradley, A. Curir and M. Francaviglia, *Gen. Rel. Grav.* **23**, 1011 (1991). *See* §5.2.

[43] C. Brans and R.H. Dicke, *Phys. Rev.* **124**, 925 (1961). *See* §5.4.3.

[44] R.R. Caldwell, M. Kamionkowski and L. Wadley, *Phys. Rev.* **D59**, 027101 (1998). *See* §4.1.

[45] M. Carmeli, Ch. Charach, *Phys. Lett.* **75A**, 333 (1980). *See* §4.1, 4.5.2.

[46] M. Carmeli, Ch. Charach, *Found. Physics* **14**, 963 (1984). *See* §4.6.1.

[47] M. Carmeli, Ch. Charach and A. Feinstein, *Ann. Phys.* **150**, 392 (1983). *See* §5.4.

[48] M. Carmeli, Ch. Charach and S. Malin, *Phys. Rep.* **76**, 79 (1981). *See* §4.1, 4.5.1, 4.5.2, 5.4, 5.4.2.

[49] J. Carot and E. Verdaguer, *Class. Quantum Grav.* **6**, 845 (1989). *See* §8.5.

[50] B.J. Carr, *Astron. Astrophys.* **89**, 6 (1980). *See* §4.1.

[51] B.J. Carr and E. Verdaguer, *Phys. Rev.* **D28**, 2995 (1983). *See* §4.4.1, 5.1.2.

[52] J. Centrella and R.A. Matzner, *Phys. Rev.* **D25**, 930 (1982). *See* §4.6.3.

[53] J. Céspedes and E. Verdaguer, *Class. Quantum Grav.* **4**, L7 (1987). *See* §4.5.

[54] J. Céspedes and E.Verdaguer, *Phys. Rev.* **D36**, 2259 (1987). *See* §4.5, 5.3.

[55] S. Chandrasekhar, *Proc. R. Soc. London* **A408**, 209 (1986). *See* §6.1.

[56] S. Chandrasekhar and V. Ferrari, *Proc. R. Soc. London* **A396**, 55 (1984). *See* §7.1.

[57] S. Chandrasekhar and B.C. Xanthopoulos, *Proc. R. Soc. London* **A398**, 233 (1985). *See* §7.1.

[58] S. Chandrasekhar and B.C. Xanthopoulos, *Proc. R. Soc. London* **A402**, 37 (1985). *See* §7.1.

[59] S. Chandrasekhar and B.C. Xanthopoulos, *Proc. R. Soc. London* **A403**, 189 (1985). *See* §7.1.

[60] S. Chandrasekhar and B.C. Xanthopoulos, *Proc. R. Soc. London* **A408**, 175 (1986). *See* §7.1, 7.2.

[61] S. Chandrasekhar and B.C. Xanthopoulos, *Proc. R. Soc. London* **A410**, 311 (1987). *See* §7.1.

[62] S. Chandrasekhar and B.C. Xanthopoulos, *Proc. R. Soc. London* **A414**, 1 (1987). *See* §7.1.

[63] A. Chodos and S. Detweiler, *Phys. Rev.* **D21**, 2167 (1980). *See* §5.4.4.

[64] C.M. Cosgrove, *J. Math. Phys.* **21**, 2417 (1980). *See* §1.2, 8.1, 8.5, 8.5.2.

[65] C.M. Cosgrove, *J. Math. Phys.* **22**, 2624 (1981). *See* §1.2, 3.1, 8.1.

[66] C.M. Cosgrove, *J. Math. Phys.* **23**, 615 (1982). *See* §1.2, 8.1.

[67] J. Cruzate, M.C. Diaz, R.J. Gleiser and J.A. Pullin, *Class. Quantum Grav.* **5**, 883 (1988). *See* §4.6.3.

[68] A. Curir and M. Francaviglia, *Gen. Rel. Grav.* **17**, 1 (1985). *See* §5.2.

[69] A. Curir, M. Francaviglia and C. Sgarra, *Gen. Rel. Grav.* **18**, 745 (1986). *See* §5.2.

[70] A. Curir, M. Francaviglia and E. Verdaguer, *Astrophys. J.* **397**, 390 (1992). *See* §4.5.2.

[71] A.D. Dagotto, R.J. Gleiser and C.O. Nicasio, *Class. Quantum Grav.* **7**, 1791 (1990). *See* §4.6.3.

[72] A.D. Dagotto, R.J. Gleiser and C.O. Nicasio, *Phys. Rev.* **D42**, 424 (1990). *See* §3.4.

[73] A.D. Dagotto, R.J. Gleiser and C.O. Nicasio, *Class. Quantum Grav.* **8**, 1185 (1991). *See* §3.1, 4.6.3.

[74] A.D. Dagotto, R.J. Gleiser and C.O. Nicasio, *Class. Quantum Grav.* **10**, 961 (1993). *See* §3.1, 3.4.

[75] P.D. D'Eath, in *Sources of gravitational radiation*, ed. L. Smarr (Cambridge: Cambridge University Press, 1979). *See* §7.1.

[76] M.C.Diaz, R.J. Gleiser, G.I. Gonzalez and J.A. Pullin, *Phys. Rev.* **D40**, 1033 (1989). *See* §5.4.2.

[77] M.C.Diaz, R.J. Gleiser and J.A. Pullin, *Class. Quantum Grav.* **4**, L23 (1987). *See* §5.4.4.

[78] M.C.Diaz, R.J. Gleiser and J.A. Pullin, *Class. Quantum Grav.* **5**, 641 (1988). *See* §5.4.4.

[79] M.C.Diaz, R.J. Gleiser and J.A. Pullin, *J. Math. Phys.* **29**, 169 (1988). *See* §5.4.4.

[80] M.C.Diaz, R.J. Gleiser and J.A. Pullin, *Astrophys. J.* **339**, 1 (1989). *See* §5.4.4.

[81] R.A. d'Inverno and R.A. Russell-Clark, *J. Math. Phys.* **12**, 1258 (1971). *See* §4.3.1.

[82] M. Dorca and E. Verdaguer, *Nucl. Phys.* **B403**, 770 (1993). *See* §7.3, 7.3.3.

[83] M. Dorca and E. Verdaguer, *Nucl. Phys.* **B484**, 435 (1997). *See* §7.1.

[84] R.K. Dodd, J.C. Eilbeck, J.D. Gibbon and H.C. Morris, *Solitons and nonlinear wave equations* (New York: Academic Press, 1982). *See* §1.1.

[85] T. Dray and G. 't Hooft, *Class. Quantum Grav.* **3**, 825 (1986). *See* §8.5.

[86] M.J. Duff, B.E.W. Nilsson and C.N. Pope, *Phys. Rep.* **130**, 1 (1986). *See* §1.5, 5.4.3.

[87] A. Economou and D. Tsoubelis, *Phys. Rev.* **D38**, 498 (1988). *See* §6.3.2.

[88] A. Economou and D. Tsoubelis, *J. Math. Phys.* **30**, 1562 (1989). *See* §5.1.1, 6.3.3.

[89] J. Ehlers and W. Kundt, in *Gravitation: an introduction to current research*, ed. L. Witten (New York: Wiley, 1962). *See* §4.3.2.

[90] A. Einstein and N. Rosen, *J. Franklin Inst.* **223**, 43 (1937); J. Weber, *General relativity and gravitational waves* (New York: Interscience, 1961). *See* §1.2, 1.3, 4.3.

[91] G.F.R. Ellis and M.A.H. MacCallum, *Commun. Math. Phys.* **12**, 108 (1969). *See* §4.1, 4.5.1.

[92] G.F.R. Ellis and M.A.H. MacCallum, *Commun. Math. Phys.* **19**, 31 (1970). *See* §4.1.

[93] F.J. Ernst, A. García-Díaz and I. Hauser, *J. Math. Phys.* **28**, 2155 (1987). *See* §7.3.4.

[94] F.J. Ernst, A. García-Díaz and I. Hauser, *J. Math. Phys.* **29**, 681 (1988). *See* §7.3.4.

[95] L.D. Faddeev and N. Yu. Reshetikhin, *Ann. Phys.* **167**, 227 (1986). *See* §1.1.

[96] H. Farhoosh and R.L. Zimmerman, *Phys. Rev.* **D21**, 317 (1980). *See* §8.5.1.

[97] A. Feinstein, *Phys. Rev.* **D36**, 3263 (1987). *See* §4.4.1, 4.5.2, 4.6.3.

[98] A. Feinstein, in *Recent developments in gravitation*, eds. E. Verdaguer, J. Garriga, and J. Céspedes (Singapore: World Scientific, 1990). *See* §7.1.

[99] A. Feinstein and Ch. Charach, *Class. Quantum Grav.* **3**, L5 (1986). *See* §4.6, 4.6.1, 4.6.2.

[100] A. Feinstein and Ch. Charach, *Gen. Rel. Grav.* **26**, 743 (1994). *See* §4.6, 4.6.1.

[101] A. Feinstein and J.B. Griffiths, *Class. Quantum Grav.* **11**, L109 (1994). *See* §5.4.4.

[102] A. Feinstein and J. Ibáñez, *Phys. Rev.* **D39**, 470 (1989). *See* §7.1, 7.3, 7.3.2.

[103] V. Ferrari, *Phys. Rev.* **D37**, 3061 (1988). *See* §7.3.4.

[104] V. Ferrari and J. Ibáñez, *Gen. Rel. Grav.* **19**, 405 (1987). *See* §7.1, 7.3, 7.3.3.

[105] V. Ferrari and J. Ibáñez, *Proc. R. Soc. London* **A417**, 417 (1988). *See* §7.1, 7.3.3, 7.3.4.

[106] V. Ferrari, J. Ibáñez and M. Bruni, *Phys. Rev.* **D36**, 1053 (1987). *See* §7.1, 7.3.4.

[107] V. Ferrari, J. Ibáñez and M. Bruni, *Phys. Lett.* **A122**, 459 (1987). *See* §7.3.4.

[108] V. Ferrari, P. Pendenza and G. Veneziano, *Gen. Rel. Grav.* **20**, 1185 (1988). *See* §7.2.2.

[109] X. Fustero and E. Verdaguer, *Gen. Rel. Grav.* **18**, 1141 (1986). *See* §4.6.3, 6.1, 6.2.3.

[110] A. Garate and R.J. Gleiser, *Class. Quantum Grav.* **12**, 119 (1995). *See* §2.3.2, 3.1, 3.4, 3.5.

[111] C.S. Gardner, J.M. Green, M.D. Kruskal and R.M. Miura, *Phys. Rev. Lett.* **19**, 1095 (1967). *See* §1.1.

[112] D. Garfinkle, *Phys. Rev.* **D41**, 1112 (1990). *See* §7.1.

[113] J. Garriga and E. Verdaguer, *Phys. Rev.* **D36**, 2250 (1987). *See* §4.6, 4.6.3, 6.1, 6.2, 6.2.3, 6.2.4.

[114] J. Garriga and E. Verdaguer, *Phys. Rev.* **D43**, 391 (1991). *See* §7.1, 7.2.2.

[115] R. Gautreau and J.L. Anderson, *Phys. Lett.* **A25**, 291 (1967). *See* §8.5.2.

[116] R. Geroch, *Commun. Math. Phys.* **13**, 180 (1969). *See* §8.5.

[117] G.W. Gibbons, *Commun. Math. Phys.* **45**, 191 (1975). *See* §7.1.

[118] R.J. Gleiser, *Gen. Rel. Grav.* **16**, 1077 (1984). *See* §4.5.2, 5.1.1,

[119] R.J. Gleiser, M.C. Diaz and R.D. Grosso, *Class. Quantum Grav.* **5**, 989 (1988). *See* §5.4.2

[120] R.J. Gleiser, C.O. Nicasio and A. Garate, *Class. Quantum Grav.* **10**, 2557 (1993). *See* §3.1.

[121] R.J. Gleiser, C.O. Nicasio and A. Garate, *Class. Quantum Grav.* **11**, 1519 (1994). *See* §4.6.3.

[122] R.J. Gleiser, A. Garate and C.O. Nicasio, *J. Math. Phys.* **37**, 5652 (1996). *See* §2.3.

[123] R.J. Gleiser and M.H. Tiglio, *Phys. Rev.* **D58**, 124028 (1998). *See* §3.1, 6.3.1.

[124] B.B. Godfrey, *Gen. Rel. Grav.* **3**, 3 (1972). *See* §8.5.1.

[125] R.H. Gowdy, *Phys. Rev. Lett.* **27**, 826 (1971). *See* §4.1.

[126] R.H. Gowdy, *Ann. Phys.* **83**, 203 (1974). *See* §4.1.

[127] J.B. Griffiths, *Colliding waves in general relativity* (Oxford: Clarendon Press, 1991). *See* §7.1, 7.3.1, 7.3.4.

[128] J.B. Griffiths, *Class. Quantum Grav.* **10**, 975 (1993). *See* §5.4.4.

[129] J.B. Griffiths, *J. Math. Phys.* **34**, 4064 (1993). *See* §5.4.4.

[130] J.B. Griffiths and S. Miccichè, *Gen. Rel. Grav.* **31**, 869 (1999). *See* §4.5.2.

[131] L.P. Grishchuk, *Ann. NY Acad. Sci.* **302**, 439 (1977). *See* §4.1.

[132] D.J. Gross and M.J. Perry, *Nucl. Phys.* **B266**, 29 (1983). *See* §1.5, 5.4.3.

[133] A. Guth, *Phys. Rev.* **D23**, 347 (1981). *See* §4.1.

[134] J.J. Halliwell and S.W. Hawking, *Phys. Rev.* **D31**, 1777 (1985). *See* §4.1.

[135] B.K. Harrison, *Phys. Rev. Lett.* **41**, 1197 (1978). *See* §1.2, 8.1.

[136] B.K. Harrison, *Phys. Rev.* **D21**, 1695 (1980). *See* §1.2, 8.1.

[137] I. Hauser and F.J. Ernst, *Phys. Rev.* **D20**, 362 (1979). *See* §8.1.

[138] I. Hauser and F.J. Ernst, *J. Math. Phys.* **21**, 1126, 1418 (1980). *See* §3.1, 8.1.

[139] I. Hauser and F.J. Ernst, *J. Math. Phys.* **30**, 872, 2322 (1989). *See* §7.1.

[140] I. Hauser and F.J. Ernst, *J. Math. Phys.* **31**, 871 (1990). *See* §7.1.

[141] I. Hauser and F.J. Ernst, *J. Math. Phys.* **32**, 198 (1991). *See* §7.1.

[142] M.A. Hausman, D.W. Olson and B.D. Roth, *Astrophys. J.* **270**, 351 (1983). *See* §5.4.4.

[143] S.W. Hawking and G.F.R. Ellis, *The large scale structure of spacetime* (Cambridge: Cambridge University Press, 1973). *See* §5.4.2, 5.4.4, 7.2.1.

[144] S.W. Hawking and I.G. Moss, *Phys. Lett.* **110B**, 35 (1982). *See* §4.1.

[145] S.A. Hayward, *Class. Quantum Grav.* **6**, 1021 (1989). *See* §7.1, 7.3, 7.3.3.

[146] R.W. Hellings, *Phys. Rev. Lett* **43**, 470 (1979). *See* §4.1.

[147] R.W. Hellings and G.S. Downs, *Astrophys. J.* **265**, L35 (1983). *See* §4.1.

[148] B.L. Hu, *Phys. Rev.* **D12**, 1551 (1975). *See* §4.1.

[149] C. Hoenselaers, *Prog. Theor. Phys.* **56**, 324 (1976). *See* §8.5.2.

[150] J. Ibáñez and E. Verdaguer, *Phys. Rev. Lett.* **51**, 1313 (1983). *See* §4.6.3.

[151] J. Ibáñez and E. Verdaguer, *Phys. Rev.* **D31**, 251 (1985). *See* §4.6.3.

[152] J. Ibáñez and E. Verdaguer, *Phys. Rev.* **D34**, 1202 (1986). *See* §5.4.3, 5.4.4.

[153] J. Ibáñez and E. Verdaguer, *Astrophys. J.* **306**, 401 (1986). *See* §5.4.4.

[154] J. Ibáñez and E. Verdaguer, *Class. Quantum Grav.* **3**, 1235 (1986). *See* §5.4.1.

[155] W. Israel, *Phys. Rev.* **D15**, 935 (1977). *See* §6.1, 8.5.2.

[156] W. Israel and K.A. Khan, *Nuovo Cimento* **33**, 331 (1964). *See* §8.5.2.

[157] R.T. Jantzen, *Nuovo Cimento* **59B**, 287 (1980). *See* §2.2, 4.6.1.

[158] R.T. Jantzen, *Gen. Rel. Grav.* **15**, 115 (1983). *See* §8.5.1, 8.5.2.

[159] P. Jordan, *Ann. Phys. (Leipzig)* **1**, 219 (1947). *See* §5.4.3.

[160] P. Jordan, J. Ehlers and W. Kundt, *Akad. Wiss. Mainz Mabk. Math. Nat. Kl. Jahrg,* No. **2** (1960). *See* §6.1

[161] K. Khan and R. Penrose, *Nature* **229**, 185 (1971). *See* §7.1, 7.3.

[162] Th. Kaluza, *Sitzungsber. Preuss. Akad. Wiss. Phys. Math. Kl.* **LIV**, 966 (1921). *See* §1.5, 5.4.3.

[163] A. Karlhede, *Gen. Rel. Grav.* **12**, 693 (1980). *See* §4.6.1.

[164] E. Kasner, *Am. J. Math.* **43**, 217 (1921). *See* §4.2.

[165] R.P. Kerr, *Phys. Rev. Lett.* **11**, 273 (1963). *See* §8.3.

[166] W. Kinnersley, *J. Math. Phys.* **10**, 1195 (1969). *See* §8.5.

[167] W. Kinnersley and D.M. Chitre, *J. Math. Phys.* **18**, 1538 (1977). *See* §8.1, 8.6.

[168] W. Kinnersley and D.M. Chitre, *J. Math. Phys.* **19**, 1926 (1978). *See* §6.3.3, 8.1, 8.6.

[169] W. Kinnersley and D.M. Chitre, *J. Math. Phys.* **19**, 2037 (1978). *See* §6.3.3, 8.1, 8.6.

[170] W. Kinnersley and E.F. Kelly, *J. Math. Phys.* **15**, 2121 (1974). *See* §8.5, 8.5.1.

[171] W. Kinnersley and M. Walker, *Phys. Rev.* **D2**, 1359 (1970). *See* §8.5.1.

[172] D.W. Kitchingham, *Class. Quantum Grav.* **1**, 677 (1984). *See* §4.5.1, 5.2, 5.3, 8.1.

[173] D.W. Kitchingham, PhD Thesis, Queen Mary College, University of London, unpublished (1986). *See* §4.5, 5.3, 7.2.3, 8.1.

[174] D.W. Kitchingham, *Class. Quantum Grav.* **3**, 133 (1986). *See* §4.6.1, 5.4.1, 8.1.

[175] O. Klein, *Z. Phys.* **37**, 875 (1926). *See* §1.5, 5.4.3.

[176] A.S. Kompaneets, *Sov. Phys. JETP* **7**, 659 (1958). *See* §1.2, 6.1.

[177] P. Kordas, *Phys. Rev.* **D48**, 5013 (1993). *See* §2.3.2, 3.5.

[178] D. Kramer and G. Neugebauer, *J. Phys.* **A14**, L333 (1981). *See* §3.1.

[179] D. Kramer, H. Stephani, M. MacCallum and E.Herlt, *Exact solutions of Einstein's field equations* (Cambridge: Cambridge University Press, 1980). *See* §1.2, 4.1, 4.2, 4.3.1, 4.3.2, 5.2, 5.4.1, 6.1, 6.2, 7.2.1, 8.1, 8.5.1, 8.5.2.

[180] A. Krasiński, *Inhomogeneous cosmological models* (Cambridge: Cambridge University Press, 1997). *See* §1.2, 4.1, 5.4, 5.4.2, 5.4.4.

[181] L.M. Krauss and M. White, *Phys. Rev. Lett.* **69**, 869 (1992). *See* §4.1.

[182] W. Kundt and M. Trümper, *Z. Phys.* **192**, 419 (1966). *See* §8.1.

[183] H. Lamb, *Hydrodynamics* (Cambridge: Cambridge University Press, 1906). *See* §4.4.1.

[184] L.D. Landau and E.M. Lifshitz, *The classical theory of fields* (Oxford: Pergamon Press, 1971). *See* §4.2, 7.1, 7.2.1.

[185] L.D. Landau and E.M. Lifshitz, *Physical kinetics* (Oxford: Pergamon Press, 1981). *See* §4.6.3.

[186] P.D. Lax, *Commun. Pure Appl. Math.* **21**, 467 (1968). *See* §1.1.

[187] P.S. Letelier, *J. Math. Phys.* **16**, 1488 (1975). *See* §5.4.2.

[188] P.S. Letelier, *J. Math. Phys.* **20**, 2078 (1979). *See* §5.4.2.

[189] P.S. Letelier, *Phys. Rev.* **D22**, 807 (1980). *See* §5.4.3.

[190] P.S. Letelier, *Phys. Rev.* **D26**, 2623 (1982). *See* §5.4.2.

[191] P.S. Letelier, *J. Math. Phys.* **25**, 2675 (1984). *See* §5.1.

[192] P.S. Letelier, *Class. Quantum Grav.* **2**, 419 (1985). *See* §2.2.

[193] P.S. Letelier, *J. Math. Phys.* **26**, 467 (1985). *See* §8.5, 8.5.1, 8.5.2.

[194] P.S. Letelier, *J. Math. Phys.* **27**, 564 (1986). *See* §5.2, 5.4.2.

[195] P.S. Letelier, *J. Math. Phys.* **27**, 615 (1986). *See* §5.4.2.

[196] P.S. Letelier and S.R. Oliveira, *Class. Quantum Grav.* **5**, L47 (1988). *See* §5.4.2.

[197] P.S. Letelier and R. Tabensky, *Nuovo Cimento* **B28**, 407 (1975). *See* §5.4.2.

[198] P.S. Letelier and E. Verdaguer, *Phys. Rev.* **D36**, 2981 (1987). *See* §5.4.2, 5.4.3.

[199] P.S. Letelier and E. Verdaguer, *J. Math. Phys.* **28**, 2431 (1987). *See* §5.4.2.

[200] P.S. Letelier and E. Verdaguer, *Class. Quantum Grav.* **6**, 705 (1989). *See* §5.4.2.

[201] T. Levi-Città, *Rend. Acc. Lincei* **28**, 3 (1919). *See* §6.1.

[202] E.P. Liang, *Astrophys. J.* **204**, 235 (1976). *See* §5.4, 5.4.3.

[203] A.D. Linde, *Phys. Lett.* **108B**, 389 (1982). *See* §4.1.

[204] E. Lifshitz, *J. Phys. (USSR)* **10**, 116 (1946). *See* §4.1, 5.4.

[205] V.N. Lukash, *Sov. Phys. JETP* **40**, 792 (1975). *See* §5.3.

[206] D. Maison, *Phys. Rev. Lett.* **41**, 521 (1978). *See* §1.2.

[207] M.A.H. MacCallum, *Commun. Math. Phys.* **20**, 57 (1971). *See* §4.5.1.

[208] M.A.H. MacCallum, in *Cargese lectures in physics*, ed. E. Schtzman (New York: Gordon and Breach, 1973). *See* §4.1.

[209] M.A.H. MacCallum, in *General relativity, an Einstein centenary survey*, eds. S.W. Hawking and W. Israel (Cambridge: Cambridge University Press, 1979). *See* §4.1.

[210] M.A.H. MacCallum, in *Lecture notes in physics: Retzbach seminar on exact solutions of Einstein's field equations*, eds. W. Dietz and C. Hoenselaers (Berlin: Springer Verlag, 1984). *See* §4.1, 5.4.

[211] M.A.H. MacCallum, in *The origin of structure in the universe*, eds. E. Gunzig and P. Nardone (Dordrecht: Kluwer Academic, 1993). *See* §4.1.

[212] M.A.H. MacCallum, in *The renaissance of general relativity and cosmology*, eds. G.F.R. Ellis, A. Lanza and J.C. Miller (Cambridge: Cambridge University Press, 1993). *See* §4.1.

[213] L. Marder, *Proc. R. Soc. London* **A244**, 524 (1958). *See* §6.1, 6.2.3.

[214] L. Marder, *Proc. R. Soc. London* **A313**, 83 (1969). *See* §6.1.

[215] B. Mashhoon, *Mon. Not. R. Astron. Soc.* **199**, 659 (1982). *See* §4.1.

[216] B. Mashhoon and L.P. Grishchuk, *Astrophys. J.* **236**, 990 (1980). *See* §4.1.

[217] B. Mashhoon, B.J. Carr and B.L. Hu, *Astrophys. J.* **246**, 569 (1981). *See* §4.1.

[218] R.A. Matzner and F.J. Tipler, *Phys. Rev.* **D29**, 1575 (1984). *See* §7.1, 7.3.3.

[219] E.A. Milne, *Quart. J. Math., Oxford* **5**, 64 (1934). *See* §5.4.1.

[220] C.W. Misner, *J. Math. Phys.* **4**, 924 (1963). *See* §8.3.

[221] C.W. Misner, *Phys. Rev. Lett.* **22**, 1071 (1969). *See* §4.1.

[222] C.W. Misner, K.S. Thorne and J.A. Wheeler, *Gravitation* (San Francisco: Freeman, 1973). *See* §4.3.3, 5.4.3, 7.2.2.

[223] V.F. Mukhanov, H.A. Feldman and R.H. Brandenberger, *Phys. Rep.* **215**, 203 (1992). *See* §4.1.

[224] G. Neugebauer, *J. Phys.* **A12**, L67 (1979). *See* §1.2, 8.1.

[225] G. Neugebauer, *J. Phys.* **A13**, L19 (1980). *See* §8.6.

[226] G. Neugebauer and D. Kramer, *J. Phys.* **A16**, 1927 (1983). *See* §3.1, 7.1.

[227] G. Neugebauer and R. Meinel, *Astrophys. J.* **414**, L97 (1993). *See* §8.3.

[228] G. Neugebauer and R. Meinel, *Phys. Rev. Lett.* **73**, 2166 (1994). *See* §8.3.

[229] G. Neugebauer and R. Meinel, *Phys. Rev. Lett.* **75**, 3046 (1995). *See* §8.3.

[230] E. Newman, L. Tamburino and T. Unti, *J. Math. Phys.* **4**, 915 (1963). *See* §7.3.4, 8.3.

[231] S. Novikov, S.V. Manakov, L.P. Pitaevsky and V.E. Zakharov, *Theory of solitons. The inverse scattering method* (New York: Consultants Bureau, 1984). *See* §1.1.

[232] Y. Nutku and M. Halil, *Phys. Rev. Lett.* **39**, 1379 (1977). *See* §7.1.

[233] S. O'Brien and J.L. Synge, *Proc. Dublin Inst. Adv. Stu.* **A9**, 1 (1952). *See* §7.3.

[234] G. Oliver and E. Verdaguer, *J. Math. Phys.* **30**, 442 (1989). *See* §4.5.3, 5.4.2.

[235] D.W. Olson and J. Silk, *Astrophys. J.* **233**, 395 (1979). *See* §5.4.4.

[236] A. Papapetrou, *Ann. Inst. H. Poincaré* **A4**, 83 (1966). *See* §8.1.

[237] P.J.E. Peebles, *Physical cosmology* (Princeton: Princeton University Press, 1971). *See* §4.1.

[238] P.J.E. Peebles, *The large scale structure of the Universe* (Princeton: Princeton University Press, 1980). *See* §5.4.

[239] P.J.E. Peebles, *Astrophys. J.* **257**, 438 (1982). *See* §5.4.4.

[240] P.J.E. Peebles, *Principles of physical cosmology* (Princeton: Princeton University Press, 1993). *See* §4.1.

[241] R. Penrose, *Rev. Mod. Phys.* **37**, 215 (1965). *See* §7.1, 7.2.2.

[242] T. Piran, P.N. Safier and R.F. Stark, *Phys. Rev.* **D32**, 3101 (1985). *See* §6.3.

[243] T. Piran and P.N. Safier, *Nature* **318**, 271 (1985). *See* §6.3, 6.3.

[244] A. Polyakov and P.B. Wiegmann, *Phys. Lett.* **B131**, 121 (1983). *See* §1.1.

[245] R. Rajaraman, *Solitons and instantons* (Amsterdam: North-Holland, 1982). *See* §2.3, 4.6.3.

[246] A.K. Raychaudhuri, *Phys. Rev.* **D41**, 3041 (1990). *See* §8.5.

[247] C. Rebbi and G. Soliani, *Solitons and particles* (Singapore: World Scientific, 1984). *See* §1.1, 1.2.

[248] M.J. Rees, *Mon. Not. R. Astron. Soc.* **154**, 187 (1971). *See* §4.1.

[249] M.J. Rees, *Gravitazione sperimentale*, eds. B. Bertotti (Rome: Academia Nazionale dei Lincei, 1977). *See* §4.1.

[250] W. Rindler, *Phys. Rev.* **119**, 2082 (1960). *See* §8.5, 8.5.1.

[251] W. Rindler, *Am. J. Phys.* **34**, 1174 (1966). *See* §8.5.

[252] I. Robinson and A. Trautman, *Proc. R. Soc. London* **A265**, 463 (1962). *See* §8.5.

[253] R.W. Romani and J.H. Taylor, *Astrophys. J.* **265**, L35 (1983). *See* §4.1.

[254] N. Rosen, *Bull. Res. Coun. Israel* **3**, 328 (1954). *See* §4.4.1.

[255] L.A. Rosi and R.L. Zimmerman, *Astrophys. Space Sci.* **45**, 447 (1976). *See* §4.1.

[256] V.A. Rubakov, M.V. Sazkin and A.V. Veryaskin, *Phys. Lett.* **115B**, 189 (1982). *See* §4.1.

[257] M.P. Ryan and L.S. Shepley, *Homogeneous relativistic cosmologies* (Princeton: Princeton University Press, 1975). *See* §4.1.

[258] K. Schwarzschild, *Sitz. Preuss. Akad. Wiss.* **189** (1916). *See* §8.3.

[259] A.C. Scott, F.Y.F. Chu and D.W. McLaughlin, *Proc. IEEE* **61** 1443 (1973). *See* §1.1, 4.6.3.

[260] S.T.C. Siklos, *J. Phys.* **A14**, 395 (1981). *See* §5.3.

[261] S.T.C. Siklos, in *Relativistic astrophysics and cosmology*, eds. X. Fustero and E. Verdaguer (Singapore: World Scientific, 1984). *See* §5.3, 5.4.1.

[262] J.J. Stachel, *J. Math. Phys.* **7**, 1321 (1966). *See* §5.3, 6.1.

[263] J.L. Synge, *Relativity: the special theory* (Amsterdam: North-Holland, 1955). *See* §4.6.

[264] P. Szekeres, *Nature* **228**, 1183 (1970). *See* §7.1, 7.3, 7.3.1.

[265] P. Szekeres, *J. Math. Phys.* **13**, 286 (1972). *See* §4.3.1, 7.1, 7.3.

[266] R. Tabensky and P.S. Letelier, *J. Math. Phys.* **15**, 594 (1974). *See* §5.4.2.

[267] R. Tabensky and A.H. Taub, *Commun. Math. Phys.* **29**, 61 (1973). *See* §5.4.2, 5.4.3.

[268] A.H. Taub, *Ann. Math.* **53**, 472 (1951). *See* §7.3.4, 8.5, 8.5.1, 8.3.

[269] A.H. Taub, *J. Math. Phys.* **21**, 1423 (1980). *See* §4.5.2.

[270] Y.R. Thiry, *C.R. Acad. Sci. Paris* **226**, 216 (1984). *See* §5.4.3.

[271] Y.R. Thiry, *C.R. Acad. Sci. Paris* **226**, 1881 (1984). *See* §5.4.3.

[272] K.S. Thorne, *Phys. Rev.* **138**, 251 (1965). *See* §4.3.3, 6.1.

[273] K.S. Thorne, in *Theoretical principles in astrophysics and relativity*, eds. N.R. Lebonitz, W.H. Reid and P.O. Vandervoot (Chicago: University of Chicago Press, 1978). *See* §4.1.

[274] K.S. Thorne, in *Particle and nuclear astrophysics and cosmology in the next millennium*, eds. E.W. Kolb and R.D. Peccei (Singapore: World Scientific, 1995). *See* §4.1.

[275] A. Tomimatsu, *Prog. Theor. Phys.* **63**, 1054 (1980). *See* §8.6.

[276] A. Tomimatsu, *Gen. Rel. Grav.* **21**, 613 (1989). *See* §6.1, 6.3, 6.3.1, 6.3.2.

[277] A. Tomimatsu and H. Sato, *Phys. Rev. Lett.* **29**, 1344 (1972). *See* §6.3.3, 8.6.

[278] A. Tomimatsu and H. Sato, *Suppl. Prog. Theor. Phys.* **70**, 215 (1981). *See* §8.5, 8.6.

[279] M.S. Turner, *Astrophys. J.* **233**, 685 (1979). *See* §4.1.

[280] N. Turok, *Nucl Phys.* **B242**, 520 (1984). *See* §4.1.

[281] T. Vachaspati and A. Vilenkin, *Phys. Rev.* **D31**, 3052 (1985). *See* §4.1.

[282] T. Vachaspati and A. Vilenkin, *Phys. Rev.* **D37**, 898 (1988). *See* §4.1.

[283] T. Vachaspati, *Nucl. Phys.* **B277**, 593 (1986). *See* §6.2.

[284] E. Verdaguer, *J. Phys.* **A15**, 1261 (1982). *See* §8.2, 8.5.1.

[285] E. Verdaguer, in *Observational and theoretical aspects of relativistic astrophysics and cosmology*, eds. J.L. Sanz and L.J. Goicoechea (Singapore: World Scientific, 1985). *See* §1.5, 4.5.1.

[286] E. Verdaguer, *Gen. Rel. Grav.* **18**, 1045 (1986). *See* §4.6.1.

[287] E. Verdaguer, *Nuovo Cimento* **100B**, 787 (1987). *See* §7.2.3.

[288] E. Verdaguer, *Phys. Rep.* **229**, 1 (1993). *See* §1.5, 4.1.

[289] G. Veneziano, *Mod. Phys. Lett.* **A2**, 899 (1987). *See* §7.3.4.

[290] A. Vilenkin, *Phys. Rev. Lett.* **46**, 1169 (1981). *See* §4.1.

[291] A. Vilenkin, *Phys. Rep.* **121**, 263 (1985). *See* §6.1, 6.2.

[292] B.H. Voorhees, *Phys. Rev.* **D2**, 2119 (1970). *See* §8.5, 8.5.2.

[293] J. Wainwright, *Phys. Rev.* **D20**, 3031 (1979). *See* §4.1, 5.3.

[294] J. Wainwright, *J. Phys.* **A12**, 2015 (1979). *See* §4.1, 4.2.

[295] J. Wainwright, *J. Phys.* **A14**, 1131 (1981). *See* §4.1.

[296] J. Wainwright and P.J. Anderson, *Gen. Rel. Grav.* **16**, 609 (1984). *See* §7.2.3.

[297] J. Wainwright, W.C.W. Ince and B.J. Marshman, *Gen. Rel. Grav.* **10**, 259 (1979). *See* §4.5.1, 5.4.2.

[298] J. Wainwright and B.J. Marshman, *Phys. Lett.* **72A**, 275 (1979). *See* §4.1, 5.3.

[299] R.M. Wald, *General relativity* (Chicago: Chicago University Press, 1984). *See* §4.3.2, 4.6.3.

[300] J. Weber and J.A Wheeler, *Rev. Mod. Phys.* **29**, 429 (1957). *See* §4.4.1.

[301] S. Weinberg, *Gravitation and cosmology* (New York: John Wiley, 1972). *See* §4.1, 5.4, 5.4.3.

[302] G.B. Whitham, *Linear and non-linear waves* (New York: John Wiley, 1974). *See* §1.1.

[303] B.C. Xanthopoulos, *Phys. Lett.* **B178**, 163 (1986). *See* §6.3.2.

[304] B.C. Xanthopoulos, *Phys. Rev.* **D34**, 3608 (1986). *See* §6.3.2.

[305] C.N. Yang, *Phys. Rev. Lett.* **38**, 1377 (1977). *See* §5.4.2.

[306] U. Yurtsever, *Phys. Rev.* **D36**, 1662 (1987). *See* §7.1.

[307] U. Yurtsever, *Phys. Rev.* **D37**, 2790 (1988). *See* §7.3.

[308] U. Yurtsever, *Phys. Rev.* **D37**, 2803 (1988). *See* §4.1, 4.2, 4.6.3, 7.1.

[309] U. Yurtsever, *Phys. Rev.* **D38**, 1706 (1988). *See* §4.1, 4.2, 4.6.3, 7.1, 7.3.

[310] U. Yurtsever, *Phys. Rev.* **D40**, 360 (1989). *See* §7.1.

[311] V.E. Zakharov and A.V. Mikhailov, *Sov. Phys. JETP* **47**, 1017 (1978). *See* §1.1.

[312] V.E. Zakharov and A.B Shabat, *Funct. Anal. Appl.* **13**, 166 (1980). *See* §1.1.

[313] Ya.B. Zeldovich, *Mon. Not. R. Astron. Soc.* **160**, 1 (1972). *See* §5.4.2.

[314] Ya.B. Zeldovich and I.D. Novikov, *Relativistic astrophysics: The structure and evolution of the Universe* (Chicago: University of Chicago Press, 1983). *See* §4.1, 5.4.4.

[315] R.L. Zimmerman and R.W. Hellings, *Astrophys. J.* **241**, 475 (1980). *See* §4.1.

[316] D.M. Zipoy, *J. Math. Phys.* **7**, 1137 (1966). *See* §8.5.2.

Index